Not by Design

Not by Design

Retiring Darwin's Watchmaker

John O. Reiss

UNIVERSITY OF CALIFORNIA PRESS

Berkeley   Los Angeles   London

University of California Press, one of the most distinguished university presses in the United States, enriches lives around the world by advancing scholarship in the humanities, social sciences, and natural sciences. Its activities are supported by the UC Press Foundation and by philanthropic contributions from individuals and institutions. For more information, visit www.ucpress.edu.

University of California Press
Berkeley and Los Angeles, California

University of California Press, Ltd.
London, England

Library of Congress Cataloging-in-Publication Data

Reiss, John O., 1961–
    Not by design : retiring Darwin's watchmaker / John O. Reiss.
       p. cm.
    Includes bibliographical references and index.
    ISBN 978-0-520-25893-8 (cloth : alk. paper)
    1. Natural selection. 2. Evolution (Biology) 3. Evolutionary genetics. 4. Intelligent design (Teleology) 5. Cuvier, Georges, baron, 1769–1832. I. Title.
    QH375.R45 2009
    576.8'2—dc22   2009004282

Manufactured in the United States of America.

16   15   14   13   12   11   10   09
10   9   8   7   6   5   4   3   2   1

The paper used in this publication meets the minimum requirements of ANSI/NISO Z39.48-1992 (R 1997) (*Permanence of Paper*). ∞

Cover image: Blue-footed booby (*Sula nebouxii*), Galápagos Islands, Ecuador. Photo courtesy of Leslie Sinclair.

For Shorty

For whatever you see that lives and breathes and thrives
Has been, from the very beginning, guarded, saved
By its trickery or its swiftness or brute strength

—LUCRETIUS, CA. 55 BC, *DE RERUM NATURA*

It is in vain, therefore, to insist upon the uses of the parts in animals
or vegetables, and their curious adjustment to each other. I would fain
know how an animal could subsist, unless its parts were so adjusted?

—PHILO, 1779, IN HUME'S *DIALOGUES CONCERNING NATURAL RELIGION*

In all these cases it should be remembered, that animals could not exist
without these adaptations.

—CHARLES DARWIN, 1838, ABSTRACT OF MACCULLOCH'S *ATTRIBUTES OF GOD*

CONTENTS

The metaphor of design, with the organism as artifact, is at the heart of Darwinian evolutionary biology.
—MICHAEL RUSE, 2003, *DARWIN AND DESIGN*

Many evolutionary biologists today are in the rather peculiar position of denying design in their battle with "intelligent design" proponents over the teaching of evolution in the schools, while at the same time they embrace a design metaphor for understanding the features of organisms. The basic structure of the approach is simple: because of past natural selection, organisms appear "as if" designed for the end of survival and reproduction, and thus we can think of them "as if" they were designed—but please don't think that they actually were designed. This position seems uncomfortable, if not absurd.

The relation of current evolutionary biology to the "design problem" is indeed rather strange. We deny evolution any teleology, any goal-directedness; we are convinced that natural selection and mutation are mechanistic evolutionary processes, which can't foresee the future. Yet we use past natural selection as a way to explain the current adaptedness of organisms, in much the same way that pre-Darwinian natural theologians invoked the past actions of the Creator—as exemplified by Paley's (1802) famous metaphor of the Creator as divine watchmaker. To put this another way: the general position of evolutionists is that Darwin destroyed the design argument. The implication is that the design argument was valid in the absence of natural selection, that natural selection was needed to fill a void that existed prior to Darwin, a void previously filled by the Creator: "Paley was correct to choose design over chance, but he did not know that there was a natural as well as a transcendent source of design" (Scott 2005, 83).

What I attempt to show in this book is that the analogy between natural selection and a designer is both pernicious and unnecessary. The design argument was not valid before Darwin: it had been destroyed by Lucretius, Hume, and a host of others, even in the absence of evolution. I argue that the design metaphor, rooted in the analogy between natural selection and the Creator, has resulted in a curiously "squishy" structure to current evolutionary theory, including the theory of mathematical population genetics, and that this "squishiness" has contributed to the persistent nature of many of the major controversies in evolutionary biology. Among these controversies are the validity of adaptationism, the relative roles of natural selection and genetic drift in evolution, the levels of selection, and even the proper definition of such fundamental terms as *adaptation, fitness,* and *natural selection* itself.

I believe that we can achieve clarity on many problematic issues by connecting evolutionary theory with the antiteleological theories of the past, rather than the natural theological ones. We can thereby achieve a more balanced structure for evolutionary theory, a structure in which natural selection is reconnected with ecology and integrated with other evolutionary "forces," in which "internal selection" is accorded a place as important as "external selection," in which mechanisms that generate novelty can play their roles alongside mechanisms that spread the novelties that arise.

For me, the bridge to making this connection, and thus to dealing with the problem of teleology, comes in the form of Georges Cuvier's (1769–1832) principle of the "conditions for existence" *(conditions d'existence)*: anything that exists must be doing so by virtue of the fact that it is satisfying its conditions for existence. Cuvier has traditionally been seen as the conservative antievolutionist, who used his power and prestige to squash the prescient Jean-Baptiste Lamarck and Etienne Geoffroy Saint-Hilaire (the evolutionists of his day). It is rather ironic, then, that his principle may help move evolutionary theory beyond the teleology implicit in Darwinian natural selection.

To explain the origins of this book, I must indulge in some personal history. When I arrived at the Museum of Comparative Zoology in fall 1984, a vigorous controversy raged the halls, between Steve Gould and Richard Lewontin, on the one hand, and E. O. Wilson, on the other. This controversy centered on the validity of what Gould and Lewontin termed "adaptationism" and the "Panglossian paradigm"—the assumption that natural selection rules evolution and that, as a result, all structures and behaviors of organisms can be understood as adaptations, evolved under natural selection for a specific function or functions. I had come to join Pere Alberch's lab, which focused on the role of developmental constraints in morphological evolution, work closely allied to Gould and Lewontin's critique of adaptationism. Yet as I thought more about the issues involved, it seemed to me that both sides of the debate had merit. This was not just a controversy to be entered into, but rather a problem I might be able to solve—a presumptuous idea, but that's how young grad students are.

I began to read widely in the historical and philosophical aspects of this debate, which in morphology has always been expressed as that over the primacy of form versus function. One day, as I rode my bicycle home from the lab, I was mulling over Cuvier's principle. It suddenly occurred to me that Cuvier's principle of the conditions for existence could be seen as *identical* to Darwin's principle of natural selection, in the sense that those organisms that don't satisfy their conditions for existence are summarily "selected against."

This book is the further development of this thought, which I believe offers a promising approach to resolving some of the persistent controversies in evolutionary biology. As I have come to realize, the principle of natural selection plays at least two different roles in evolutionary biology. On the one hand, it can be conceived of as the principle of differential survival and reproduction of genetic variants due to their differential adaptedness—that is, as a *mechanism* of evolutionary change in populations. I call this the principle of evolution by "narrow sense" selection; it is the common conception and is clearly not the same as Cuvier's principle.

On the other hand, the principle of natural selection can be conceived of as a *principle of interpretation or understanding* applied to the evolutionary process; a principle that can be applied at any level. This conception (when teleological associations have been removed) is indeed structurally equivalent to Cuvier's principle. I call this the principle of "broad sense" selection. Whatever the path of evolution from time A to time B, all organisms (as well as genes, cell lineages, characters, species, etc.) participating in this path must have been satisfying their conditions for existence. We can use this knowledge to guide our investigation. This is what I mean when I say that the principle of the conditions for existence functions as a principle of interpretation or understanding. The conditions for existence are the boundary conditions within which evolution must occur.

I argue that both conceptions of the principle of natural selection are valid, but that confusion between them frequently gives interpretations of evolution by natural selection a teleological flavor. Unlike the conditions for existence, which admit no exceptions, the mechanism of natural selection is only one factor in evolution, however important it may be. By considering the *mechanism* or *process* of narrow sense natural selection responsible for the entire path of evolution, one in effect treats the end result achieved as the *goal* of the process. This is the standard view that since the eye is adapted for seeing, its structure can be explained as the end result of a process of selection for seeing.

Gould himself was one of the worst in this respect: by asking us to consider each trait as an "adaptation" (originating by natural selection for its proper function), a "nonaptation" (originating somehow else), or an "exaptation" (originating by natural selection of a distinct adaptation or nonaptation), he accepted the division of the organism into discrete traits, some of which were the goal of past natural selection, some of which weren't. Gould had no difficulty in considering the eye an adaptation for seeing. Nevertheless, given that other evolutionary factors must have played some role in the origin of the eye over millions of years (though how great a role can certainly be debated), it simply can't be true that selection for seeing was responsible for all the change that occurred along the path to the present form.

The conditions for existence, unlike past selection, are subject to empirical investigation. For example, whatever mechanisms were responsible for the changes involved in the evolution of the eye—and I suspect that selection for seeing played a prominent if not exclusive role—it is clear that all the organisms along this path were able to see well enough to obtain the information necessary to satisfy their conditions for existence. This observation leads naturally into studies on the role of vision in the life of current organisms, with inference back to ancestors of living taxa. Such inference may not seem profound, because it does not provide a *causal* account of the evolution of vision, but it does enable us to *understand* how it was that these organisms left descendants at all, and thus stayed in the evolutionary game.

Although I am trying to avoid some of the subtleties of the issues involved here (there will be time enough for them in the body of the book), it is important to note another, related use of the principle of the conditions for existence—a use that is more closely connected with the *mechanism* of narrow sense natural selection. This use of the principle deals with the *conditions for continued existence* (persistence) *of characters in populations*. What both Darwin and Wallace pointed out was that for any feature ("variation") or variant type ("variety") to persist in a population, that feature or type must satisfy *its* conditions for existence. They assumed in particular that a feature or type will generally persist only if it is associated with greater success in the struggle for existence than its alternatives. This might be called the Darwin-Wallace principle; it is here that their thought clearly went beyond Cuvier's focus on the individual organism.

A classic example of such reasoning is the constraints we accept on our thinking about the dorsal plates of *Stegosaurus*. The function of these plates has been debated, but the assumption that they must have had *some* function can be justified on the basis that if they did not, they would not have persisted in the population. Of course, we can also assume that the individual stegosaurs must have been satisfying their conditions for existence *as organisms*.

I do not claim originality for any of the ideas in this book but seek only to put existing ideas in a more rigorous conceptual framework, a framework that helps clarify the interrelationships among the basic concepts of evolutionary biology. I am all too aware of my own limitations and how they have affected my ability to produce the book I wanted to. Nevertheless, once I got started, I had to write this book, if only to get it out of my system—and to raise questions that I think have not been satisfactorily answered by the debate as it stands today. I will be gratified if it encourages people to reexamine for themselves the relation among functional explanation, adaptedness, and Darwinian natural selection.

This book owes an intellectual debt to far too many people to mention individually, and no doubt to some I have now forgotten. Nevertheless, I must acknowledge the importance of my adviser, the late Pere Alberch, and my fellow Typologists. I owe to them my introduction to the historical and philosophical issues involved and, indeed, to "philosophical natural history" in general. Steve Gould and Richard Lewontin opened my eyes to the problem of neo-Darwinian teleology. Leon Croizat and Dov Ospovat helped me see just how this teleology enters the theory of natural selection and where this teleology came from, thus pointing the way to how to get beyond it. Finally, E. S. Russell introduced me to Cuvier and to a functionalism not rooted in past natural selection. Without all of these people, this book could never have been written.

I am also grateful to the many people have read some or all of the book at various stages of its preparation and given me comments. I would like to thank in particular Alan de Queiroz, Megan Donahue, A. W. F. Edwards, Richard Lewontin, Amy McCune, Peter McLaughlin, Will Provine, Karen Reiss, Phillip Sloan, Philippe Taquet, and J. Scott Turner. Michael Wade and Robin Waples helped by their criticism of a draft of chapter 8 submitted to (and rejected by) *Evolution*.

Two chapters previously appeared in print in different form: material in chapter 2, previously published as Reiss (2005), is included with kind permission from Keith R. Benson, editor in chief, *History and Philosophy of the Life Sciences*; material in chapter 7, previously published as Reiss (2007), is included with kind permission from Springer Science+Business Media.

A key aspect of my approach has been to include some lengthy quotes from historical sources. Excerpts from the following works are reprinted by kind permission of the publishers listed. Xenophon (1965), *Recollections of Socrates and Socrates' Defense before the Jury,* translated by A. S. Benjamin, excerpts reprinted by permission of Pearson Education, Inc., Upper Saddle River, NJ; Aristotle (1952), *Physica,* translated by R. P. Hardie and R. K. Gaye, excerpts reprinted by permission of Oxford University Press; Lucretius (1995), *On the Nature of Things: De rerum natura,* translated by A. M. Esolen, excerpts reprinted by permission of The Johns Hopkins University Press; Descartes (1985), *The Philosophical Writings of Descartes, Vol. 2,* translated by J. Cottingham, R. Stoothoff, and D. Murdoch, excerpts reprinted by permission of Cambridge University Press; Stewart and Kemp (1963), *Diderot, Interpreter of Nature: Selected Writings,* translated by J. Stewart and J. Kemp, excerpts reprinted by permission of International Publishers Co., New York; Kant (1987),

Critique of Judgment, translated by W. S. Pluhar, excerpts reprinted by permission of Hackett Publishing Company, Inc.; Rudwick (1997), *Georges Cuvier, Fossil Bones, and Geological Catastrophes: New Translations and Interpretations of the Primary Texts,* translated by M. J. S. Rudwick, excerpts reprinted by permission of University of Chicago Press; Bernard (1927), *An Introduction to the Study of Experimental Medicine,* translated by H. C. Greene, excerpts reprinted by permission of Dover Publications, Inc.; Fisher (1930a), *The Genetical Theory of Natural Selection,* excerpts reprinted by permission of Oxford University Press; Letteney (1999), *Georges Cuvier, Transcendental Naturalist: A Study of Teleological Explanation in Biology,* excerpts from this unpublished PhD dissertation included by kind permission of Michael Letteney. In all cases, copyright still obtains and all rights are reserved to the copyright owner.

I would like to thank Julia Graham and the rest of the Interlibrary Loan staff at the Humboldt State University library for going beyond the call of duty in helping me obtain some of the older literature. Valerie Budig-Markin graciously reviewed and corrected my French translations. Chuck Crumly, my editor at University of California Press, has been supportive and patient beyond any reasonable limit. Francisco Reinking and Kate Hoffman at the Press, and Aline Magee and Laura Larson of Michael Bass Associates, were a great help with the logistics of preparing the final manuscript. A grant from Humboldt State University Sponsored Programs Foundation helped with publication costs. Finally, my wife, Karen Reiss, has been there for me throughout. I'd especially like to thank her for confiscating the nascent manuscript to prevent me from working on it, thus allowing me to finish my doctoral dissertation—and for later being willing to give it back.

PART ONE

Prolegomena

1

The Problem

More attention to the History of Science is needed, as much by scientists as by historians, and especially by biologists, and this should mean a deliberate attempt to understand the thoughts of the great masters of the past, to see in what circumstances or intellectual milieu their ideas were formed, where they took the wrong turning or stopped short on the right track.

—R. A. FISHER, 1959, *NATURAL SELECTION FROM THE GENETICAL STANDPOINT*

Adaptation, natural selection, and *fitness* are the trinity of terms that form the core of the explanatory framework of modern evolutionary biology. However, unlike fundamental terms in physics, such as *mass, energy,* or *velocity,* these terms currently have no generally agreed-on meaning, either empirical or theoretical (see Ridley 2004; Futuyma 2005; Freeman and Herron 2007). This is obviously a problem; it is by no means a new one.

In 1866, seven years after the publication of the *Origin of Species,* Alfred R. Wallace wrote to Charles Darwin, proposing that Darwin eliminate the term *natural selection* from his work, replacing it with Herbert Spencer's *survival of the fittest.* As the independent elucidator of the principle, Wallace certainly did not object to the theory itself. His grounds were rather that "this term is the plain expression of the *fact;* Natural Selection is a metaphorical expression of it, and to a certain degree *indirect* and *incorrect,* since, even personifying Nature, she does not so much select special variations as exterminate the most unfavourable ones.... Natural Selection is, when understood, so necessary and self-evident a principle that it is a pity that it should be in any way obscured" (Marchant 1916, 141–142).

As the result of Wallace's suggestion, Darwin did work the "survival of the fittest" into subsequent editions of the *Origin;* nevertheless, he retained "natural selection," arguing "As in time the term must grow intelligible, the objections to its use will grow weaker and weaker" (Marchant 1916, 144). The continuing debate, 150 years after *The Origin of Species,* makes one wonder whether the term has in fact "grown intelligible."

TELEOLOGY AND NATURAL SELECTION

In this book, I try to show that the concept of natural selection is often invoked to explain evolutionary transformations for which we have no evidence that the *mechanism* of natural selection, as currently understood, was wholly or even partially responsible for the transformation. I argue that we have never been able to overcome the major weakness of the metaphor of natural selection, the one that led Wallace to object so strongly to its use. This weakness is the implication that there is, in nature, an agent with actions analogous to those of the breeder in artificial selection, a teleological (goal-directed) agent that intentionally, and with foresight, "selects" variations directed toward the improvement of the organism.

The teleology implicit in the metaphor of natural selection subtly permeates many of the most basic concepts of evolutionary biology, including, most prominently, the concepts of adaptation and fitness. I believe that this implicit teleology has led to many of the numerous objections to the theory over the years. Moreover, and most importantly, this teleology is entirely unnecessary, and does not contribute anything to our understanding of the evolutionary process; instead, it often makes us think we have explained a phenomenon when we have in fact merely restated the case in different terms.

This book is far from an argument against natural selection, since the process elucidated by Darwin and Wallace is clearly an essential part of evolution. Nevertheless, I believe that many of our conceptions of the role that the process of natural selection plays in evolution (which generally come directly from Darwin's) are unjustifiable additions to the basic process. These conceptions are often derived not from the *mechanism* of natural selection but only from the *metaphor*. Consequently, they partake of the teleology of the metaphor.

The general mode of thinking that I object to goes as follows: "character x plays a useful ('adaptive') role in the life of organism y; therefore, character x must have evolved by natural selection for this role." The definitional equation of adaptations and past natural selection is fairly standard in evolutionary biology today: "a feature is an adaptation for a particular function if it has evolved by natural selection for that function" (Futuyma 2005, 265). When combined with the assumption that useful features or characters are in fact adaptations by this definition, a teleological role for natural selection results. In this role, natural selection is inferred to have directed evolution from an unimproved (poorly adapted) past state toward an improved (well-adapted) present state, merely on the basis that the present state exists and is well adapted.

For example, consider the following passage from Ernst Mayr, one of the founders of the modern synthesis: "Pelagic marine invertebrates have a great diversity of mechanisms by which to stay afloat in the water: gas bubbles, oil droplets, or an enlargement of the body surface. In each case natural selection, which is always opportunistic, made use of that part of the available variation that led most easily to the needed adaptation" (Mayr 1982, 590). We are presented with an image of poor invertebrate larvae sinking to the bottom, desperately in need of some way to stay afloat, only to be saved by the improving force of natural selection. As pointed out by Croizat (1962), among others, this mode of thinking is not only teleological but also Lamarckian: a need gives rise to an organ to fulfill that need.

While the best solution might be to do what Wallace suggested so long ago—to completely extirpate the term *natural selection* from the lexicon of evolutionary biology—the term is by now too well established to replace. Instead, I would only propose that its use be restricted to those situations that are known to meet the criteria for the mechanism to act. These situations almost always involve current populations under study, general situations, or (possibly) molecular evolution, in contrast to particular historical transformations of phenotypes.

However, none of this discussion is particularly new or original. Many authors have argued against the historical definition of adaptation and function, and the simple equation of present adaptedness and past natural selection. Moreover, current textbooks (e.g., Ridley 2004; Futuyma 2005; Freeman and Herron 2007) are certainly sensitive to the issues involved. I believe that the continued teleological use of the concept of natural selection, in spite of the obvious problems involved, is due primarily to the absence of another evolutionary principle that can be used to interpret patterns of macroevolutionary transformation. Fundamental to my restriction of the term *natural selection* will be the reintroduction of another principle, related to and often confused with that of natural selection. This principle is founded on the concept of the necessary conditions for an organism's (or other evolutionary entity's) continued existence; it states that (by definition) the existence of any organism is contingent on its satisfaction of these conditions. The usefulness of the principle of the *conditions for existence* was first insisted on more than two hundred years ago by the great comparative anatomist, vertebrate paleontologist, and arch antievolutionist, Georges Cuvier.

A ROLE FOR HISTORY

Because of the nature of the subject matter, the approach I take in this book is at times critical, at times theoretical, but first and foremost, historical. In fact, as the book emerged, I was surprised to see how much time needed to be spent on the historical background to the story. Or rather, the historical background turned out to be a major part of the story. The issue of teleological explanation has been with us since the dawn of Western philosophy, to an extent presumably realized by all philosophers, but certainly not by most working biologists. To understand the depth of the problem of teleology, it is not enough to begin with Darwin or even with the battles of the Enlightenment. It is necessary to recognize that the problem of teleological explanation in Western science goes back to the debate between the ancient atomists, on the one hand, and Socrates, Plato, and Aristotle, on the other.

Only by understanding the long history of the debate can one hope to understand the circumstances that produced the intellectual *milieux* within which Cuvier and Darwin developed their ideas. Only through the lens of history can one see how other options might have been available, options overlooked due to the predominant beliefs of the times. Most importantly, only through history can one understand how a naturalistic, mechanistic theory of natural selection came to be associated with the design metaphor, when for two thousand years naturalistic philosophers had opposed any intimation of design in nature.

I follow the debate over design from the Greeks, through Darwin, to the modern synthesis. I show how Darwin's perspective, founded on a natural theological view in which adaptedness is something more than mere existence, influenced the formalization of his theory in the structure of mathematical population genetics, most notably at the hand of Sewall Wright. I show that the issues at stake in a number of key controversies of twentieth-century evolutionary biology—including that over Wright's "adaptive landscape" and Fisher's "fundamental theorem," the meaning and significance of genetic load, and the relative roles of natural selection and genetic drift in evolution—can be clarified by viewing these controversies in light of the principle of the conditions for existence. Furthermore, by examining this history, I show how Cuvier's principle can be used to help resolve some of the most troublesome conceptual issues in evolutionary biology today. These include the meaning of key terms such as *fitness, adaptation,* and *natural selection,* the distinction between natural selection and genetic drift, and the role of constraints in evolution.

My aim throughout is to examine, as Fisher advised, "the thoughts of the great masters of the past," to attempt to see "where they took the wrong turning or stopped short on the right track." The method I follow, which might thus be called "historico-critical," is bound to seem at times like history, at times like philosophy, and at times like biology. These days, history and philosophy of science are fields of their own, fields with their own training and specialist journals. Yet if history and philosophy of science are to do anything for science itself, one would hope that they might inform our understanding of the persistent issues of science, the issues that will not go away. The problem of biological "design," of teleology, is clearly one such issue. In its modern guise, this *is* the problem of the proper definition of function, adaptation, and natural selection. History can help us here. It can do so because the conceptual problems we face (unlike the empirical ones) are mostly of our own making, coming from uncritical acceptance of thought structures handed down to us from our predecessors.

OVERVIEW OF THE BOOK (AND HOW TO READ IT)

The book is divided into four parts. In the first part, I provide the necessary philosophical background by reviewing the forms that teleological explanations of the structure of the universe take, and examining which of these are acceptable to science. In particular, I distinguish conditional teleological explanations, which are acceptable, from purposive and deterministic teleological explanations, which are not. I show that Cuvier's principle of the conditions for existence is a conditional teleological principle.

In the second part, I pursue the debate over design from its origins in ancient Greek philosophy through the Enlightenment to Cuvier and Darwin. I show that the "Epicurean hypothesis" was an ever-present threat, a threat the design argument was intended to counter. I argue that Cuvier, with his broad training in philosophy and his distrust of "systems," was able to find the common ground between the materialists and the teleologists in his recognition of the principle of the conditions for existence. By contrast, I argue that Darwin, coming from a background in British natural theology, bequeathed us an evolutionary theory preserving some of the contradictions in natural theology, an implicitly teleological evolutionary theory in which the adaptedness of organisms is only loosely connected to their existence.

In the third part of the book, I show how the teleological aspect of Darwin's theory was translated into the mathematical language of population genetics, particularly by Sewall Wright. This teleology is exemplified by Wright's metaphor of the adaptive landscape; it is absent from Fisher's fundamental theorem. I examine the debate over genetic load, showing that confusion over the meaning of fitness arising out of Wright's (and Haldane's) teleological view of selection led to confusion between two distinct issues: the total variance in relative rate of increase available for evolutionary change, and the mean absolute rate of increase (mean "fitness") of a population. Finally, I examine the fundamental distinction between natural selection and genetic drift, starting from a consideration of how they are measured, and suggest that it may not be as fundamental as Wright suggested.

In the fourth part of the book, I step back to take a larger view and survey evolutionary biology from the perspective I have been developing. I first consider the meaning of adaptedness. I examine in particular its relation both to natural selection and to the conditions for existence, bringing in some empirical examples to ground the discussion. Next, I ask how we can talk about macroevolution without lapsing into teleology, and show that the principle of the conditions for existence can play a key role here. I end by considering some of the ways in which the conditions for existence can serve as a unifying concept in evolutionary biology, tying together such disparate fields as quantitative genetics, the levels of selection, evo-devo, ecology, physiology, and conservation biology.

Many of the reviewers of the manuscript asked me, Who is this book for? This is a valid question, as my approach and subject shift from history to philosophy to biology and back throughout the book; moreover, some parts involve mathematics (if only simple algebra), which can put off many readers. I would answer that this book is written, first and foremost, for *biologists*. Biology is the scientific study of life, and a key question for all biologists must therefore be, How do we treat life, which is an extremely complicated physical phenomenon, scientifically? That said, this book may perhaps be read differently by different audiences. Philosophers of biology may want to merely skim the material in the first half of part 2, as it is no doubt familiar to them already. Working biologists may want to turn directly to part 4, where the practical implications of the views I develop are worked out. Finally, general readers (if any are left out there) may want to read the whole book but skip the mathematical sections of part 3, a task I have made easier by putting most of the equations in boxes so as not to interfere with the main flow of text. I think that each of these groups, despite their diverse interests, may find something of value here.

· · ·

In concluding this introductory chapter, I wish to stress again that I make no claim for originality of thought in this book. My goal is certainly not to promote any new theory of evolution, but rather to argue for rigorous and clear definitions of key terms that will provide a sound basis for investigating evolution as a phenomenon. To this end, I have provided a glossary at the end of the book and done my best to adhere to the definitions therein. The paths I have explored in the last twenty years of following the principle of the conditions for existence where it would lead me have led to much clarification of my own thoughts, and have shown me many surprising and satisfying connections. I can only hope that the theoretical framework provided by the principle of the conditions for existence seems as promising to others as it does to me.

2

Philosophical Background

Natural History nevertheless has a rational principle that is exclusive to it and
which it employs to great advantage on many occasions; it is the conditions
for existence, *or, vulgarly,* final causes.

—G. CUVIER, 1817, *LE RÈGNE ANIMAL*

This book is about how evolutionary biology can rid itself of illegitimate teleological thinking. Such teleological thinking—rooted in the metaphor of natural selection as the designer of biological systems—is widespread, perhaps even more
so among philosophers of biology than biologists. In this chapter, I provide the
necessary philosophical background by defining just what I mean by teleological
thinking and why I think it is a problem.

TELEOLOGICAL EXPLANATION: INTENTIONAL,
REPRESENTATIONAL, AND CONDITIONAL

Human intentional action is perhaps the psychological source of teleological thinking in general. We commonly say, for example, "I went to the store to buy bread"
and consider this an explanation of our actions. Almost everyone agrees that from
a scientific standpoint this is not problematic and involves no reverse causation, or
cause that comes after its effect (a common concern of critics of teleology). My desire for bread, combined with my belief that I can get bread at the store (both of
which occur prior to my going to the store), explains the act of getting there and
buying the bread. This is *intentional teleological explanation*, explanation of future-
directed action based on positing an *intentional system* with a prior mental representation of the goal of the action and the means to get there, as well as the actual
means to carry out the act, as part of the efficient cause of the action.

Intentional teleology is also involved in explaining the production (and use) of
artifacts. If I make something, say, a clock, my desire for a device to tell time (end)
together with my belief that a device built in a certain way will allow me to tell the

time (means), and my ability to actually implement these means, explain the production (and thus the existence) of the clock.

If we treat the intentional teleology involved in acts and the making of artifacts as the paradigm for teleology in general, the issue for teleological conceptions in biology is whether nature can have intentions, or at least something analogous to intentions, and thus to what degree we can unite explanations of intentional acts and artifacts with explanations of instinctive behaviors of organisms and of organisms themselves.

There are really two distinct questions here. One is the question of whether representations of goals and the means to achieve them exist outside the human mind. If such representations exist, and we can determine what they are, then we can explain the behavior of systems containing them with reference to such representations. Such explanations might be called *representational teleological explanations*. Intentional explanations are merely a special case of this general type of explanation—the case in which the representations are made to an internal consciousness. Explanations based on internal representations of the goal state (souls, vital powers, entelechies, developmental programs, etc.) have been common in the history of biology.

The second question, and the one I am interested in here, is entirely different from considerations of such representational teleology operating within organisms. It is the question of to what degree the evolutionary process that resulted in the organisms we see around us today can be used to explain features of those organisms— in particular, their apparent complexity and adaptedness—in a way analogous to the use of the representation (intention) as efficient cause in representational (intentional) explanation. These questions are distinct, though they have often been confused, no doubt because part of organismal complexity is of course behavior that appears to us "adaptive" or "goal directed," and thus dependent on the existence of some internal representation of a goal state.

However, in the same examples used to illustrate intentional or representational teleology, another sort of teleological explanation is possible. Given that I do go to the store and buy bread, my going to the store (means) can be explained as a necessary condition* for buying bread (the end result). Reciprocally, my buying bread (end result) can be explained by the prior occurrence of the necessary conditions (means). Moreover, the means can be broken up into parts, and the partial actions that make up this complex action (getting in my car, starting it, following a particular route, etc.) can all also be explained as necessary conditions for the completion of the specific act of which they were a part and thus for the result. Given that

---

* By "necessary condition," I don't mean that there was no other possible way to get bread, but rather that in the context of the actual act, going to the store was instrumental in getting the bread in a way that other activities of mine, such as chatting with a friend I ran into there, were not.

I do make a clock (means) that has hands that move in a certain way (end), my making it was a necessary condition for its having hands that move that way. Moreover, my giving the parts of the clock a certain shape and certain relationships can be explained as necessary conditions for the movement of the hands in just that way. This is what I will call *conditional teleological explanation*. Conditional teleological explanation is a form of causal explanation, but it is not causal in the same way that explanations based on physical laws are—a specific end result must be proposed for the significance of the preceding actions to be clear.

Importantly, once an overall goal, end, or result is given or assumed, conditional teleological explanation does not depend on any intention or representation. If I go to the store and buy bread, my going to the store can be used to explain my buying bread, whether I intended to buy it or not. If I make a clock that has hands that move in a certain way, their movement can be explained by my making it that way, even if they don't move the way I intended. Such explanations of actions do not seem terribly informative, but this is because they are explanations of entire actions. What makes conditional teleological explanations useful is that the means-ends relationship can be broken up into smaller parts, thus making them *functional explanations*. The partial actions that make up a complex action can each be explained as means to the complex action as an end, and the parts of an artifact can be explained as means to the action of the artifact as a whole as an end. Getting in my car, starting the engine, and so forth, can be explained as means to the end of buying bread; having gears of a certain size and relation, and so forth, can be explained as means to the end of the clock's movement. In the case of artifacts, the independence from the intentions of an agent in such functional explanations is especially clear; given a clock that has hands that move in a certain way, the contribution of the parts to the overall motion of the clock can clearly be examined with no knowledge of its creator's intentions (though some end, e.g., a specific type of motion, must be assumed). As Peter McLaughlin (2001) has pointed out, the problem of functional explanation is not so much the problem of future direction as it is the problem of the relation of the part to the whole, of reductionism versus holism. In conditional teleological functional explanations, the whole is taken as explanatorily prior to the parts; the parts are explained as conditions for the whole, and the whole as conditioned on the parts.

In fact, it is the means-end relationship involved in conditional teleology that informs intentional teleology. If I desire bread, it is my belief that a certain set of conditions must be satisfied as means to that end that informs my actions. If I desire a clock, my belief that a certain structure will work to tell the time informs my actions. In intentional action we want something to happen, and we do what we think is necessary (we supply what we think are necessary conditions) to bring it about. A mental representation of conditional teleological relations thus plays a role in intentional action (see Rosen 1985). However, such representations are not the same as the actual conditional teleological relations. Shipwrecked on an island, we can pray to God, believing that this is a necessary condition for our salvation, but this doesn't mean that when we are saved, our prayers were of any use.

More generally, any system that behaves purposively may encode a representation of the goal state and have the means to get there—for example, the set point of a thermostat and the mechanism that turns on the furnace when the temperature falls below it. But again, there is no necessary connection between the representation and achieving the goal: if the thermostat sends a current to turn on the furnace, but the furnace is broken and the house warms up by the sunlight coming through the windows, the representation had no actual role in the event.

Thus, while conditional teleological explanation focuses on future ends, it is free from any assumptions about desires, beliefs, or representations. Moreover, because the ends are known to have occurred, no problem of the "missing goal object" occurs. When a prior representation of the end and the means to get there agrees with the end achieved and the means actually used, representational and conditional teleological explanations are both possible, but this is not always the case. When the goal represented is frustrated (the case of the "missing goal object"), only representational teleological explanation is possible. When a set of events results in conditions necessary to a purpose being fulfilled with no role of a corresponding representation, only conditional teleological explanation is.

TELEOLOGY AND NECESSITY

What is it that makes an explanation of a system or process teleological rather than simply nomological—that is, occurring in accordance with natural laws? This distinction turns out to depend on the type of necessity involved. As Aristotle first recognized, teleological explanations (both representational and conditional) involve only a "hypothetical necessity," rather than the "absolute necessity" involved in deterministic or lawlike explanations (Randall 1960; see chap. 3). In other words, teleological explanations must be explanations of events that are contingent (for us), because if the event is determined by known laws—and associated boundary conditions—nothing is left for the end to explain that hasn't already been accounted for by these laws. Hypothetical necessity invokes only necessary conditions for a process to occur, whereas absolute necessity invokes sufficient conditions. Thus, in order to buy bread, it is necessary that I go to the store, but it is not necessary that I go to the store to buy bread in any absolute sense; I am not determined (at least by laws we are familiar with) to go. Confusion between these two modes of necessity, in addition to confusion between representational and conditional teleology, has also resulted in scientifically objectionable teleological explanations.

If one doesn't recognize the distinction between absolute and hypothetical necessity, one effectively treats teleological explanations as if they were deterministic.

This fallacy might be called *deterministic teleology*. Such deterministic teleology can occur in two ways. First, explicit representations of ends may be considered to determine the result, although the causal connection between the representation and the end is not clear; this is deterministic teleology proper. For example, I might treat my representation of buying bread as the deterministic cause of my going to the store, although I can't trace the deterministic causal chain from my mental representation through nerve impulses to muscular movements that result in my getting to the store.

A second, more subtle deterministic teleology can occur when known mechanisms are considered to account for a particular result, even though it can't be derived from them; this might be called *teleological determinism*. To give an historical example, the young, precritical Kant presented a cosmogony in which he explained the origin of the structure of the solar system on the basis of Newton's laws, although he did not derive this structure from the laws (Kant 1981; see chap. 4).* Such a false determinism is a representational teleological explanation in disguise; when we view an achieved end as necessary (in an absolute sense), even though we can't derive that necessity from natural laws, we use an implicit representation to complete the determination of the system. The location of this representation is often rather vague; it may be conceived of as in the laws themselves, the matter, or the mind of God. Such teleological determinism confuses the absolute necessity of the laws with the hypothetical necessity of the results.

In summary, we see that there are two primary dangers to be avoided in the use of teleological explanations in science: (1) the danger of attributing explicit intentions (or at least representations of the goal state) to nature where they don't exist and (2) the danger of confusing a teleological with a deterministic explanation and thus promoting a false, teleological determinism.

A TAXONOMY OF TELEOLOGY

Two types of teleology are commonly recognized: purposive and functional teleology (Beckner 1967; Woodfield 1998). This distinction has been generalized from the two major situations where teleological explanations have been considered appropriate: intentional behavior and the production and use of artifacts. *Purposive* or *goal-directed teleology* involves the behavior of a system with respect to a goal of the system as a whole. As is widely recognized, purposive or goal-directed teleology is *intrinsic* or *internal* teleology, in that the end that we recognize is a future state of the system itself. It is thus monadic (Boylan 1986), in the sense that it is unitary and incapable of further division.

---

* In contrast to Kant, Newton himself wisely refrained from considering his laws adequate even to account for the stability of the solar system, because he could not demonstrate that stability from them.

In contrast to purposive teleology, *functional teleology* involves behaviors of parts that contribute to some activity of a larger system to which they belong. Thus, we say that the spark plugs function to ignite the gasoline within the larger system of the turning engine or driving car, or the heart functions to pump blood within the larger system of the living organism. Functional teleology is *external* or *extrinsic* teleology, in that the function of the part is always relative to an activity of a larger containing system. For this reason, Boylan (1986) called this type of teleology *systemic teleology*.

The distinction between purposive and functional teleology can be summarized by noting that purposive teleology focuses on ends; functional teleology, on means. However, all teleology, whether functional or purposive, necessarily involves reference both to an end (goal) and the means to that end, and it is clear that more must be going on here. To clarify this situation, we must make another important distinction that cuts across the distinction between purposive and functional teleology in general. This is the distinction between *teleological systems* and *teleological explanations*. One needs to distinguish between positing the actual existence (in some sense) of ends in the world and our use of ends as principles of understanding.

These combined distinctions can be most easily summarized in the form of a table, relating teleological (functional and purposive) systems and teleological (conditional and representational) explanations (table 2.1). We can see that five types of explanations are possible for the behavior of such systems, though not all are valid:

- *Deterministic explanation.* In a deterministic/reductionistic explanation, the behavior of the overall system can be inferred from the properties of the parts and the laws of their interaction, together with appropriate boundary conditions. Such explanation is obviously valid, given that we can define the parts and their properties in a way that the system properties can indeed be inferred from them and can specify the relevant boundary conditions. For example, we can construct a theoretical model of a pendulum clock, which derives its operating behavior from the laws of gravity, together with initial and boundary conditions such as the position of the pendulum, the mass of the pendulum, the ratios of the gears, and the local gravitational field. Similarly, we can construct a theoretical model of a system composed of a thermostat, an electrical circuit controlling a furnace, and a house being heated, in which the operating behavior is derived from properties of the metal, laws of electrical conduction, and so forth, together with initial and boundary conditions such as ambient external temperature and the rate of heat gain or loss.
- *Teleological deterministic explanation.* As already discussed, when one attempts to give a deterministic explanation but does not specify the relevant laws and boundary conditions in a way that allows determination of the

TABLE 2.1 A Taxonomy of Teleology

Explanation	System	
	Functional Interacting parts with characteristic behavior	*Purposive/representational* Contains internal representation of goal
Deterministic Behavior explained by properties of parts and laws of interaction	Valid—if laws and properties can be found and boundary conditions specified	Valid—if laws and properties can be found and boundary conditions specified
Teleological deterministic Properties of parts used to explain behavior, but they don't	Invalid—implicit representation used to complete determination	Invalid—implicit representation used to complete determination
Representational (internal) Internal representation of behavior/goal explains it	Invalid—no such representation exists	Valid—though may be difficult to establish *which* behavior/goal is represented internally; purposive functions determined with respect to overall goal
Representational (external) External representation of behavior/goal explains both behavior and system structure	Valid—if system has been designed; design functions determined with respect to overall goal	Valid—if system has been designed; design functions determined with respect to overall goal
Conditional/functional System behavior explained by necessary conditions of parts and their interaction	Valid—conditional functions determined with respect to a particular behavior	Valid—conditional functions determined with respect to a particular behavior

overall system behavior, one is guilty of a false, teleological determinism. The example mentioned earlier was Kant's verbal model of the origin of the solar system on the basis of Newton's laws. Such an "explanation" is at best a sketch of the possibility of a deterministic/reductionist explanation, a hypothesis, not such an explanation itself. This type of pseudodeterministic explanation is invalid both for functional and for purposive systems. Historically, many evolutionary explanations have been of this type

· *Internal representational explanation.* In this mode of teleological explanation, the behavior of the system is explained on the basis of an internal representation of the behavior or goal within the system. This makes no sense for systems that are merely functional and have no such representation of a goal, but for purposive (representational) systems, it does make some sense to explain their behavior with respect to this goal. Thus, it makes sense to say that the thermostat turned the furnace on in order to bring the house back up to temperature or that I went to the store in order to buy bread. Within such an explanation, the *purposive function* of parts of the system can be determined with respect to the overall goal. Thus, the function of my pulling the car out of the driveway is to head toward the store, with respect to my overall goal of going to the store to buy bread. As with all teleological explanations, such an explanation is hierarchical, in that functions can be considered purposes with respect to lower-level (smaller) parts of the system: my turning the key to start the car has a purposive function not only with respect to the overall purpose of buying bread but with respect to the subsidiary purpose (function) of pulling the car out of the driveway.

· *External representational explanation.* In this mode of teleological explanation, the behavior and structure of a system are explained on the basis of an external representation of this behavior and structure. This is the explanation based on intentional design. It is valid both for functional and purposive/representational systems, given that they are in fact designed (i.e., we can explain the behavior of both a car engine and a guided missile by the relevant intention, together with the assumption that the designer accurately understood the behavior that would result from the design). With respect to the overall representation of the intended behavior of the system, *design functions* are the intended functions of the parts. Again, these are hierarchical explanations: the design function of the engine is to provide power for the car to move down the road (intended system behavior or goal); the spark plugs have the design function of igniting the gas not only within the larger system of the car but within the smaller system of the engine.

· *Conditional teleological (functional) explanation.* In this mode of explanation, the overall goal or system behavior one is attempting to explain must be given; given this behavior, the necessary conditions for the behavior are objective facts that explain its occurrence. Such an explanation is valid both for merely functional systems and for purposive/representational systems. For example, given that the engine is running, and this is the behavior of the system we are interested in, with respect to that behavior, the (conditional) function of the spark plugs is to ignite the gas. The fact that the spark plugs are igniting the gas helps explain the running of the engine. Likewise, given that the thermostat does turn on the furnace, and the relatively constant temperature of the house is the system behavior we want to explain, with respect to this behavior the (conditional) function of the thermostat is to turn on the furnace. Such conditional explanation has often been confused with explanation based on external representations (design), presumably because in the case of artifacts, the system behavior we choose to focus on is generally the design function of the system for humans.

What is clear from this overview is that both functional and purposive systems are generally subject to the same sort of explanations. This is none too surprising because purposive systems are merely a type of functional system. The major difference, of course, is that only purposive systems can have their actions explained on the basis of internal representations of the goal state. Much of the confusion in discussions of teleology in biology has come from confusion between purposive and functional explanations of purposive systems, because the same system undergoing the same behavior can be understood both purposively (based on an internal representation of its goal) and functionally. In fact, the behavior of a designed purposive system like a homing torpedo can be explained in four distinct but valid ways: (1) deterministically, (2) based on the internal representation of the goal state, (3) based on the external representation of the goal state (design), and (4) functionally.

I have focused on artifacts so far, rather than organisms, because understanding organisms is more difficult. Given that we are trying to achieve a naturalistic explanation, and that a deterministic explanation of organisms is beyond our ability, only internal representational and conditional/functional explanations are possible. I have already noted that internal representational explanation of organisms is problematic, because it is difficult to determine what goals are internally represented; in any case, such explanation does not really concern me in this book. Instead, we need to focus on functional explanations. The big problem for functional explanation of organisms is, of course, determining an overall goal to which the functions of the parts can be relativized. In the case of artifacts, this goal is derived from the intention/purpose of the designer or user, but such relativism to a goal is clearly antinaturalistic in the case of organisms—we no longer want to base functions on God's design. Alternatively, we can just relativize the functions to any activity we see the organism carrying out. But doing so creates the problem that there is no way to distinguish between "true" functions and false ones, outside of a given assumed purpose. There would be no way to say that pumping blood is a function of the heart, but making heart sounds is not. Thus, functions become entirely relative to our interests (as in the approach of Cummins 1975).

Instead, I will argue that one can make functional/conditional explanations of biological systems nonarbitrary by grounding them on the continued existence of the system itself. An organism is a complexly organized system that depends on the activity of its parts for its continued existence (survival). The activities of the parts can thus be relativized to that continued existence as their end result or "goal." This is the principle of the conditions for existence.

THE PRINCIPLE OF THE CONDITIONS FOR EXISTENCE

For Georges Cuvier (1769–1832), the founder of vertebrate paleontology and the leading comparative anatomist of his generation, the principle of the conditions for existence (*conditions d'existence*) was the fundamental guide for work in natural history. Cuvier's emphasis on this principle was in keeping with his entire approach to zoology, which always involved an Aristotelian (and Kantian) stress on the functional unity of the organism. I will examine Cuvier's life and thought in more detail in chapter 5; for now, I will focus on understanding the principle itself.

The principle was first broadly introduced in his *Leçons d'anatomie comparée* (Cuvier 1800–1805, 1: 47, translation adapted from several sources):

> Of what use would sensation be to us, if muscular force did not help it, even in the most trifling circumstances? What use could we make of touch, if we could not carry our hands toward the palpable object? And what should we behold if we could not turn our eyes or head at pleasure? It is on this mutual dependence of the functions, and this reciprocal help they lend each other, that the laws that determine the relations of their organs are founded. . . . For it is evident, that an appropriate harmony among the organs acting upon one another is a necessary condition for the existence *(condition nécessaire de l'existence)* of the being to which they belong, and that if one of these functions were modified in a way incompatible with the modifications of others, that being could not exist.

And, as quoted in the epigraph to this chapter, Cuvier elaborated further on this principle in his *Règne animal* (Cuvier 1817, 1: 6):

> Natural History nevertheless has a rational principle that is exclusive to it and which it employs to great advantage on many occasions; it is the *conditions for existence*, or, vulgarly, *final causes*. Since nothing can exist that does not fulfill the conditions that render its existence possible, the different parts of each being must be coordinated in such a way as to render possible the existence of the being as a whole, not only in itself, but also in relation with other beings, and the analysis of these conditions often leads to general laws which are as certain as those which are derived from calculation or from experiment.

As Cuvier recognized, this is a teleological principle, but if one considers it in light of the distinctions drawn earlier here among representational, deterministic, and conditional teleology, it is clear that it involves only a conditional teleology. If we observe that an organism exists, then it must be *possible* for it to exist, but this does not mean it was *designed* to exist or that it *had* to exist. The problem for a naturalistic teleology has always been an objective standard of value, for all teleological thinking involves not just ends but valued ends. What Cuvier recognized is that existence itself is an end that objectively exists in nature (see chap. 5). The objective existence of organisms is a solid fact on which to ground our (conditional) teleological explanations of that existence. When the conditional functions of the parts of a complex system are relativized to the continued existence of the whole, we might call them *natural functions*; biological functions are clearly of this type.

It is worth noting that my translation of the *conditions d'existence* as "conditions *for* existence" is unorthodox; almost all English translations have instead rendered it literally, as the "conditions *of* existence." However, notable exceptions have occurred, including Lee (1833), who uses "condition to existence"; Corsi (1988), who uses "prerequisites of existence"; Grene and Depew (2004), who follow Corsi and use "prerequisites of existence"; and, most recently, Rudwick (2005), who uses "conditions necessary for existence." As with other attempts to convey a more accurate idea of Cuvier's meaning, I have translated it thus because this gives the term less ambiguity: the "conditions *of* existence" can in fact mean two quite different things.

The first of these meanings, and that of Cuvier, is that of the necessary conditions *for* the existence of an organism. These conditions are a characteristic of the organism. For an animal, *obtaining enough food* is a condition of existence in this sense, where "enough" is obviously relative to the particular organism in question. The second possible meaning is that of the environmental conditions, or circumstances, *in which* an organism exists (e.g., Semper 1881). The *types of other organisms present* in the environment of an animal are conditions of existence in this sense. The reason I have preferred to translate the phrase as "conditions for existence" is that this latter meaning, entirely different from Cuvier's (Russell 1916, 34; see chap. 5), is thereby excluded.

It is interesting to note that Herbert Spencer (1864, 445), in his discussion of the "survival of the fittest," implicitly recognized the distinction between the conditions for existence (his "conditions *to* life") and the environmental conditions (his "conditions *of* life"); the destruction of the "diseased and feeble" through their "failure to *fulfil* some of the conditions *to* life, leaves behind those which are able to *fulfil* the conditions *to* life, and thus keeps up the average *fitness* to the conditions *of* life" (my emphasis).

THE CONDITIONS FOR EXISTENCE
AND THE WEAK ANTHROPIC PRINCIPLE

To conclude this discussion, it is worth noting certain parallels between the principle of the conditions for existence and the weak anthropic principle (WAP), which has had some vogue in astrophysics (see especially the interesting review by Barrow and Tipler 1986). This principle was defined by Hawking (1988) as follows: "The weak anthropic principle states that in a universe that is large or infinite in space and/or in time, the conditions necessary for the development of intelligent life will be met only in certain regions that are limited in space and time. The intelligent beings in these regions should therefore not be surprised if they observe that their locality in the universe *satisfies the conditions that are necessary for their existence*" (124, my emphasis).

Even in astrophysics, this view is closely tied in many minds to the strong anthropic principle (SAP), which holds that intelligent life must evolve somewhere, and even the final anthropic principle (FAP), which holds that life progresses to the highest possible consciousness. Such extensions have justifiably led to ridicule, as in Martin Gardner's (1986) invocation of the "completely ridiculous anthropic principle" (CRAP). These alternative anthropic principles obviously make the mistake of teleological determinism—just because it happened, doesn't mean it had to have happened (unless we can show that it did!). Nevertheless, it is hard for anyone to argue with the weak anthropic principle, which only admits a conditional teleology.

This general approach embodied in the weak anthropic principle has been classified as based on recognition of an "observer selection effect" (Bostrom 2002). The principle of the conditions for existence is just such a principle, for it merely states that we should not be surprised if organisms are constructed in such a way as to allow them to satisfy the conditions for their existence, since if they could not, they wouldn't be around for us to observe (see McClellan 2001). With organisms, however, unlike the universe as a whole, we are in the strong position of being able to observe multiple similar systems and even to manipulate them to determine just what those conditions are. This gives us a path to actually doing science. In astrophysics, with only one observable universe and one known example of life, we can only use the weak anthropic principle to avoid surprise at certain facts.

NATURAL SELECTION AND THE ARGUMENT FROM DESIGN

The classical *argument from design* posited an external representational explanation for the structure and behavior of organisms: God designed organisms like we design a clock, and his desire to make just that organism explained the existence of particular organisms. Likewise, specific features of the organism (e.g., valves in the veins) could be explained in terms of their design function—that is, by their intended contribution to the overall effect they were designed to produce (life). That God's plans worked just as intended—that intention and actuality coincided—could not be doubted. After all, an omniscient and omnipotent Creator could not have failed to achieve his purpose, or only accidentally hit on something that works.

We know now that organisms were not designed; they have evolved. The existence, structure, and behavior of organisms are thus to be explained by the evolutionary processes that led up to it. But this knowledge leaves the biologist in a real muddle over teleology. Did the advent of an evolutionary perspective somehow eliminate teleology from biology? Or did it naturalize it? Does it make sense to talk of function (or even purpose) without a designer?

Many people, beginning with Darwin himself, have tried to answer this question by metaphorically comparing natural selection to the designer. Thus, one might say that the valves in the veins were put there by natural selection. This point of view is common in biology and biological philosophy, where an adaptation or function is usually defined as the product of past natural selection. Recall the quote from Michael Ruse which formed the epigraph to the preface: "The metaphor of design, with the organism as artifact, is at the heart of Darwinian evolutionary biology" (Ruse 2003, 266). I would agree—but while Ruse embraces this metaphor, I see it as a serious problem.

Metaphors may have heuristic value in science, but they are dangerous if taken literally. This is especially true of the metaphor of intentional design when used for natural systems. Nothing in nature corresponds to the designer of artificial systems; nature has no intentions. Instead, the teleology we can posit in nature corresponds only to the *conditional teleology* involved in an intentional act. There is no intention or design, but given a result (e.g., survival), the conditions necessary for that result clearly happened to occur and can be used to explain that result. Legitimate (conditional) teleological analysis in biology consists of the following: We see some result, and we try to understand that result by the conditions that make it possible. When we see a living organism, we try to understand its life by the conditions that make that life possible. One of these conditions, perhaps, is the presence of valves in the veins. These valves (or, rather, their effects) can thus be understood as conditions for the existence of the organism.

In discussing possible modes of explanation, I pointed out that a false, teleological determinism results when one explains a result based on deterministic laws, without being able to derive the result from those laws. I believe that the metaphor of design by natural selection has fostered such a false determinism in biology. While natural selection is indeed a mechanism of evolution, one can't just ascribe achieved results of the evolutionary process to natural selection. If one does so, natural selection effectively becomes a (representational) teleological agent, aiming toward the achieved results.

Why has this been permitted to occur? To understand, we must look at the other side of the classical argument of design. Not only can one explain organismal features *from* God's design, or intention (the heart was put there to pump blood), one can argue from the order, end-directedness, adaptedness to function, and beauty of the world *to* design, and thus to a designer.* Order, end-directedness, and adaptedness appear to be empirical features of the world and, as such, to call for some explanation. Currently the customary explanation for these organismal properties is, of course, a historical process of natural selection, viewed as analogous to intentional design. It has been the assumed necessity for a historical explanation for these features that has given natural selection its teleological role in Darwinian evolutionary theory.

---

* The distinction I am making between the argument *of*, *to*, and *from* design is not standard, but neither is any other usage—all of these phrases have been used interchangeably by various authors. For a similar distinction, see Ruse (2003).

However, the argument to design fails, even without natural selection. This is because the problem that the argument to design—and Darwin—set out to solve does not really exist. In particular, the argument ignores the fundamental difference between artifacts and organisms. The teleology of artifacts like a clock is externally imposed and relative to the goal of some assumed agent. Given the goal, conditional teleology can explain how that goal is realized, but it will not get you the goal. By contrast, the conditional teleology of organisms is intrinsic, in that the only "goal" we need recognize is the continued existence of the system itself. When the only end we posit is continued existence, or survival, the order, end-directedness, and adaptedness to function of the world can be explained not as the product of design but as a chance result that is merely a condition for its continued existence. This point has been recognized by antiteleologists since Democritus and Epicurus (see chap. 3). What is most surprising about the historical path that led to the mess we are currently in is that Darwin's natural selection stood in for the creationists' designer, rather than the antiteleologists' chance subject to the conditions for existence.

THE CONDITIONS FOR EXISTENCE
AND EVOLUTIONARY EXPLANATION

How, then, is Cuvier's principle—developed in a nonevolutionary context—relevant to present-day biology, which is evolutionary? Spencer's passage quoted earlier gives us a clue. If Cuvier was wrong in thinking that evolution was impossible because the functional integration of the organism was such that any change in one part would render it unable to satisfy its conditions for existence, he was right that any evolutionary change must be subject to these conditions. The conditions for existence of organisms are in fact part of the range of boundary conditions within which the evolutionary process must occur. The unadapted can't satisfy their conditions for existence and die; the adapted satisfy their conditions for existence and live. Adaptedness from this perspective is not a product of evolution, it is a condition for evolution. In this broad sense of natural selection, then, the principle of natural selection may be considered identical with the principle of the conditions for existence.

However, lest we carry this identification too far, it is important to note another aspect of the principle of natural selection, which has to do with the conditions for the continued existence of traits in populations. This aspect is what Jean Gayon (1998) has called the "hypothesis of natural selection," which is fundamentally a hypothesis about the genetic physiology of populations. The basic postulate is that for a trait to continue to exist, organisms characterized by that trait must survive and reproduce at a higher rate than alternatives (must be "naturally selected"), and the trait must be heritable. As I noted in the preface, it was in the fundamental recognition that for any feature ("variation") or variant type ("variety") to persist in a population, that feature or type must satisfy *its* conditions for existence that Darwin's (and Wallace's) principle went beyond anything in Cuvier's. This aspect of natural selection provides the basis for much of modern evolutionary theory, from evolutionarily stable strategies to sex ratios. Nevertheless, it does not give us a reason to compare natural selection to a designer. Rather, it invites us to provide conditional teleological explanations for the existence of specific traits, which—as I will try to show in later chapters—is in fact what most successful evolutionary explanations have been.

Returning to the question of functional explanation, the principle of the conditions for existence helps explain why we feel that functions should explain the existence of parts of organisms—they do, given that the system itself continues to exist, and the function helps explain that continued existence (compare Wouters 1995, 2007; Asma 1996; McLaughlin 2001). The beating of the heart in fact explains its own continued existence in two ways: (1) *as an individual organ*, because if it stopped beating, the organism would die and the heart would, too; and (2) *as a trait in a population*, because if organisms with beating hearts died more frequently than those without, hearts would disappear from the population (given certain assumptions about their genetic basis). However, this view is not the same as saying that organisms can be assimilated to artifacts, where the design function can explain the existence of the part, because the causal basis is entirely different.

In organisms, where only conditional teleology applies, organs and behaviors have functions only to the extent that they do actually contribute to the continued existence of the organism. Many have felt that functional ascriptions should be normative—that the statement "The function of the heart is to pump blood" should imply, for example, that a heart that is not pumping blood still has that function. There is no reason to accept such normativity (Wouters 1999; Amundson 2000; Davies 2001). Normative concepts are derived from intentional teleology and the goal that was aimed at. Nature doesn't aim at anything and thus can't miss. In other words, it makes sense to say that the heart is malfunctioning when a heart attack occurs and the individual dies. However, an arrhythmia is only a malfunction if it impacts survival—normative concepts based on the average or common condition are not compatible with evolution. Thus, while for Cuvier normative concepts made sense, for us they don't. Functional explanation in biology must always be relative to the continued existence of the organism (or some other evolutionary entity).

Nevertheless, as I noted in discussing the weak anthropic principle, an important distinction can be made among systems that are amenable to conditional teleological explanation. What makes functional explanations in biology testable is that the systems to which we apply functional explanations are members of classes of similar systems. If we have many organisms with beating hearts, we can stop the heart from beating in one and examine the effect, then apply the knowledge gained to other similar organisms. However, if we had only one organism, we would be severely limited in the methods we could use to establish the functions of parts. This is almost the situation we are in with respect to the overall evolutionary process, which happens only once. In a recent discussion, Francisco Ayala (1999) distinguished determinate from indeterminate natural teleology, contrasting the individual development of a chicken with the evolutionary development of a species. Now, most importantly, these processes can be distinguished as the contrast between a representational system and a nonrepresentational one. However, if we remove considerations of intrinsic representational teleology (developmental programs, etc.), what distinguishes these processes is primarily that we have multiple examples of chicken development to study, with similar initial and final conditions, but only one example of the evolutionary process:

THE FUNCTION DEBATE

In the previous sections of this chapter, I have promoted the distinction between representational and conditional teleology and use of the principle of the conditions for existence as a way to solve some of the persistent problems with function and teleology in biology. Of course, this is by no means a virgin field—there is a large literature on the scientific interpretation of function within the field of the philosophy of science. Of all the recent work, it will be most helpful to look at that of Arno Wouters (1995, 2003a, 2005a, 2005b, 2007), both because his approach is closely allied to mine, and because he considers a broad range of possible definitions of function and understanding of functional explanation. In a recent review covering the field as a whole (Wouters 2005a), he distinguishes five main philosophical approaches to the analysis of biological function:

- The "causal role" or "systemic approach" of Cummins (1975), in which a function is a contribution to a capacity of a complex system
- The "goal contribution approach" of Nagel (1961) and Boorse (1976, 2002), in which a function is a contribution to achieving or maintaining a goal state of the overall system
- The "life chances approach" of Canfield (1964, 1965), Ruse (1971, 1973), Wimsatt (1972), and Bigelow and Pargetter (1987), among others, in which a function is an effect that increases the life chances of the function bearer
- The "etiological or historical approach" of Wright (1973), Millikan (1989), Neander (1991), Griffiths (1993), Godfrey Smith (1994), and others, in which functions are past effects that explain the current presence of the function bearer
- The "non–historical selection approach" of Walsh (1996) and Walsh and Ariew (1996) (Wouters also includes Kitcher 1993 here), in which functions are seen as effects for which the function bearer is selected (in certain circumstances)

Where does the view I have promoted here fall in this classification? A key distinction made by Wouters is that between individual- and population-level definitions of function. Both the "etiological" and "non–historical selection" approaches are population-level definitions; functions are defined as those features that have been positively selected in the past or are positively selected in the present, where selection is defined as a population-level process. Even more importantly, as I have argued at greater length elsewhere (Reiss 2005), Larry Wright's "etiological" views also contain an implicit teleology. Because Wright grounds his definition of function by ascribing achieved results to a selective process he knows only by these results, he is implicitly supplying an end for the process. This is the same problem we saw exemplified by the standard definition of "adaptation."

By contrast, the view I am promoting here based on Cuvier's principle of the conditions for existence is clearly an individual-level definition and centered in the present. And while one could define function in terms of intrinsic purposiveness of biological systems, such purposive functions are not the sort I am interested in here, so that the goal contribution approach is not applicable. The two that are left are the systemic and the life chances approaches.

The systemic approach of Robert Cummins (1975, 2002), also called the "causal role" approach (Amundson and Lauder 1994), has much to recommend it. By focusing on systems as chains of "capacities," it shows a clear recognition that functional understanding is based on chains of conditional necessity, and it is accompanied by an incisive critique of the teleological underpinnings of the etiological or selected effects view. However, from the perspective of the principle of the conditions for existence, it can be faulted for not recognizing the continued existence of the system itself as a valid end to which the functions of all parts of the system can be relativized, thus resulting in a purely subjective or relative system of functions.

Likewise, the life chances (or what might equally well be called the "individual fitness") approach to function (Canfield 1964, 1965; Ruse 1971, 1973; Wimsatt 1972; Bigelow and Pargetter 1987) also has much to recommend it. In particular, it focuses on continued existence or survival as the key to function. Thus, for Canfield a function is an effect of a part that is "useful" to the containing system. It is useful "by aiding to *preserve the life* of things having the item, or else what it does is useful by aiding to *preserve the species* of things having it" (1964, 291, my emphasis). Yet the formulation by Bigelow and Pargetter (1987), in which functions are explicitly tied to the "propensity" to survive and reproduce, makes it especially clear what is wrong with the life chances or individual fitness approach to functions from the perspective I have been developing here. I have argued that functions are used to explain the *actual* existence of an organism; they are effects that are necessary conditions for its continued existence. But in the life chances view, functions are dispositions, with a certain probability of conferring survival in a specific environment. Thus, one is effectively treating functions as effects that (probabilistically) *determine* the survival or not of the organism. This is exactly the mistake of teleological determinism: since we can't really *determine* the survival of the organism on the basis of the functions of its parts (even probabilistically), we can't embed functions in a forward-looking theory as the life chances approach attempts to do. Instead, functional explanations must start with the real, existing organism and look backward, asking, "What were the necessary conditions for this organism to make it to where I can observe it today?" This is what makes functional explanation a form of conditional teleological explanation.

As it turns out, Wouters's views on these issues are quite similar to my own, though he has developed them from a different perspective and with a different terminology (Wouters 1995, 1999, 2003a, 2003b, 2005a, 2005b, 2007). In a recent paper (Wouters 2007), he calls attention to the fact that both functional ascriptions themselves, and functional explanations with respect to them, are a key part of biology, one not touched by those concerned with "etiological" or "current selection" approaches to function.

> The state of being alive critically depends on the individual's organization, that is, on the spatial arrangements of its parts and the timing of their activities. As the diversity of living organisms testifies, there are many ways ("forms of organization" or "designs") to construct mechanisms that are alive. However, not all arbitrary combinations of parts result in a living mechanism. Certain ways of life . . . are possible only if certain constraints . . . are satisfied. This makes it possible to explain the structure and activity of a living mechanism's parts on the basis of the constraints imposed on it by other characteristics of this mechanism and the conditions under which it works. (77)

Wouters's notion of functional dependencies as constraints on being alive is quite closely related to Cuvier's principle of the conditions for existence. This connection has not been lost on him. In a 2005 paper promoting the independence of "the functional perspective of organismal biology," Wouters explicitly tied his views to Cuvier's: "A design explanation is primarily concerned with what kinds of organisms can viably exist. More precisely, it is concerned with how matter must be organized to obtain an organism that is able to maintain itself, to grow, to develop and to produce offspring (that is with what Cuvier, the founding father of the discipline of functional animal morphology at the end of the 18th century, would call 'the conditions of existence')" (2005b, 54).

I consider Wouters' terminology unfortunate, because a "design explanation" certainly suggests ties to the argument of design (his earlier "viability explanation" is far better). Nevertheless, I can't help but enthusiastically endorse his general approach to function, especially his emphasis on the primacy of the individual level in functional explanations, his stress on the organizational requirements of the life state, and his defense of the autonomy of functional biology, points that have tended to be lost in the ongoing debate over the relative merits of systemic versus selected effect definitions of function. He and I are clearly looking toward the same sort of resolution to the function debate.

. . .

In this chapter, I have tried to show that the distinction between representational and conditional teleological explanation is fundamental, and that it can help resolve many of the difficulties in the literature over the nature of teleological explanation. In particular, I have argued that to conceive of past selection as the efficient cause of organismal traits leads necessarily to a teleological (and implicitly representational) determinism; there is simply no basis for the analogy between natural selection and an intentional designer. In biology, continued existence, survival, or simply life is the phenomenon we are trying to explain; we can explain it only through a conditional teleology, since determinism fails us and we reject intentional (external representational) teleology. Thus, the functions of parts of organisms should be understood as Cuvier understood them, as conditions for the continued existence of those organisms.

Armed with these conceptual distinctions, I next turn to a review of the historical debate over teleological thinking in science. This debate is both much older and more complex than often represented. Understanding this historical background is essential for understanding Cuvier's position, articulated in the context of Enlightenment rationalism; for understanding Darwin's position, articulated amid the resurgence of British natural theology in the wake of the French Revolution; and, finally, for understanding the weaknesses of much modern thinking on the relation of biology and teleology.

PART TWO

How Did We Get into This Mess?
From Socrates and Lucretius
to Cuvier and Darwin

3

Design versus the Epicurean Hypothesis

Two Thousand Years of Debate

"The front teeth in all living beings are designed to cut; the molars to receive food from the incisors and to grind it. The mouth, through which enter the things that living beings desire, was placed near the eyes and nose. Since excrements are unpleasant, the ducts that get rid of them are as far as possible from the sense organs. When these have been made with so much foresight, are you at a loss to say whether they are the work of chance or of design?"

"No, by Zeus!" replied Aristodemus. "When I look at it that way, they do appear to be the work of a wise and loving creator."

—SOCRATES, CA. 399 BC, IN XENOPHON'S *MEMORABILIA*

Nothing is born in us that we might use it,
But what is born produces its own use.

—LUCRETIUS, CA. 55 BC, *DE RERUM NATURA*

Having examined some of the basic distinctions in teleological thought and met the principle of the conditions for existence, we need to go back a bit and examine some of the forms that teleological and antiteleological thinking has taken in Western thought from the Greeks through the Enlightenment. This review is by no means intended to be a definitive history; Ruse (2003), Grene and Depew (2004), and Barrow and Tipler (1986) provide much fuller treatments, the former two giving a good survey of teleology in its biological context, the latter in a much broader one. Sedley (2007) provides an excellent overview of the ancient debates. Other useful sources include Hicks (1883), Osborn (1929), Lennox (1992), Amundson (1996), and Spaemann and Löw (2005).

Here I want to focus on contrasting teleological views with those of the atomist or Epicurean school. The atomists do not enter the standard historiography of evolution;

they are missing, for example, from the accounts of Ridley (2004), who begins with the Enlightenment, and Futuyma (2005), who discusses Plato and Aristotle, but not Democritus or Lucretius. Even Bowler's (2003) textbook on the history of evolution begins with the Enlightenment, although—as we will see—there is no way to understand the rise of evolutionary theory in the Enlightenment except in the context of an antiteleology derived directly from Lucretius.

Only by understanding the ever-present specter of the Epicurean hypothesis can we hope to understand the long history of the argument over design. And only by understanding this history can we hope to understand how to resolve it. Cuvier himself was certainly aware of much of this history, though Darwin was not, nor is the modern biologist likely to be. We must not make the same mistake Darwin did and think that this ancient debate is irrelevant to our science. As we will see, it was by doing so that Darwin came to incorporate elements of the design argument into what was supposed to be a purely mechanistic theory of evolution. To grow beyond Darwin's teleology, we must reach back beyond Paley to the ancient atomists and connect our evolutionary theory with its proper roots in the antiteleological, naturalistic theories of the past.

This chapter and the next thus provide a crash course in the history of the debate over teleology. I have made extensive use of quotes from the original sources, because my paraphrases could never convey their feel. In painting the contrast between the design view and that of the Epicureans, I will also have the opportunity to examine the prehistory of the principle of the conditions for existence itself. I will show that many related ideas have been expressed throughout history, on both sides of the debate. It is no doubt this feature of the history that made Cuvier less conscious of any originality on his part. I begin with Socrates.

THE TELEOLOGISTS: SOCRATES, PLATO, AND ARISTOTLE

In Plato's *Phaedo*, Socrates in prison explains his disappointment on buying a copy of Anaxagoras's book, which he had heard proclaimed mind or intelligence (*nous*) as the origin of all things:

> I saw that he made no use of Mind, nor gave it any responsibility for the management of things, but mentioned as causes *aēr* and *aithēr* and many other bizarre things. This seems to me like saying that Socrates' actions are all due to his mind, and then, in trying exhaustively to enumerate the causes of my actions, to say that the reason I am sitting here is that my body consists of bones and sinews, because the bones are hard and separated by joints, that the sinews are such as to contract and relax.... [B]ut he would neglect to mention the true causes, namely that, when the Athenians thought it best to condemn me, it seemed best to me to sit here for this reason and better to remain and endure whatever penalty they order. For, by the dog, I think these sinews and bones

would have been in Megara long since, impelled by my belief in what is best, if I had not thought it better and more honourable to endure whatever penalty the city ordered rather than escape and run away. (98b–99a, quoted in Hankinson 1998, 85)

Thus does Socrates argue for the superiority of an intentional teleological view of human behavior, and by analogy, the order of the universe.

That these views were those of Socrates himself, and not only Plato, is made more likely by the testimony of Xenophon (a contemporary of Plato and another disciple of Socrates). He presents a conversation between Socrates and Aristodemus that represents the earliest full explication of the argument of design (ca. 399 BC):

"Tell me, Aristodemus, aren't there any men whom you admire for their wisdom?"

"Yes, there are."

"Who, tell me their names," said Socrates.

"In epic poetry, I admire Homer most of all; in dithyramb, Melanippedes; in tragedy, Sophocles. In sculpture, Polyclitus; in painting, Zeuxis."

"Who do you think deserves more admiration? Those who create images without sense or power to move, or those who create living beings able to think and act?"

"By Zeus, of course those who create living beings, provided they come into being through design, not by some chance."

"Suppose there are things which give no hint as to the purpose of their existence, and also things which clearly serve a useful purpose. Which do you judge to be works of chance and which works of design?"

"Works that serve a useful end must be works of design."

"Don't you think that, from the very first, the creator of men endowed us with senses for a useful purpose? With eyes to see visible objects, and ears to hear sounds? What use would odors be to us if we had no nose? Would we taste sweet, bitter, and all the delights of the palate if our tongue had not been made to taste them? Besides, don't you think that other things, too, are likely to be the result of forethought? The sense of sight is weak, for example, and therefore the eyes are given eyelids which, like doors, open wide when we have to use our eyes and close when we sleep. Eyelashes grow, like screens, so that the wind does not hurt our eyes. Above the eyes, eyebrows project like cornices to prevent harm from the sweat of the brow. There is also the fact that ears receive all sounds, but never are clogged up. The front teeth in all living beings are designed to cut; the molars to receive food from the incisors and to grind it. The mouth, through which enter the things that living beings desire, was placed near the eyes and nose. Since excrements are unpleasant, the ducts that get rid of them are as far as possible from the sense organs. When these have been made with so much foresight, are you at a loss to say whether they are the work of chance or of design?"

"No, by Zeus!" replied Aristodemus. "When I look at it that way, they do appear to be the work of a wise and loving creator."

"What of the desire to beget children, the mother's desire to raise her children, the children's longing to live, and the great fear of death?"

"To be sure, they seem to be the contrivances of someone who planned for living creatures to exist." (Xenophon 1965, 23–24)

We see here the language that we will recognize in the design argument for the next 2,400 years: "design," "contrivance"; indeed, many of these examples of the foresight and care of the Creator (the eyes, the teeth, the reproductive instinct) will recur repeatedly. Note that the fundamental argument is that things come about in only two ways, by chance and by design, and things that serve a purpose must have been made with design (intention).

Plato subsequently greatly elaborated this view in his late dialogue *Timaeus*, which gives us a teleological cosmogony that was to prove extremely influential through the Middle Ages and beyond (Taylor 1928; Cornford 1937; Wright 2000; Sedley 2007). As Sedley (2007) points out, Plato's emphasis is on the argument *from* design, not *to* design, unlike Xenophon's Socrates or the later Stoics. Yet this difference, in the end, is not fundamental, because Plato certainly assumed that the world must be viewed as the product of design.

As a variety of the design argument, Plato's creation myth follows the general pattern. The intention ("reason") of the Creator explains both the existence and the function of the parts of organisms (and indeed of the whole universe). Interestingly, however, for Plato reason must work within the constraints imposed by the material: reason can "overrule and persuade" necessity only by conforming to it (Cornford 1937). It would take us too far afield to examine Plato's views at greater length. I would only like to point out that this contrast between reason and material presages the well-known distinction made by his student, Aristotle, between the formal and final causes, on the one hand, and the material and efficient causes, on the other. It is to Aristotle we now turn.

Of this illustrious lineage, Aristotle is by far the most important figure for our history. Not only was he the founder of zoology as a science (his *History of Animals* and *Parts of Animals* provided the fundamental basis for the development of zoology through the Renaissance), but he is well known for maintaining a teleological view of nature, a view at least partly driven by his zoological studies, especially his observations on development. For Aristotle, natural ends or final causes ("that for the sake of which") are causes for both animate and inanimate bodies. Moreover, as we will see in chapter 5, there are important connections between Cuvier and Aristotle.

Aristotle's account of biological development explicitly takes on the materialistic account of Empedocles and the atomists. How, according to Aristotle, do we know that nature acts "for the sake of" something?

> Why should not nature work, not for the sake of something, nor because it is better so, but just as the sky rains, not in order to make the corn grow, but of necessity? What is drawn up must cool, and what has been cooled must become water and descend, the result of this being that the corn grows. Similarly if a man's crop is spoiled on the threshing floor, the rain did not fall for the sake of this—in order that the crop might be spoiled—but that result just followed. Why then should it not be the same with the

parts in nature, e.g. that our teeth should come up *of necessity*—the front teeth sharp, fitted for tearing, the molars broad and useful for grinding down the food—since they did not arise for this end, but it was merely a coincident result; and so with all other parts in which we suppose that there is a purpose? Wherever then all the parts came about just what they would have been if they had come to be for an end, such things survived, being organized spontaneously in a fitting way; whereas those which grew otherwise perished and continue to perish, as Empedocles says his "man-faced ox-progeny" did. (Aristotle 1952, *Physica*, 198b)

At the heart of his argument for teleology in nature (and against Empedocles), Aristotle invokes an empirical argument: teeth and other natural things come about always or for the most part in the same way, and this is not characteristic of chance or spontaneous events.* "If, then, it is agreed that things are either the result of coincidence or for an end, and these cannot be the result of coincidence or spontaneity, it follows that they must be for an end" (199a). Again, it helps to think of development. Since a chicken egg almost always produces a chicken, it is impossible to believe that chicken eggs just happen, by chance, to make chickens. In the terms used in the previous chapter, the chicken egg is clearly a purposive system, directed toward the form of an adult chicken.

The developmental link is clear in his further justification of this position:

If then in art there are cases in which what is rightly produced serves a purpose, and if where mistakes occur there was a purpose in what was attempted, only it was not attained, so it must also be in natural products, and monstrosities will be failures in the purposive effort. Thus in the original combinations the "ox-progeny" if they failed to reach a determinate end must have arisen through the corruption of some principle corresponding to what is now the seed. . . . Again, in plants too we find the relation of means to end, though the degree of organization is less. Were there then in plants also "olive-headed vine progeny," like the "man-headed ox progeny," or not? An absurd suggestion; yet there must have been, if there were such things among animals.

Moreover, among the seeds anything must have come to be at random. But the person who asserts this entirely does away with "nature" and what exists "by nature." For those things are natural which, by a continuous movement originated from an internal principle, arrive at some completion: the same completion is not reached from every principle [read seed]; nor is any chance completion, but always the tendency in each is towards the same end, if there is no impediment. (199b)

Aristotle concludes this section with a forceful statement of the teleological nature of the universe: "It is absurd to suppose that purpose is not present because we do not observe the agent deliberating. Art does not deliberate. If the ship-building art

* This is the sole ancient passage referred to by Darwin in his "Historical Sketch" added to later editions of the *Origin of Species,* and here Darwin appears to think that Empedocles' views were Aristotle's own. See Gotthelf (1999).

were in the wood, it would produce the same results *by nature*. If therefore, purpose is present in art, it is present also in nature. The best illustration is a doctor doctoring himself: nature is like that. It is plain then that nature is a cause, a cause that operates for a purpose" (199b). Thus, although Aristotle begins cautiously enough, he is led to believe in the existence of natural purposes (as Kant was later to call them) by the regularity and specificity of development, which truly appears directed toward the very end achieved.*

This much is well known. That Aristotle held a teleological view of the universe is certainly not news. However, it is less widely remembered that Aristotle introduces an important qualification to his teleological view of physics, centered on the definition of necessity. This is the distinction between absolute and hypothetical (sometimes translated "conditional") necessity, noted in the previous chapter. Once again Aristotle is worth quoting in full:

> As regards what is "of necessity," we must ask whether the necessity is "hypothetical," or "simple" as well. The current view places what is of necessity in the process of production, just as if one were to suppose that the wall of a house necessarily comes to be because what is heavy is naturally carried downwards and what is light to the top, wherefore the stones and foundations take the lowest place, with earth above because it is lighter, and wood at the top of all as being the lightest. Whereas, though the wall does not come to be *without* these, it is not *due* to these, except as its material cause: it comes to be for the sake of sheltering and guarding certain things. Similarly in all other things which involve production for an end; the product cannot come to be without things which have a necessary nature, but it is not due to these (except as its material); it comes to be for an end. For instance, why is a saw such as it is? To effect so-and-so and for the sake of so-and-so. This end, however, cannot be realized unless the saw is made of iron. It is, therefore, necessary for it to be of iron [or some similar hard material], *if* we are to have a saw and perform the operation of sawing. What is necessary then, is necessary *on a hypothesis*; it is not a result necessarily determined by antecedents. (200a)

In other words, natural processes, to the extent they are "for the sake of something," do not necessarily occur; there are only certain necessary conditions for their occurrence.

This view is a logical corollary of the belief in purposive behavior; for if seemingly purposive behavior is in fact necessary, then there appears to be no point in a teleological view. Teleology only makes sense if the outcome of the process is in some way contingent in the absence of purpose; otherwise, there is nothing left to explain. In other words, if one observes a certain result to have been achieved, one can say, "It had to be so," but the "had to" in such a statement is not the same as that

---

* Rather strangely, we also find him appearing to argue that rain in winter is for a purpose, because it occurs "always or for the most part." Sedley (2007) has used this and some similar passages to argue that Aristotle was a cosmic teleologist after the manner of Plato.

in "the circumference of a circle has to be π times the diameter." This stricture on the type of necessity involved in teleological explanations is often overlooked in treatments of Aristotle's teleology.

Two further aspects of Aristotle's philosophy that will be relevant to us later are worth mentioning here. The first is Aristotle's promotion of the supposition "Nature does nothing in vain," which was to reverberate through the ages down to Harvey, Newton, and beyond. Aristotle's use of this phrase has recently been reviewed by Lennox (2001). From our perspective, this phrase appears to be Aristotle's proposal of an *inductive generalization*, which, once made, can then be used as a *principle of reasoning*. Aristotle's examples are explanations of why a set of milk teeth are present in mammals, why both tusks and sawteeth are not found together, why both gills and lungs aren't found together (he obviously didn't know of lungfish!). This is clearly an argument of adaptedness, of appropriateness of means for ends. Nature "always, given the possibilities, does what is best for the substantial being of each kind of animal; accordingly, if it is better a certain way, that is also how it is by nature" (*On Animal Locomotion* 2:704b, quoted in Lennox 2001, 206).

The second point to be noted is his actual approach to zoological explanation, which bears some resemblance to Cuvier's (see Grene and Depew 2004). This he lays down clearly in the introduction to his *Parts of Animals*: "Hence we should if possible say that because this is what it is to be a man, therefore he has these things; for he cannot be without these parts. Failing that, we should get as near as possible to it: we should either say altogether that it cannot be otherwise, or that it is at least good thus" (640a33–b1, Balme 1972, Balme's translation). Here, at the outset of scientific zoology, we find a variety of modes of explanation for the parts of animals proposed: (1) the inclusion of the part as part of the definition or essence of that *type* of animal, (2) the (hypothetical) necessity of the part for the existence of that *particular* animal, and, finally, (3) that having the part is better (in some sense) than not having it (Balme 1972; Cooper 1987; Gotthelf 1987). The third type of explanation is clearly what Aristotle refers to by the supposition that nature "does nothing in vain." Even for Aristotle, the explanation based on the assumption of adaptedness was the least justifiable explanation.

Both Socrates and Plato looked to an external creator for an explanation for organic complexity; they stand at the head of a long design tradition in Western philosophy. Aristotle, in contrast looked to an inner perfecting principle to explain the apparent goal-directedness of development, but he also imparted powers to nature such as "always doing what is best" that, if they don't imply consciousness, certainly don't preclude it. By introducing the concept of hypothetical necessity, and by his level-headed discussion of modes of explanation in zoology, Aristotle moved the debate forward, beyond the natural theology of Socrates and Plato. Yet in later times, these subtleties of his approach tended to be forgotten, and his views were assimilated to Neo-Platonic, Stoic, and Christian thought. As we will see, this amalgam provided the basis for the development of the argument of design from the later Middle Ages through the nineteenth century.

EMPEDOCLES AND THE ATOMISTS

The teleological strain coming down from Socrates through Plato to Aristotle did not arise in a vacuum. We have seen Aristotle invoke an argument from the end-directedness of development to counter Empedocles' assertion that the order we see among organisms results from a winnowing of preexistent diversity of "monsters." Socrates' teleology was developed in response to the ateleological or antiteleological theories of Empedocles and the atomists, who developed explicitly naturalistic (if not necessarily well-articulated) theories of the origin of life.

Empedocles (ca. 495–435 BC) was not only a philosopher but a shaman of sorts. His interest for us is his primitive doctrine of a process something like natural selection as explanatory of the order of the world. Empedocles' Universe was composed of the four elements (earth, air, fire, water) and two forces, love and strife. He proclaimed a cosmic cycle, in which a period of increasing love, during which there was a universal mixture of the elements and the generation of our world, was followed by a period of increasing strife, in which elements separated out of the mixture and the world was destroyed. During the period of increasing love, our current world was generated. We are now in the period of increasing strife.

How did life come to be? As love expanded from its lowest ebb, there was a period of "creative tension" (Inwood 2001, 54) during which

> things which had previously learned to be immortal grew mortal,
>   and things previously unblended were mixed, interchanging their paths.
> And as they were mixed ten thousand tribes of mortals poured forth,
>   fitted together in all kinds of forms, a wonder to behold.
> (61/35.14–17, Inwood translation)

But these tribes were not perfectly formed at first, but instead were represented only by miscellaneous parts:

> as many heads without necks sprouted up
>   and arms wandered naked, bereft of shoulders,
>   and eyes roamed alone, impoverished of foreheads.
> (64/57)

As we have seen reported by Aristotle (*Physica*, 198b29–32), there next followed a period of combination of parts and survival of the fit among these combinations: "Where all came together as though for a purpose, these survived, being formed automatically in a fitting way. As many as did not perished and are perishing, as Empedocles says of the 'ox-like animals with men's faces'" (Inwood translation). Whether this theory is in fact "a distant ancestor of the enormously powerful and successful explanatory paradigm of natural selection" (Hankinson 1998, 49) will be discussed later. For the moment, I wish to note only that this first protoevolutionary theory was nonteleological, if not necessarily antiteleological (see Sedley 2007).

Unlike Empedocles, the early atomists Leucippus (fl. 450–420 BC) and Democritus (ca. 460–ca. 370 BC) conceived of the world as composed of an infinite number of atoms, moving in the void. In their cosmogony, the universe emerged as the result of a "whirl" within the cosmic soup (Bailey 1964). Moreover,

> the atoms did not form themselves into a whirl in order that a cosmos might result: there is no design either on their part or on the part of any extraneous force or power. They fall into the whirl "accidentally" and the result by a process of strict necessity is a world. Democritus' intention was no doubt anti-teleological: this, as has been seen, was one of his great aims in establishing "necessity" as the basis of his system, and it remained always a cardinal point in the Atomic tradition that the creation of a world was not the result of design. (Bailey 1964, 141)

The antiteleological views of the early atomists on the origin of the universe were in later times elaborated into an atomic theory of the survival of the fittest by Epicurus (ca. 341–271 BC) and his Roman follower Lucretius (ca. 99–55 BC), combining Democritus's views on the origin of the universe with those of Empedocles on the origin of life.

The young Epicurus first visited Athens in 323 BC, the year that Alexander died and Aristotle left Athens in fear for his life. Epicurus subsequently wandered the Near East, studying philosophy and eventually founding his own school. In 307, he returned to Athens and bought a house and his famous Garden. Here he lived with a community of disciples. His cult, popularly associated with overindulgence and hedonism, in fact focused on human happiness as peaceful contentment and freedom from worry. It was to last some four hundred years. Much of our knowledge of Epicurus's teachings comes from Lucretius's poem *De rerum natura (On the Nature of Things)*. This ode to Epicurus was to prove extremely influential, if heretical, in the Renaissance and Enlightenment.

We have seen Aristotle proclaiming the inadequacies of his predecessors' views on the role of chance in the origin and continued existence of the universe; for him nature is at root purposive. Epicurus and Lucretius will have none of this. Not only does the universe emerge by chance from the primeval whorl, but so does life:

> For surely the atoms did not hold council, assigning
> Order to each, flexing their keen minds with
> Questions of place and motion and who goes where.
> But because many atoms in many ways,
> Spurred on by blows through the endless stretch of time,
> Are launched and carried along by their own weight
> And come together and try all combinations,

Whatever their assemblies might create,
In just this manner, scattered through the ages,
Chancing upon all combinations, motions,
They at last tossed together and suddenly fused
Into the origin of mighty things,
Of the earth and the sea and sky and all that live.
(Lucretius 1995, V: 419–431)

First the Earth gave the shimmer of greenery
And grasses to deck the hills; then over the meadows
The flowering fields are bright with the color of springtime,
And for all the trees that shoot into the air
It's a growing contest—and the reins are free!
As feathers or bristles or hair are the first to form
On birds' strong wings or the limbs of galloping horses,
So the new earth sent the grass up first, and the brush,
Then made the many mortal animals,
Various kinds arising in various ways.
(V: 780–789)

Yet not all these first-formed creatures were perfect—as with Empedocles, Epicurus proposed a period of weeding out of the imperfect, nonfunctional forms:

For the earth, way back then, tried to bring forth many
Prodigies, strange of feature, monster-limbed,
The halfway not-this not-the-other hermaphrodite,
Some orphaned of feet, some widowed of both their hands,
Many even lacking a mouth, or blind and eyeless,
Or manacled up by the members clung together,
Unable to do a thing, go anyplace,
Dodge harm, grab hold of anything they'd need.
Other such freaks and monsters Earth created—
In vain, for Nature frightened off their growth,
They couldn't attain the wished-for flower of adulthood,
Or seek out food or join in the act of love.
For many things, we see, must coincide
So as to forge the generations; first
There must be food; next, ways for the genital seeds
To stream through the body from the slack-fallen members,
And, so that males and females may unite,
A way to exchange the mutual joys of love.

And many kinds of creatures must have died
Unable to plant out new sprouts of life.
For whatever you see that lives and breathes and thrives
Has been, from the very beginning, guarded, saved

By its trickery or its swiftness or brute strength.
(V: 834–858)

This view of the mechanical, chance origin of life, subject to winnowing at the hands of functional requirements for survival, was closely associated with a denial of teleological explanations for structure:

> Here I most violently want you to
> Avoid one fearful error, a vicious flaw.
> Don't think that our bright eyes were made that we
> Might look ahead; that hips and knees and ankles
> So intricately bend that we might take
> Big strides, and the arms are strapped to the sturdy shoulders
> And hands are given for servants to each side
> That we might use them to support our lives.
> All other explanations of this sort
> Are twisted, topsy-turvy logic, for
> Nothing is born in us that we might use it,
> But what is born produces its own use.
> (IV: 820–831)

A clearer denial of design cannot be imagined.

How did Epicurus view the relation between chance, which he credits with the formation of the world, and necessity? The early atomists had preached a universal determinism; all chance is only relative to our ability to understand the true causes of events.* This was not true of Epicurus, who "admitted the existence of real contingency in nature, an element of 'chance' which at times worked in contravention of necessity" (Bailey 1964, 326). Whether or not this element of chance can ultimately be traced back to the atomic swerve, this admission of the reality of chance stands in contrast both to the views of the earlier atomists, who considered chance to be an illusion, and to those of Aristotle, who considered all chance relative to purpose, to final causes.

THE END OF THE CLASSICAL ERA
AND THE RISE OF CHRISTIANITY

In later Roman times, a variety of philosophies coexisted: Epicureans, Skeptics, Stoics, Peripatetics, Neo-Platonists, and Christians mingled in the streets of Rome and Athens. The Epicureans, as we have just seen, were fundamentally antiteleological in their approach to the world. The Skeptics, who included adherents of the

---

* Interestingly, Jacques Monod's (1971) excellent book on *Chance and Necessity* derives its title from a quote he ascribes to Democritus: "Everything existing in the Universe is the fruit of chance and of necessity." Democritus does not seem to have made such a statement; instead, he typically viewed chance as only our subjective view of necessity.

"New Academy" that followed on Plato, argued against the certainty of all knowledge and thus endorsed neither antiteleology nor teleology. Stoics, Peripatetics, Neo-Platonists, and Christians, however, were all teleologists, in spite of their differences. As the classical era ended and the Middle Ages began, these traditions all combined to help create the now-ascendant Christian ethic.

The Stoic view, which was transmitted to the Christians of the Middle Ages by Cicero and Seneca, involved a deterministic universe ruled by a rational creator (logos), who has arranged all things for the best. It is in many ways akin to that view expressed by Plato in the Timaeus, but shorn of its complicated and mystical metaphysics, and combined with a causal determinism as rigorous as that of the early atomists. By this determinism, the Stoics allowed for God's purpose in creating the universe as it is but in a sense disallowed any other purpose in the world; if all is preordained by God, then, contra Aristotle, there are no "natural purposes," only apparent ones. Thus, free will was a real problem for them (Bobzien 1998).

The Stoics, like the Christians who were to follow, conceived of the entire universe as God's plan. They thus made full use of the argument of design, both proving the existence of a rational God by the evidence of the rationality of the universe, and inferring the purpose of the parts of the universe (to the best of their ability) on the assumption that the Creator was rationally aiming at the best (Long 1986). As we will see, their appeal to the design argument was the direct ancestor of the now more familiar one of Paley some 1,800 years later.

For the men of the Renaissance and the Enlightenment, the Stoic philosophy was readily accessible in Cicero's dialogue on The Nature of the Gods (ca. 45 BC). The dialogue represents a conversation among an Epicurean, Velleius; a Stoic, Balbus; and a Skeptic of the New Academy, Cotta. Cicero has his Stoic Balbus proclaim:

> If then the products of nature are better than those of art, and if art produces nothing without reason, nature too cannot be deemed to be without reason. When you see a statue or a painting, you recognize the exercise of art; when you observe from a distance the course of a ship, you do not hesitate to assume that its motion is guided by reason and by art; when you look at a sun-dial or a water clock, you infer that it tells the time by art and not by chance; how then can it be consistent to suppose that the world, which includes both the works of art in question, the craftsmen who made them, and everything else besides, can be devoid of purpose and of reason? (Cicero 1961, II: 87, Rackham translation)

And in a passage on the order of the universe parallel to that of Paley so many years later, he continues: "Suppose a traveler to carry into Scythia or Britain the orrery recently constructed by our friend Posidonius, which at each revolution reproduces the same motions of the sun, the moon and the five planets that take place in the heavens every twenty-four hours, would any single native doubt that this orrery was the work of a rational being?" (II: 88). Like Plato's, the Stoic teleology was thoroughly anthropocentric: "For whose sake then shall one pronounce the world to have been created? Doubtless for the sake of those living beings which have the use of reason; these are the gods and mankind, who assuredly surpass all other things in excellence, since the most excellent of all things is reason. Thus we are led to believe that the world and all the things that it contains were made for the sake of gods and men" (II: 133).

Thus, both in Plato (and thence the Neo-Platonists) and in the Stoics, the early church fathers were presented with what from our point of view are extremely similar metaphysical schemes. It is of interest that the only work of Plato generally known to the early Middle Ages was the *Timaeus*. The rational creator of Plato and the Stoics fused (however uncomfortably) with the stern, jealous God of the Old Testament and the meek, loving God of the New Testament to give rise to a strange amalgam that surprisingly still holds together, however tenuously, today. The design argument thus passed directly into the Christian metaphysics that dominated Western thought for the next millennium and a half. Rather than pursuing this transformation any further, however, I want to examine the revolt against the dominant teleological worldview that arose in the course of what has come to be known as the scientific revolution.

THE SCIENTIFIC REVOLUTION
AND THE REVOLT AGAINST TELEOLOGY

Because it is so well known, it is easy to forget the significance of the fact that the rise of modern science in the Renaissance was coupled to a vociferous rejection of Scholasticism, the Aristotelianism of the later Middle Ages. The Renaissance scholars had rediscovered and promoted the diffusion of such classical works as Lucretius's *De rerum natura* and Cicero's *De natura deorum*, which made the Scholastic Christian/Aristotelian synthesis appear as only one among many possible ways of thinking. The Copernican revolution, completed by Kepler (1571–1630) and Galileo (1564–1642), made it clear that relatively easily gathered facts could contradict centuries of tradition, thought, and argument. This empirical orientation of the new science was associated with a rejection of the validity of teleological explanations and the substitution of a revived mechanical atomism, in which material bodies move according to mathematical laws.

Galileo is, of course, the great representative here; his mathematical account of the universe in terms of geometrical space and time left no room for purpose in science. As expressed by E. A. Burtt (1954) in his classic work on *The Metaphysical Foundations of Modern Science*: "In place of the teleological categories into which scholasticism had analyzed change and movement, we now have these two formerly insignificant entities given new meanings as ultimate mathematical continua and raised to the rank of ultimate metaphysical notions. The real world, to repeat, is a world of mathematically measurable motions in space and time" (93). From Galileo's conception of mechanical, efficient causality as the only causality in nature, "the conception of the world as a perfect machine is thus rendered inevitable, and it is no accident that first in Huyghens and (in a more philosophical form) in Leibniz we have this position unequivocally proclaimed" (101).

Galileo accepted God's grace and providence in general, but he effectively eliminated them from his science. He refused even to discuss what the cause of motion might be; his task was only to describe its form. Moreover, although Galileo felt that nature and the Bible could not disagree when properly interpreted, he insisted on the priority of observation and reason over biblical text in determining the structure of the universe (it is for this attitude, as much as the specific doctrines he endorsed, that he was condemned).

Galileo's biology is of particular interest here. In his *Dialogues Concerning Two New Sciences* (Galileo 1954), we find an interesting discussion of the scaling of bones to support the skeleton of small and large animals. He nowhere brings in the Creator's design but rather considers the mechanical limits involved: "nor can nature produce trees of extraordinary size because the branches would break down under their own weight; so also it would be impossible to build up the bony structures of men, horses, or other animals so as to hold together and perform their normal functions if these animals were to be increased enormously in height" ("Second Day," 130). Such mechanical constraints on organic form (which in our terminology could be called mechanical conditions for the existence of these forms) were all he would allow into his science.

Around the same time that Galileo was composing his *Dialogues* and showing that science could proceed quite effectively in willful ignorance of final causes, Francis Bacon (1561–1626) and René Descartes (1596–1650) were denying their utility on theoretical grounds. Bacon famously stated that "inquiry into final causes is sterile, and, like a virgin consecrated to God, produces nothing" (quoted in Barrow and Tipler 1986, 49), although he clarified that this was not because they were not true but because "their excursions into the limits of physical causes hath bred a vastness and solitude in that track" (quoted in Barrow and Tipler 1986, 52). Similarly, Descartes declared, "I consider the customary search for final causes to be totally useless in physics; there is considerable rashness in thinking myself capable of investigating the purposes of God" (*Meditations on the First Philosophy*, 1641, "Fourth Meditation"; Descartes 1985, II: 39). Instead, he preferred to focus on clear and simple truths of reason, such as the famous *cogito ergo sum*, from which he begins (*Discourse on the Method*, 1637, pt. 4; 1985, I: 127).

For the development of thought on teleology, two aspects of the Cartesian philosophy are especially relevant. First is his famous mind-body dualism. For Descartes, like Plato, the mind or rational soul was preeminent. Nevertheless, the very separation he proposed tended to promote increased recognition of the purely mechanical aspects of the universe, and living organisms in particular. This tendency was strengthened by his insistence that only man had a mind; beasts were mere automata, as was man's body. This mechanistic view had its effects, though—as a perusal of Descartes' physiology in his *Treatise on Man* makes clear; his physiological ideas were not only mistaken but almost laughably crude. He never even fully understood the greatest triumph of mechanistic physiology of the time: Harvey's discovery of the circulation of the blood (see Grene and Depew 2004).

The second mechanistic influence of Descartes, in addition to his postulate of animal automata, was his equally mechanistic model for the origin of the universe based on his famous "vortices" (Descartes 1983). In Descartes we can clearly see the downfall of the mechanistic philosophy. With both his animal automata and his model for the origin of the universe and our world, Descartes fell into the trap of what I have called teleological determinism: to attribute absolute necessity to an achieved result, though the necessity is not derived from the laws proposed to account for the result. Thus, according to Descartes, he had explained the origin of the universe on the basis of laws: "I showed how, in consequence of these laws, the greater part of the matter of this chaos had to become disposed and arranged in a certain way, which made it resemble our heavens; and how, at the same time, some of its parts had to form an earth, some planets and comets, and others a sun and fixed stars" (Descartes 1985, *Discourse on the Method*, I, pt. 5: 132–133). Of course, although we now adopt a similar model for the origin of the universe, we consider the formation of the precise universe we know contingent, not necessary, and only ascribe such features of the universe to laws as can be derived from those laws.

RATIONAL THEOLOGY AND THE ARGUMENT OF DESIGN: THE LATER SEVENTEENTH CENTURY

In the later seventeenth century, we see the emergence of the modern argument of design in full form (for an excellent survey of this period, see Willey 1934, 1940). Only a few years after the work of Harvey, Galileo, Bacon, and Descartes had promoted the mechanical point of view, Thomas Hobbes (1588–1679) adopted a thorough mechanistic materialism. Hobbes was both a friend and a promoter of Harvey and Galileo. Descartes had admitted the human soul, at the price of making all animals soulless. Hobbes argued vigorously against an immaterial soul, thus controverting both Catholicism and Aristotelianism; he nevertheless retained the Creator (most felt disingenuously) as the divine artificer. In the introduction to his *Leviathan* (1651), he uses mechanism to promote his analogy of Man the machine (designed by God) and the machine of State (designed by man); in fact, he compares both to a watch:

> NATURE (the Art whereby God hath made and governes the world) is by the *Art* of man, as in many other things, so in this also imitated, that it can make an Artificial Animal. For seeing life is but a motion of Limbs, the begining whereof is in some principall part within; why may we not say, that all *Automata* (Engines that move themselves by springs and wheeles as doth a watch) have an artificiall life? For what is the

Heart, but a *Spring*; and the *Nerves*, but so many *Strings*; and the *Joynts*, but so many *Wheeles*, giving motion to the whole Body, such as was intended by the Artificer? *Art* goes yet further, imitating that Rationall and most excellent worke of Nature, Man. For by Art is created that great LEVIATHAN called a COMMON-WEALTH, or STATE, which is but an artificial Man. (Hobbes 1973, 1)

Such a mechanistic physiology easily lent itself, as can be seen in Descartes' discussion of the senses, to a focus on the wise contrivance of this machine by its creator (though this was certainly not Descartes' or Hobbes's main point). In the early seventeenth century, a number of factors conspired to promote the adoption of such a rationalist, mechanistic natural theology. This natural theology had, as we have seen, been anticipated by Socrates, Plato, and especially the Stoics.

A key figure in the rise of a modern natural theology is Pierre Gassendi (1592–1655), a French Catholic priest, physicist, and philosopher who was a friend of both Galileo and Hobbes, and a correspondent of Descartes. Gassendi is chiefly known for promoting Epicurean atomism. However, as shown in the excellent recent study by Osler (1994; see also Spink 1960), Gassendi in fact combined a mechanistic and atomistic view of the world with a voluntarist God and providentialism based on the argument of design; he "baptized" Epicurean philosophy. In 1632, he wrote in a letter to a friend: "As for the Epicurean Commentaries, you seem to hesitate lest I go wrong in religion. But I am opposed to anything which could conflict with it. You insist on Providence: truly I defend it against Epicurus" (quoted in Osler 1994, 36). Most importantly for our theme, against the Epicurean hypothesis that the present order can be understood as an unintentional result of the accidental coming together of infinite atoms in infinite space in infinite time, Gassendi specifically argued for the necessity of design. Yet, like Descartes and Galileo, "he rejected any kind of immanent finality in the Aristotelian sense. What Gassendi called final causes are actually divine intentions reflected in the design of the creation. Thus, the purposiveness found in nature is, for Gassendi, externally imposed by God. The natural order itself is ruled only by efficient causes, including God" (Osler 1994, 49).

In an extended debate with Descartes, beginning with Gassendi's *Fifth Set of Objections* to Descartes' *Meditations* in 1641, Gassendi championed the argument to design and the contingency of the Creator's will against the rationalist, a priori argument for the existence of God proposed by Descartes. Gassendi's argument is worth paying close attention to. "Your rejection of the employment of final causes in physics," he says to Descartes,

might have been correct in a different context, but since you are dealing with God, there is obviously a danger that you may be abandoning the principal argument for establishing by the natural light the wisdom, providence and power of God, and indeed his existence. Leaving aside the entire world, the heavens and its other main parts, how or where will you be able to get any better evidence for the existence of such a God than from the function of the various parts in plants, animals, man, and yourself (or your body), seeing that you bear the likeness of God? We know that certain great thinkers [presumably Galen] have been led by a study of anatomy not just to achieve a knowledge of God but also to sing thankful hymns to him for having organized all the parts and harmonized their functions in such a way as to deserve the highest praise for his care and providence.

You will say that it is the physical causes of this organization and arrangement which we should investigate, and that it is foolish to have recourse to purposes rather than to active causes or materials. But no mortal can possibly understand or explain the active principle that produces the observed form and arrangement of the valves which serve as the openings to the vessels in the chambers of the heart. Nor can we understand the source from which this active principle acquires the material from which the valves are fashioned, or how it makes use of them, or what it requires to ensure that they are of the correct hardness, consistency, fit, flexibility, size, shape and position. Since, I say, no physicist is able to discern and explain these and similar structures, why should he not at least admire their superb functioning and the ineffable Providence which has so appositely designed the valves for this function? Why should the physicist not be praised if he then sees that we must necessarily acknowledge some first cause which arranged these and all other things with such supreme wisdom and in such precise conformity with his purposes?

You say that it is rash to investigate the purposes of God. But while this may be true if you are thinking of the purposes which God himself wished to remain hidden or ordered us not to investigate, it surely does not apply to purposes which he left on public display, as it were, and which can be discovered without much effort. (Descartes 1985, II: 215)

In other words, as long as we can't explain the organization and function of organisms as the deterministic result of natural laws, why not just admire God's handiwork? The purpose (function) of body parts is clear and must be assumed to be the purpose of their Creator. Descartes' response to this is interesting:

The points you make to defend the notion of a final cause should be applied to efficient causation. The function of the various parts of plants and animals etc. makes it appropriate to admire God as their efficient cause—to recognize and glorify the craftsman through examining his works; but we cannot guess from this what purpose God had in creating any given thing. In ethics, then, where we may often legitimately employ conjectures, it may admittedly be pious on occasion to try to guess what purpose God may have had in mind in his direction of the universe; but in physics, where everything must be backed up by the strongest arguments, such conjectures are futile. We cannot pretend that some of God's purposes are more out in the open than others; all are equally hidden in the inscrutable abyss of his wisdom. Nor should you pretend that none of us mortals is incapable of understanding other kinds of cause; they are all much easier to discover than God's purposes, and the kinds of cause which you put forward as typical of the difficulties involved are in fact ones that many people consider they do know about. (Descartes 1985, II: 258)

Thus, Descartes is quite willing to go along with Gassendi's use of the argument *of* design in general. What he is arguing against is instead the argument *from* a designer to any particular design—indeed, any focus on the function, or conditional teleological role—of animal parts. As we have seen, he wants to explain animals not functionally but mechanically. He wants to know not how they manage to live but how they move and act. In fact, he completely rejects the scientific validity of a functional rather than a mechanical understanding of organisms. Unfortunately, Descartes' mechanisms were very crude and certainly didn't account for the phenomena; instead, they led him to a deterministic teleology.

For Gassendi, in contrast, the organism must be interpreted functionally, because we don't understand it mechanically. From his perspective, it was reasonable to assume that a function we see being fulfilled (and fulfilled well) was intended to be fulfilled; he thus assumed an intentional teleology in nature and used this to explain the functional role of parts and the adaptedness of parts for their function. The adaptedness of the heart valves for their function is evidence for a creator who created them for just that function. For Gassendi, the argument to design is "the royal road, smooth and easy to follow, by which one comes to recognize the existence of God, his power, his wisdom, and his goodness and his other attributes, which is nothing other than the marvelous working of the universe, which proclaims its author by its grandeur, its divisions, its variety, its order, its beauty, its constancy, and its other particularities" (quoted in Osler 1994, 161). Furthermore, he supported an anthropocentric teleology: all was made for man, and man was made to worship God (as in Plato and the Stoics). Against the deterministic materialism of Hobbes, the modern representative of Democritus, Gassendi argued for the reality of human free will (though not on the grounds of the Epicurean "swerve") and the immortality of the soul (against Epicurus). As we will see, Gassendi's views are quite similar to those later expressed by the chemist Robert Boyle, whom he likely influenced. Before we turn to Boyle, however, we must examine some further developments in philosophy that likewise contributed to the rise of natural theology.

THE DEISTS, THE PLATONISTS, AND
THE REBIRTH OF NATURAL THEOLOGY

In the mid–seventeenth century, natural theology was promoted by two related movements that arose at the boundary between theology and philosophy, both aiming to ground theology on a rational basis, and both originating in England (Tulloch 1874; Willey 1934; Patrides 1980). The first was the Deists, of whom Lord

Herbert of Cherbury (1583–1648) is commonly styled as the "father of Deism." The Deists wanted to found a religion entirely on rational grounds; this "natural religion" was supposed to lie at the root of all actual ones. Deism found few followers, however, until the late seventeenth and early eighteenth centuries.

The second movement was a more conservative one, based not on the mere exercise of reason but on the appeal to classical tradition, and was far more influential at the time. This was the so-called Cambridge Platonists, who in the 1650s and 1660s opposed the materialism (and feared atheism) of Descartes and Hobbes with a revival of Platonic idealism (Tulloch 1874; Willey 1934; Patrides 1980). This was a conservative but not a reactionary movement; the Cambridge Platonists, all ordained ministers, proposed to combat the dangers of atheism with a classically based, rational religion accepting of doctrinal differences. Their most prominent members were Henry More (1614–1687), who published an *Antidote against Atheism* in 1653, and Ralph Cudworth (1617–1688), whose major work was entitled *The True Intellectual System of the Universe: The First Part; wherein, All the Reason and Philosophy of Atheism Is Confuted; and Its Impossibility Demonstrated* (1678).

Book II of More's *Antidote* is in many ways the type of all later natural theologies, leading up to Paley's of 1800. Here More argues explicitly against the idea that the world could be the effect of the blind motion of matter, and he asks his reader smitten by atheism "diligently to attend to those many and most manifest marks and signes that I shall point him to in this outward frame of things that naturally signify unto us that *there is a God*" (Patrides 1980, 245). These signs include the cycle of the seasons and of day and night, the mountains that give rise to useful rivers, the provision for man of building materials and fuel, and for navigation lodestone that he might navigate the oceans, the steady direction of the Earth's axis to keep his compass true, and strange countries that he might exercise his mind. With respect to animals and plants, he cites their form, beauty, structure, and use to man. He assumes that most plants and animals were made for man's use (noting that horse's hoofs are well designed to fit horseshoes!). However, he also has some more astute observations of adaptedness, or fitness to function, in natural history, including the instincts of animals, the aerodynamic shape of birds, the reduction of the eyes, good hearing and sharp claws of moles, the eyes of hares that allow them to look backward at a pursuing enemy, the feet of ducks that function as oars for swimming, and the long neck of the heron used in feeding.

Most critically for our story, More notes (quite rightly) that "Counsell most properly is there implyde where we discerne a variety and possibility of being otherwise, and yet the best is made choise of" (266). In other words, the argument to design depends on the assumption not only that the world works but that it might have existed yet worked worse. This is the assumption of *superabundant providence* or *grace*. As we will see, understanding this assumption is critical to understanding Darwin's path to natural selection. More does not hesitate to make this assumption:

> I demand first in generall concerning all those Creatures that have *Eyes* and *Eares*, whether they might not have had onely *one Eye* and *one Eare* a piece . . . and subsisted though they had been no better provided for then thus. But it is evident that their having *two Eyes* and *two Eares*, so placed as they are, is more safe, more sightly, and more usefull. Therefore that being made so constantly choice of, which our own Reason deemeth best, we are to inferr that the choice proceeded from *Reason* and *Counsell*.
>
> Again I desire to know why there be no *three-footed Beasts*, (when I speak thus, I doe not meane *Monsters*, but a constant *Species* or kind of Animalls) for such a Creature as that would make a limping shift to live as well as they that have *foure*. Or why have not some beasts more than foure-feet, suppose *sixe*, & the two middlemost shorter then the rest, hanging like the two legges of a Man a horse-back by the horse sides? For it is no harder a thing for Nature to make such frames of Bodies than others that are more elegant and usefull. But the works of Nature being neither uselesse nor inept, she must either be wise her self, or be guided by some higher Principle of *Knowledge*. (273–274)

This higher principle is, of course, that an "*Eternall Mind* (that put the universall Matter upon Motion, as I conceive most reasonable, or if the Matter be confusedly mov'd of its self, as the Atheist wilfully contends) . . . takes the easy and naturall results of this general Impresse of *Motion*, where they are for his purpose, where they are not he rectifies and compleats them" (247–248). The echoes of the *Timaeus* are here quite loud; remember Plato's "Reason overruling and persuading Necessity."

The English Deists and Platonists were not the only philosophers of the time who tried to respond to the challenge to provide a rational basis for religion. Much greater philosophers, such as Spinoza, Leibniz, and Locke, did so as well. Gottfried Leibniz (1646–1716) is most important to our story. As More and Cudworth had been provoked by Descartes and Hobbes, Leibniz had been provoked by Spinoza. Like More, like Cudworth, he attempts to defend the traditional designing God against such atheism. Nevertheless, he accepts a mechanical world, in which a complete causal determinism exists. A good overview of his attitude is found in Section XIX of his *Discourse on Metaphysics* (1686), entitled "The Utility of Final Causes in Physics" (Leibniz 1951, 318–320). Here he directly attacks Descartes and Spinoza: "As I do not wish to judge people in ill part I bring no accusation against our new philosophers who pretend to banish final causes from physics, but I am nevertheless obliged to avow that the consequences of such a banishment appear to me dangerous." After all, "all those who see the admirable structure of animals find themselves led to recognize the wisdom of the author of things." "When one seriously holds such opinions which hand everything over to material necessity or to a kind of chance . . . it is difficult to recognize an intelligent author of nature." But the "effect should correspond to its cause"; if we see order, it must have been produced by intelligence (he here refers admiringly to Plato's passage in the *Phaedo* quoted earlier, in which

Socrates defends the use of teleology). For Leibniz, the solution was the same one adopted by the Stoics: all occurs deterministically by mechanical necessity, but all was also designed by the Creator.

For Leibniz, this reconciliation is expressed in his doctrine of *pre-established harmony*: "Souls act according to the laws of final causes, by appetitions, ends and means. Bodies act in accordance with the laws of efficient causes or of motion. And the two realms, that of efficient causes and that of final causes, are in harmony with each other" (*Monadology*, section 79, Leibniz 1951, 549). Thus, unlike Descartes, who argued that the human soul could influence matter, Leibniz confined the actions of souls and matter to completely different realms. Of course, he considered the preestablished harmony between them to be more evidence for a designing God.

Famously, Leibniz also argued that this world that God created is the best of all possible worlds, based on his principle of sufficient reason (see Lovejoy 1936):

> Now, as there is an infinity of possible universes in the ideas of God, and as only one of them can exist, there must be a sufficient reason for the choice of God which determines him to select one rather than another.
>
> And this reason can only be found in the *fitness*, or in the degrees of perfection, which these worlds contain, each possible world having a right to claim existence according to the measure of perfection which it possesses.
>
> And this is the cause of the existence of the Best; namely, that his wisdom makes it known to God, his goodness makes him choose it, and his power makes him produce it. (*Monadology*, sections 53–55, Leibniz 1951, 543–544)

Finally, it is worth noting that for Leibniz, the validation of final causes was connected to the doctrine of preformation in embryology (which denied spontaneous generation): "I am therefore of the opinion of Cudworth (the greater part of whose excellent work pleases me extremely) that the laws of mechanics alone could not form an animal where there is nothing yet organized" ("Considerations on the Principles of Life," Leibniz 1951, 196).

Leibniz's views, which removed God from the governance of the universe after its initial formation, took rationalism to the extreme. At the same time, something more akin to the older viewpoint was especially to be promulgated by three leading British scientists of the time, Boyle, Ray, and Newton, to whom we now turn.

THE MECHANICAL PHILOSOPHY AND THE ARGUMENT OF DESIGN: BOYLE, RAY, AND NEWTON

Robert Boyle (1627–1691) was the founder of modern chemistry, the first to base chemistry on atomism—or the corpuscular philosophy, as it was called in his time. He was also a devout Christian, having experienced in his youth a religious conversion during a violent thunderstorm in the Swiss Alps (Crowther 1960). Boyle's work most relevant to us is his "Disquisition about the Final Causes of Natural Things: Wherein It Is Inquired, Whether, and (If at All) with What Cautions, a

Naturalist Should Admit Them?" This work was first published in 1688. Boyle considers that

> an inquiry of this kind is now the more seasonable, because two of the chief sects of the modern philosophizers do both of them, though upon differing grounds, deny, that the naturalist ought at all to trouble or busy himself about final causes. For *Epicurus*, and most of his followers (for I except some late few ones, especially the learned *Gassendus*) banish the consideration of the ends of things; because the world being, according to them, made by chance, no ends of any thing can be supposed to have been intended. And, on the contrary, Monsieur *des Cartes*, and most of his followers, suppose all the ends of God in things corporeal to be so sublime, that it were presumption in man to think his reason can extend to discover them. So that, according to these opposite sects, it is either impertinent for us to seek after final causes, or presumptuous to think we may find them. (Boyle 1772, 5: 392–393)

After discussing many examples of the use of final causes in science ("natural philosophy"), he summarizes his conclusions as follows:

> That all consideration of final causes is not to be banished from natural philosophy; but that it is rather allowable, and in some cases commendable, to observe and argue from the manifest uses of things, that the author of nature pre-ordained those ends and uses.
>
> That the sun, moon, and other coelestial bodies, excellently declare the power and wisdom, and consequently the glory of God; and were some of them, among other purposes, made to be serviceable to man.
>
> That from the supposed ends of inanimate bodies, whether coelestial or sublunary, it is very unsafe to draw arguments to prove the particular nature of those bodies, or the true system of the universe.
>
> That as to animals, and the more perfect sorts of vegetables, it is warrantable, not presumptuous, to say, that such and such parts were pre-ordained to such and such uses, relating to the welfare of the animal (or plant) itself, or the species it belongs to: but that such arguments may easily deceive, if those, that frame them, are not very cautious, and careful to avoid mistaking, among the various ends, that nature may have in the contrivance of an animal's body, and the various ways, which she may successfully take to compass the same ends. And,
>
> That, however, a naturalist, who would deserve that name, must not let the search or knowledge of final causes make him neglect the industrious indagation of efficients. (444)

As noted by Crowther (1960, 83), Boyle embodies a tendency that emerged in post-Restoration England, which he calls "accommodation" or "trimming." Rather than confront religion with free thinking based on a mechanical philosophy, British scientists emphasized the compatibility of religion and science. Natural theology helped justify science by showing that it could be of service to religion.

Against that part of the "Epicurean hypothesis" that "ascribes the origin of things to chance," Boyle urges that "there seems to have been care taken, that the body of an animal should be furnished, not only with all things, that are ordinarily necessary and convenient, but with some superabundant provision for casualties. Thus, though a man may live very well, and propagate his kind, (as many do,) though he have but one eye; yet nature is wont to furnish men with two eyes, that, if one be destroyed or diseased, the other may suffice for vision" (428). This doctrine of "superabundant provision" has already been noted in Henry More. Thus, we see that Boyle, like Gassendi (who probably influenced him here), combined an Epicurean atomism and mechanism with a somewhat cautious use of the Stoic argument from design.

Boyle's influence on later ages extended not only from his own work but from a series of lectures he endowed "for proving the Christian religion against notorious infidels, viz. Atheists, Theists, Pagans, Jews, and Mahometans, not descending lower to any controversies, that are among Christians themselves" (quoted in Crowther 1960, 57–58). William Derham's Boyle lectures of 1711 and 1712 were published as *Physico-Theology, or a Demonstration of the Being and Attributes of God from His Works of Creation* (1713), which was to become one of the most popular works on natural theology through the following century.

Just a few years after Boyle published his *Disquisition about Final Causes*, John Ray (1627–1705), the great naturalist and systematist of plants and animals, and who is considered by many the founder of modern botany and zoology, published his *Wisdom of God Manifested in the Works of the Creation* (1691; see Willey 1940 for excellent summaries; Raven 1986). This work, like Derham's *Physico-Theology*, was to be popular throughout the following century, going through four editions in Ray's lifetime and many more after his death.

Finally, this peculiarly British tradition of natural theology was continued by the great Newton. Like his contemporaries Boyle and Ray, Newton (1643–1727) did not allow natural theological speculation to affect his science. In the preface to the first edition of his *Mathematical Principles of Natural Philosophy* (1687), he states that "the whole burden of philosophy seems to consist in this—from the phenomena of motions to investigate the forces of nature, and then from these forces to demonstrate the other phenomena." In the third book of this work, on the "System of the World," he demonstrates this approach "for by the propositions mathematically demonstrated in the former books in the third I derive from the celestial phenomena the forces of gravity with which bodies tend to the sun and the several planets. Then from these forces, by other propositions which are also mathematical, I deduce the motions of the planets, the comets, the moon, and the sea" (Newton 1952a, 1–2). The independence of science from any religious hypothesis could seem no clearer.

Nevertheless, in the "General Scholium" following Book III (added in the second edition of 1713), Newton concludes that "this most beautiful system of the sun, planets, and comets, could only proceed from the counsel and dominion of an intelligent and powerful Being" (1952a, 369). Because God is incorporeal, we can't know him directly. "We know him only by his most wise and excellent contrivances of things, and final causes; . . . a god without dominion, providence, and final causes, is nothing else but Fate and Nature. Blind metaphysical necessity, which is certainly the same always and everywhere, could produce no variety of things." Moreover, "to discourse of [God] from the appearances of things, does certainly belong to natural philosophy" (1952a, 371). In other words, while the assumption of God cannot ground any demonstrations in physics, which must be based on reasoning from phenomena, there is ample justification to reason from these phenomena to the existence and attributes of God. Thus, against Descartes' mechanical model for the origin of the solar system, he argues that "though these bodies [planets and comets] may, indeed, continue in their orbits by the mere laws of gravity, yet they could by no means have at first derived the regular position of the orbits themselves from these laws" (369).

Likewise, in his *Optics* (1706, Bk. III, Pt. I, query 28) in a much-quoted passage discussing the existence of a continuous medium by which light could be propagated by mechanical vibrations, he argues against those (i.e., Descartes, Bacon) who would deny nonmechanical causes:

> Whence is it that Nature doth nothing in vain; and whence arises all that order and beauty which we see in the world? To what end are comets, and whence is it that planets move all one and the same way in orbs concentric, while comets move all manner of ways in orbs very eccentric; and what hinders the fixed stars from falling upon one another? How came the bodies of animals to be contrived with so much art, and for what ends were their several parts? Was the eye contrived without skill in Optics, and the ear without knowledge of sounds? (Newton 1952b, 529).

Newton's views were taken up by his disciples, Bentley and Clarke, who each delivered a series of the Boyle lectures in the early 1700s using Newton's principles to help prove the existence of God. These views were to incense Leibniz, who had been engaged in a controversy with Newton since 1705, beginning with a debate over priority in inventing the calculus and including an accusation by Leibniz that Newton's gravity involved "occult qualities" like those of the Scholastics by invoking action at a distance. In 1715, Leibniz sent a letter to Caroline, Princess of Wales, accusing the Newtonians of weakening religion:

> Sir Isaac Newton, and his followers, have also a very odd opinion concerning the work of God. According to their doctrine, God Almighty wants to wind up his watch from time to time: otherwise it would cease to move. He had not, it seems, sufficient foresight to make it a perpetual motion. Nay, the machine of God's making, is so imperfect, according to these gentlemen; that he is obliged to clean it now and then by an extraordinary concourse, and even to mend it, as a clockmaker mends his work; who must consequently be so much the more unskillful a workman, as he is oftener obliged

to mend his work and to set it right. According to my opinion, the same force and vigour remains always in the world, and only passes from one part of matter to another, agreeably to the laws of nature, and the beautiful pre-established order. And I hold, that when God works miracles, he does not do it in order to supply the wants of nature, but those of grace. Whoever thinks otherwise, must needs have a very mean notion of the wisdom and power of God. (Alexander 1956, 11–12)

Leibniz here refers to Newton's remark that irregularities in the solar system due to the interaction of planets and comets "will be apt to increase, till this system wants a reformation" (*Optics*, query 28). Caroline asked Clarke to reply, initiating a wide-ranging debate with Leibniz (Alexander 1956). In his first reply to Leibniz, Clarke responded to the comparison of Newton's God to an inept watchmaker: "The notion of the world's being a great machine, going on without the interposition of God, as a clock continues to go without the assistance of a clockmaker; is the notion of materialism and fate, and tends, (under pretence of making God a *supramundane intelligence*,) to exclude providence and God's government in reality out of the world" (Alexander 1956, 14). This contrast between the British view (derived from Gassendi and the Cambridge Platonists), emphasizing God's voluntarism and providence, and the Continental view (derived from Descartes), invoking God as the great mechanician, was to continue for many years and, as we will see, has profound consequences for our story.

. . .

In spite of their varying views and internecine battles, all of the founders of modern science—from Galileo and Descartes through Boyle, Ray, and Newton—agreed that the world was created by God according to his design and that certain features of the world can only be understood as part of that design. None appear to have been willing to go anywhere as far as Epicurus in denying design, even Gassendi, who was the foremost promoter of Epicureanism. Moreover, for all of them, final causes are identified with God's intentions; none thought of an intrinsic finality apart from God. What varied among them was merely the degree to which they thought us capable of discerning God's design. And all agreed that—whatever the role of final causes in biology—they had no place in physics proper ("experimental philosophy"), which was the study of world mechanism.

4

Materialism, Teleology, and Evolution in the Enlightenment

According to the expression of Kant, the reason for the manner of being of each part of a living body resides in the whole, whereas in inorganic bodies, each part has it in itself.

—G. CUVIER, 1800, *LEÇONS D'ANATOMIE COMPARÉE*

The Enlightenment has traditionally been seen as the age of reason, when time honored traditions of belief and social structure fell under the withering assault of the free-thinking *philosophes*. The reality was, of course, far more complex than this caricature would indicate. Nevertheless, what is most important for us here is that with the rise of free thought and free expression, Lucretius could once again have his say. Not only did the design argument now come under increasing attack, as in the writings of Holbach and Hume, but heretical alternatives could be more or less openly considered, as witnessed by the early evolutionary speculations of Buffon, Maupertuis, and Diderot. Even the more conservative thinkers were forced to critically examine the ground for traditional views, as Kant did in developing his conception of the nature of the organism, and his associated reevaluation of the role of teleological reasoning in science. As we will see in the next chapter, Cuvier was in every sense a child of the Enlightenment. To understand his theoretical position, we must thus understand the Enlightenment context in which he developed it.

THE ORIGINS OF THE ENLIGHTENMENT: BAYLE

The publication of Newton's *Principia* ushered in the Age of Enlightenment, and the French Revolution ended it. But although Newton's work may have begun the age, its light first and most brightly shined in France, and its first representative is commonly taken to be Pierre Bayle (1647–1706), whose massive *Historical and Critical Dictionary* (1697) cast doubt on all established religion and philosophy by a skeptical juxtaposition and examination of opposing views. As Gibbon commented,

"In reviewing the controversies of the times, he turned against each other the arguments of the disputants; successively wielding the arms of the Catholics and Protestants, he proves that neither the way of authority nor the way of examination can afford the multitude any test of religious truth. . . . He balances the *false* religions in his sceptical scales, till the opposite quantities (if I may use the language of algebra) annihilate each other" (*Autobiography*, 89, quoted in Gay 1967, 292).

Because we know that Cuvier in his Academy days was an avid reader of Bayle (see chap. 5), it is of special interest to examine Bayle's attitude toward the design argument. In the article on Anaxagoras, Bayle appears to accept it: "It is incontestably true, that every Philosopher, who will give a good Account of the Order, which appears in the several parts of the Universe, must suppose an Intelligence as the Author of this beautiful Regularity" (Bayle 1734, Vol. I, "Anaxagoras" [G] VIII: 305). Nevertheless, he discredits the criticism of Anaxagoras by Socrates in Plato's *Phaedo* (quoted here in chap. 3). Socrates had complained that while a divine mind acting for the best should be considered the source of the order of the universe, Anaxagoras instead called on material principles such as "*aēr* and *aithēr* and many other bizarre things." Bayle insists that we have no way of knowing that all is for the best:

> By what particular Reasons can it be proved, that the Perfection of Man, and That of the Universe, require, that our two Eyes be situated as they are, and that six Eyes, placed round the Head, would occasion a Disorder in our Body, and in the Universe? . . . We must stick with this general Reason: The Wisdom of the Artificer is infinite; there-fore the Work is such as it ought to be; the Particulars are out of our reach; and they, who pretend to engage in a Detail of them, generally expose themselves to ridicule. (Vol. I, "Anaxagoras" [R] I: 311)

Similarly, in the article on Leucippus, he notes, "The epithets of madman, dreamer, visionary, are due to whoever imagines, that a fortuitous concourse of infinite corpuscles produced the world" (Vol. III, "Leucippus" [C] II: 789). Nevertheless, he objects that Mind should be considered to create matter or at least give it its movement, not just to arrange it; otherwise, one is using Mind for what is less difficult and ascribing what is more difficult to blind necessity (Vol. II, "Epicurus" [S], II–VII: 786–789). Imagining Epicurus's conversation with a Platonist, he has Epicurus object: "Now you *Platonics* agree, that there has been in Matter a real Defect, which was an Obstacle to God's Project; an Obstacle, I say, which has not permitted God to make a World free from those Disorders we perceive in it: And it is certain, on the other side, that those Disorders render the Condition of Matter infinitely more unhappy than that eternal, necessary, and independent State in which it had been before the Generation of the World" (IV). He continues:

> The most intimate, general, and infallible Notion we have of God is, that God enjoys a perfect Felicity. Now this is incompatible with the Supposition of Providence: For if he governs the World, he has created it; if he has created it, he has either foreseen all

the Disorders that are in it, or he has not foreseen them. If he has foreseen them, it cannot be said that he made the World out of a Principle of Goodness, which destroys the best Answer of the *Platonic*. If he has not foreseen them, it is impossible that, seeing the ill success of his Work, he should not have been extreamly grieved at it. (VI)

Furthermore:

If you suppose afterwards, that, instead of destroying such a Work, he obstinately resolves to preserve it, and continually to be employed either in mending it's Faults, or preventing their Increase; you give us the Idea of the most unhappy Nature that can be conceived. He designed to build a magnificent Palace for the Accommodation of animated Creatures, which were to come out of the shapeless Bosom of Matter, and there to bestow Felicity upon them; but it happened that those Creatures did but eat one another, being incapable to continue alive, if the Flesh of some did not serve as Food to others. It happened that the most perfect of those Animals did not spare even the Flesh of those of his Kind; there happened to be Cannibals among them: And those, who abstained from that Brutality, did not forbear persecuting one another, and were a prey to Envy, Jealousy, Fraud, Avarice, Cruelty, Diseases, Cold, Heat, Hunger, &c. Their Author struggling continually with the Malignity of the Matter productive of these Disorders, and obliged to have always the Thunderbolt in his Hand, and to pour down upon the Earth Pestilence, War, and Famine; which, with the Wheels and Gibbets with which Highways abound, do not hinder Evil from maintaining itself: Can their Author, I say, be looked upon as a happy Being? Can one be happy, when at the end of four thousand Years Labour he has made no further Progress in his Work than the first Day he undertook it, and which he passionately desires to finish? (VII)

Such cogent criticism was to serve as a model for the *philosophes*.

THE *PHILOSOPHES*, MATERIALISM, AND LUCRETIUS (1744–1750)

The rise of materialism in the French Enlightenment is a well-known story. Aram Vartanian (1953) has argued that this materialism derived directly from the Cartesian dualism. By postulating that only the human soul was immaterial and that the development of the world, including the formation of our solar system, the Earth, and plants and animals, could proceed solely from mechanical causes after God created the initial matter and put it in motion, Descartes effectively removed God from his creation. The Cambridge Platonists More and Cudworth had seen the danger of this view for religion (which Cudworth called that of "Mechanick Theists") and promoted the argument of design and vitalism against Descartes' ateleology and mechanism. The physico-theological point of view was perhaps never as popular in France as in England, but it was maintained by a number of writers, including, most prominently, Réaumur in his work on insects (1734–1742), and the abbé Pluche in his *Spectacle de la nature* (1732–1750, both cited in Roger 1997).

Stimulated by Newton's success in providing mathematical laws governing natural phenomena, the rising generation of *philosophes* had good reason to believe that they were indeed living in a century of greater illumination. The forces of reason, toleration, and freedom were on the increase. It seems that the restrictive religious climate of France encouraged a more radical response than the permissive one of England. Although religious restrictions in Catholic France prevented the publication of heretical works, free discussion could take place in the salons, and heretical views could be published if they were anonymous, properly disguised by pious statements, or circulated privately. All of these stratagems were employed.

The rise of atheistic mechanistic materialism among the French *philosophes*, including Diderot, La Mettrie, Buffon, and d'Holbach, is a fascinating episode. By combining an Epicurean antifinalism with Newtonian mechanics, these *philosophes* created a new, completely materialistic worldview. As part of this worldview, they also developed the first modern evolutionary theories. These theories, like their ancient progenitors, were—at least at first—vehemently antiteleological. A series of closely interrelated events over the critical period of 1744–1755 encouraged the growth of these theories. The rise of materialism and evolutionism could perhaps have led to a modern evolutionary theory some hundred years before Darwin, but it didn't. As we will see, this is at least partly because these pioneer materialists abandoned the Epicurean notion of order as merely a necessary condition for existence; instead, they came to explain order by a hylozoism or panpsychism, in which matter itself took on some of the properties of life or mind (see Reill 2005).

Denis Diderot (1713–1784), founder and editor of the famous *Encyclopédie*, well exemplifies this trend (see Vartanian 1953; Crocker 1954; Stewart and Kemp 1963). Traditionally, Diderot is seen as passing from theism to deism to an outright materialism and atheism as his thought matured. However, arguments as to his commitment to a complete materialism and atheism persist in the literature. This is a common problem for this time; the disclaimers needed to distract the censors are often hard to separate from legitimate doubts of the authors, and the common technique of putting words into the mouth of characters in a dialogue (echoing Bayle's technique) makes it difficult to know which views are the author's own. Such cleverness was, in fact, widely admired at the time. Nevertheless, Diderot's remarks in his later *Conversation with the Abbé Barthélemy* (ca. 1772–1773) may perhaps be taken at face value:

> *Diderot:* What is God but a word, a simple vocable to explain the existence of the world? And note well that after all, this word explains nothing; for if you object that no clock has ever been made without a clock-maker, I shall ask you who made the clock-maker, so that we are back at the same mark—at the same mark of interrogation.

> *Barthélemy:* However, haven't you yourself, Diderot, sometimes proclaimed the existence of this clock-maker?

Diderot: Proclaimed! That is saying a great deal.

Barthélemy: A certain letter of yours has been circulated in which you clearly say: "I believe in God, although I get along very well with the atheists."

Diderot: A letter to Voltaire.... That was to please him.... Ah! so you saw that scribble. This is what happens Abbé: when I am with atheists, since there are atheists, all the arguments in favour of the existence of God spring up in my mind; when I happen to be with believers, it's the opposite; I see rise up before me, and in spite of me, everything that combats, saps, and demolishes the Divinity. (Stewart and Kemp 1963, 202–203)

In his first original work, the *Philosophic Thoughts* (1746), Diderot examines the debate among deists, skeptics, and atheists "using Bayle's method of 'impartial' presentation of both sides of controversial questions" (Crocker 1954, 63). Here Diderot recounts his attempt to convince an atheist of God. Diderot warns against granting too much to the atheists:

I open the pages of a celebrated professor and I read: "Atheists, I grant you that movement is essential to matter; what conclusion do you draw from that? That the world is the result of a fortuitous throw of atoms? You might as well tell me that Homer's *Iliad* or Voltaire's *Henriade* is the result of a fortuitous throw of written characters." I would take care not to use that argument against an atheist; he would make quick work of the comparison. According to the laws of the analysis of chances, he would say, I ought not to be surprised that a thing happens when it is possible and when the difficulty of the result is compensated for by the number of throws. (Sec. XXI, Mason 1982, 37–38)

Even at this stage of his development, Diderot is aware that the logical force of the argument of design is vulnerable to the Epicurean hypothesis.

In 1744, only two years before Diderot published his *Philosophic Thoughts*, the Swiss naturalist Abraham Trembley had published his monograph on his experiments on the freshwater polyp, *Hydra*. In showing that the polyp could regenerate complete individuals from cut pieces, Trembley's work supported the materialist notion that the animal soul was coterminous with its body, since it was clearly divisible with the body. Trembley's work made a huge impression on Parisian society of the day (Vartanian 1950; Lenhoff and Lenhoff 1986).

Another important influence on Diderot came from Maupertuis. Pierre-Louis Moreau de Maupertuis (1698–1759) was primarily a physicist. He gained fame in the 1730s by promoting Newtonian physics in Paris, at a time when many in France still upheld Cartesian vortices. He organized an expedition to Lapland that proved that—as Newton had predicted—the Earth was flattened at the poles (see the excellent biographical study of Terrall 2002). In 1745, Maupertuis had published (anonymously) his *Vénus Physique*, in which he adduced evidence from monsters (including an albino Negro boy who had been exhibited in Paris in the previous year), from the resemblance of offspring to both parents, and from Harvey's observations on development to argue against the doctrine of preexistence of germs and for epigenesis. He proposed a theory of pangenesis, of hereditary particles derived from the organs of both parents, to account for the facts of generation. These hereditary particles swim in the seminal fluids of male and female, and they combine by laws of "affinity" to produce the embryo (Terrall 2002, 215–218). He extended his thinking to the origin of human races, suggesting that giants, dwarves, and Negroes might have originated as mutants and been driven out of the temperate zones to found races in less habitable parts of the Earth.

In 1747, Julien de La Mettrie's infamous *Man a Machine* appeared, extending Descartes' hypothesis of animal automatism to humans. La Mettrie was a physician who had studied with the great medical teacher Hermann Boerhaave in Leyden. Two years before he had published his quasi-materialist *Natural History of the Soul*. This was condemned to be burned by Parliament; to wait out the storm, he fled Paris for Holland. Here he composed what is commonly regarded as the first tract of modern materialism and atheism. He drew heavily on recent scientific discoveries to support his argument (opposed to that of merely two years before) that man and animal are both nothing more than machines.

In the following year (1748), de Maillet's speculative *Telliamed, or Conversations between an Indian Philosopher and a French Missionary on the Diminution of the Sea* appeared, though it had been written more than forty years before (de Maillet 1968). This is considered by many the first truly transformist work (Bowler 2003). In the words of Telliamed (his own name backward), De Maillet argued for a geological system in which the continents had all gradually emerged out of the ocean (thus explaining marine fossils on mountains). As part of this system, he advocated the transformation of marine into terrestrial creatures, such as flying fish into birds and sea-men into men.

Thus, when Diderot wrote his *Letter on the Blind, for the Use of Those Who See* (1749), there was much more support for a materialist, atheist, and transformist position than there had been just three years before. Diderot puts his argument in the mouth of the blind Saunderson, a Cambridge mathematician and—according to Diderot—an atheist. Dying, Saunderson is visited by a Mr. Holmes, a minister who attempts to convince him of God's existence using the argument to design. Saunderson replies:

> "But if the animal organism is as perfect as you say, and as I should like to believe . . . what has it in common with a sovereign intelligent being? If it amazes you, perhaps that is because you are in the habit of treating as a miracle everything that appears to be beyond your own capacity. I have so often been an object of admiration for you, that I have a poor opinion of what surprises you. I have drawn here from all parts of England people who cannot conceive how I could do geometry; you must agree that these people had no very clear ideas about the possibilities of things. If a phenomenon is in our opinion beyond the power of man, we say at once: 'It is God's handiwork';

our vanity is content with nothing less. Why cannot we put into our discussion a little less pride and a little more philosophy? If nature offers us a difficult knot to unravel, let us leave it for what it is; do not let us introduce, in order to untie it, the hand of a Being who then at once becomes an even more difficult knot to untie than the first one. Ask an Indian why the world stays suspended in space, and he will tell you that it is carried on the back of an elephant . . . and the elephant on a tortoise. And what supports the tortoise? . . . You pity the Indian; and yet it might be said to you, as to him: Mr. Holmes, my friend, confess your ignorance and spare me your elephant and your tortoise." (Stewart and Kemp 1963, 27–28)

And he goes on:

"I see nothing; I admit, however, an admirable order in everything; but I trust you not to expect anything more of me. I grant it you about the present state of the universe, in order to get from you, in return, the liberty of thinking what I like about its ancient and primitive state, about which you are no less blind than I. You have no evidence to oppose me here, your eyes are of no use to you. You may imagine, if you wish, that that order which impressed you has always existed. But leave me free to think it has done no such thing, and that if we went back to the birth of things and of time, and perceived matter in motion and chaos becoming unravelled, we should encounter a multitude of shapeless beings instead of a few highly organized beings. If I have no objections to offer you about the present condition of things, I can at least question you about their past condition. I can ask you, for example, who told you, Leibnitz, Clarke and Newton, that at the first moment of the formation of animals, some were not without heads, others without feet? I can maintain to you . . . that all the defective combinations of matter have disappeared, and that there have only survived those in which the organization did not involve any important contradiction, and which could subsist by themselves and perpetuate themselves. . . .

"If there had never been any shapeless creatures, you would not have failed to claim that none will ever appear and that I am plunging into fantastic hypothesis; but the order is not yet so perfect that monstrous productions do not appear from time to time. . . . I conjecture then, that in the beginning, when matter in fermentation was hatching out the universe, blind creatures like myself were very common. But why should I not believe about worlds what I believe about animals? How many worlds, mutilated and imperfect, were perhaps dispersed, reformed and are dispersing again at every moment in distant space, which I cannot touch and you cannot see, but where motion continues, and will continue, to combine masses of matter until they shall have attained some arrangement in which they can persist. O philosophers, transport yourselves with me to the confines of the universe, beyond where I can touch and where you can see organized being; move over that new ocean, and seek among its irregular movements some trace of the intelligent Being whose wisdom so astounds you here! But what is the good of taking you out of your element? What is this world? A complex whole subject to revolutions which all indicate a continual tendency to destruction; a swift succession of beings which follow each other, thrust forward and disappear; a transient symmetry; a momentary order." (28–29)

This vision, clearly Lucretian in origin, rules all design out of the universe. Yet it does not go beyond Lucretius to posit an actual evolution or transformation of life; order is a condition for existence, but the combinations are thrown up individually, not in a long, continuous chain of inheritance (Crocker 1954, 1968).

A similar point of view is found in Maupertuis's *Essay on Cosmology*, published the following year (1750). As had Diderot's "Saunderson," he proposed an explanation of biological order and adaptation drawing directly on Lucretius's antiteleology:

> May we not say that, in the fortuitous combination of the productions of Nature, since only those creatures *could* survive in whose organization a certain degree of adaptation was present, there is nothing extraordinary in the fact that such adaptation is actually found in all those species which now exist? Chance, one might say, turned out a vast number of individuals; a small proportion of these were organized in such a manner that the animal's organs could satisfy their needs. A much greater number showed neither adaptation nor order; these last have all perished. . . . Thus the species which we see today are but a small part of all those that a blind destiny has produced. (quoted in Glass 1968, 57–58)

But this Epicurean perspective did not prevail. An alternative point of view, which was to dominate the future debate, had just been published the previous year. Its author was Buffon.

BUFFON, MAUPERTUIS, AND THE BIRTH
OF EVOLUTIONARY THEORY (1749–1755)

Georges-Louis Leclerc, Comte de Buffon (1707–1788), ruled French natural history for almost fifty years (Roger 1997). In his early career, he was a mathematician and physicist; he did some work in probability theory and translated Newton's *Method of Fluxions* and Stephen Hales's *Vegetable Staticks* into French. He was appointed to the Royal Academy of Sciences in 1734 as an astronomer, but he later transferred to the section of Botany due to his increasing interest in the mechanical properties of wood. It was not entirely unreasonable, then, that he was appointed the *intendant* (keeper) of the Royal Botanical Garden *(Jardin du Roi)* in 1739, a position he retained almost until his death. He was to take great advantage of this position; he not only enlarged both the menagerie and the "Cabinet" or museum of natural history but also used it as a base from which to launch his project in natural history.

In 1749, the first three volumes of his great *Natural History, General and Particular* appeared. His project, as he explained it, was both to describe particular facts and to generalize from these, to "grasp distant relationships, fit them together, and form a body of rational ideas, after having justly appreciated their plausibilities and weighed their probabilities" (I: 38–39, quoted in Roger 1997, 83–84).

These volumes contained the "First Discourse" on the aims of the project, a "Theory of the Earth," that attempted to explain the origin of the Earth through natural causes, a consideration of "Reproduction in General," and the first natural history of a particular species, man. The details of the "Theory of the Earth" are not important to us, but it is important to note that in contrast to English writers, such as Thomas Burnet in his *Sacred Theory of the Earth*, Buffon left God out of his account of the formation of the solar system and the Earth. Newton had supposed that the regular motion of the planets in concentric circles around the sun, lying in the same plane and moving in the same direction, provided evidence for a Creator; Buffon tried to explain the same facts as the result of a comet hitting the sun. Likewise, he attempted to account for the main facts of geology through natural processes operating at present intensities; that is, he was a uniformitarian before Hutton or Lyell.

His theory of generation likewise did not require a Creator, at least for its daily operation. Buffon was a friend of both Maupertuis and Diderot, and presumably familiar with their ideas. However, he developed his theory along somewhat different lines. As a civil servant, Buffon also had to be more careful in his expression than either Diderot or Maupertuis, not to say La Mettrie—but even taking this into consideration, his thought often appears contradictory. Nevertheless, he was quite clear on what he did not want to allow in his theory:

> We ought to reject every hypothesis which supposes the thing to be already accomplished, such, for example, as that which supposes the first germ to contain all the germs of the same species, or that every reproduction is a new creation, an immediate effect of the will of the Deity; for all hypotheses of this kind are mere matters of fact, concerning which it is impossible to reason. We must likewise reject every hypothesis which is founded on final causes, such as, that reproduction is ordained in order to replace the living for the dead; that the earth may always be covered with vegetables and peopled with animals; that men may be supplied with abundance of nourishment, &c.; for such hypotheses, in place of explaining the effect by physical causes, stand on no other foundation than arbitrary relations and moral affinities. (Buffon 1791, II: 30)

In place of such hypotheses, Buffon proposed his famous theory of "organic molecules" and "internal molds." He begins with the phenomena of nutrition and growth. If animals maintain their size or grow in size, yet maintain their form, this must be because the matter they take in can be properly distributed to the organs of the body. To do this, he supposes, requires that the "organic molecules" of the food that correspond to those of each organ have some affinity for those organs, comparable to the attraction of gravity (based on mass) that penetrates to the interior of objects. These organs thus form a sort of "internal mold" by which the appropriate organic molecules are directed to their appropriate place.

Likewise, he supposes that in regeneration and asexual reproduction, it is these internal molds that are reproduced:

> For, in an organized and expanded body, nothing farther is necessary for the reproduction of a new body similar to itself, than that it should contain some particle every way similar to the whole. This particle, at its first separation, will not present to our eyes a sensible figure by which we can compare it with the whole body. But, when

separated from the body, and put in a situation to receive proper nourishment, this similar particle will begin to expand and to exhibit the form of an entire and independent being, of the same species with that from which it was detached. Thus, a willow or a polypus, as they contain a larger proportion of particles similar to the whole, than most other substances, when cut into any indefinite number of pieces, each segment becomes a new body similar to the parent from which it was separated. (II: 43–44)

It is but a short step for him from this hypothesis of asexual reproduction to sexual reproduction by a mechanism of pangenesis, similar to that proposed by Maupertuis: "the organic particles sent from all parts of the body into the testicles and seminal vessels of the male, and into the ovarium of the female, compose the seminal fluid, which, in either sex . . . is a kind of extract from the several parts of the body" (51). He considers that the "spermatic animals" seen in semen, and similar particles that he and Needham (erroneously) claimed to see in follicular fluid, may in fact be these "organic particles" or some compound of them.

Buffon's theory is at root vitalistic, in that it supposes a distinction between dead matter that makes up rocks and such and living matter that makes up organisms. Moreover, his theory is teleological; the "internal mold" is effectively the same as the "substantial form" of the Scholastics that traces back to the "formal cause" of Aristotle, and his "organic molecules" strive to realize this form. Order is not, as in the vision of Lucretius, merely a condition for existence of living beings; it is inherent in these organic molecules themselves. The forces he supposes are just those that result in the form actually realized. Nevertheless, he considers his theory mechanistic along the lines of Newton's gravitational theory (rather than the old, crude mechanism of Descartes):

Every attempt to explain the animal oeconomy, and the various motion of the human body, by mechanical principles alone, must be vain and ineffectual: for it is evident, that the circulation of the blood, muscular motion, and other functions of an animated body, cannot be accounted for by impulsion, or by any of the common laws of mechanism. It is equally evident, that growth and reproduction are effects of laws of a different nature. Why, then, do we refuse the existence of penetrating forces which act upon the whole substances of bodies, when we have examples of such powers in gravity, in magnetic attraction, in chemical affinities? Since, therefore, we are assured by facts, and by a number of constant and uniform observations, that there are powers in nature which act not by impulsion, why are not these powers ranked among mechanical principles? (53)

In Buffon, we see developed a tendency already intimated in Maupertuis's *Physical Venus* and La Mettrie's *Man a Machine*: in place of the old (and now-discredited) theory of Epicurus, in which chance produces order, order is now attributed to the laws of nature. To attribute order to the self-organizing properties of nature is not incompatible with recognition of the conditions for existence as a principle of order, but, in their enthusiasm for the powers of matter, the materialists tended to forget the Epicurean hypothesis. In this way, rather ironically, the new materialism also became a sort of vitalism, not so far removed from that of the Cambridge Platonist Cudworth and his "Plastick Powers" of Nature, though of course material, not spiritual (again, see Reill, 2005).

An important source of empirical confirmation of such ideas was the microscopical work of the Englishman John Turberville Needham in collaboration with Buffon, first published in 1750 in French translation. This work supposedly demonstrated the spontaneous generation of animalcules in boiled infusions of mutton broth. Redi had shown in the seventeenth century that maggots do not spontaneously generate in rotting meat, and Malpighi's (faulty) observations on chicks had supported the doctrine of preformation, obviously inconsistent with spontaneous generation. By claiming to show that animalcules (protozoa) could be generated from dead matter, Needham's work encouraged the materialists even more toward a vitalistic viewpoint.

This work certainly had a major impact on Maupertuis; in the summer of 1750, he read Needham's book and was overwhelmed. "What a new universe!" he exclaimed (quoted in Terrall 2002, 318). In his *System of Nature* (1752), Maupertuis went beyond his stance in the *Physical Venus* and argued that simple chemical affinities couldn't account for the organization of plants and animals. Instead, "we must have recourse to some principle of intelligence, to something similar to what we call desire, aversion, memory" in matter (quoted in Terrall 2002, 328). This does not conflict with religion, because "if we admit without peril some principle of intelligence in large collections of matter; such as the bodies of animals, what greater peril will we find in attributing such a principle to the smallest particles of matter?" (Terrall 2002, 330). This property of matter can explain embryonic development, in which each material particle "retains a kind of memory of its previous situation and will resume it whenever it can, so as to form the same part in the fetus" (328). Mistakes (mutants) occur when memory is imperfect. Maupertuis thus went beyond Buffon's division of matter into organic and dead, to propose that all matter has some of the properties of mind—a panpsychism.

Maupertuis distinguished his view of active, sentient matter from that of Lucretius, "an impious philosopher" admired by "the libertines of our day" (331). Instead, he said, "God, in creating the World, endowed each particle of matter with this property, by which he wished all the individuals he had formed to reproduce themselves. And since intelligence is necessary for the formation of organized bodies, it seems greater and more worthy of the Divinity that they form themselves by the properties that [God] distributed all at once to the elements, than if these bodies were in every instance the immediate productions of his power" (quoted in Terrall 2002, 331). Maupertuis here indeed sounds rather like Cudworth (and not that different from Darwin, some hundred years later).

However, the same essay proposing sentient matter contains further development of Maupertuis's pioneer observations on genetics and speculation on evolution. In particular, he traced the pedigree of six-digitism in a Berlin family, showing clearly that both males and females could transmit the trait and that it was inherited with what we would now call variable penetrance. He also recounted his breeding experiments with dogs, in which a coat color variant reappeared in two successive generations. "These varieties, once they are confirmed by a sufficient number of generations where they appear in both sexes, found a species; and it is perhaps thus that species have multiplied" (quoted in Terrall 2002, 338). In fact, "could we not explain in this manner how the multiplication of the most dissimilar species could have sprung from just two individuals? They would owe their origin to some fortuitous productions in which the elementary parts deviated from the order maintained in the parents. Each degree of error would have created a new species, and as a result of repeated deviations the infinite diversity of animals that we see today would have come about" (338). This evolutionary theory, perhaps the first to rightly deserve the name, is not too distant from Darwin's; it is especially close to that later formulated by de Vries under the name of the "Mutation Theory" (Glass 1968; Terrall 2002).

Also appearing around the same time (1753) was the fourth volume of Buffon's *Natural History*, containing his "Discourse on the Nature of Animals" and the "Natural History of the Horse and the Ass." In the "Discourse," Buffon upheld the Cartesian distinction between man and animals, and attacked the natural theology of Réaumur: "Who, indeed, has the grander view of the Supreme Being, he who sees Him as the creator of the Universe, the orderer of all things, founder of Nature on invariable and eternal laws, or he who seeks Him and wants to find Him occupied with leading a republic of flies and very busy with the way a beetle's wing should fold?" (quoted in Roger 1997, 241). More important for the development of evolutionary theory, in the "Natural History of the Ass," Buffon explicitly raises the question of evolution: could it be that the horse and the ass, though "essentially different" (Buffon 1791, III: 399), spring from the same stock (as the classification of them in the same genus by Linnaeus would suggest)? Daubenton, Buffon's collaborator, had described the homologies between the horse's limb and the human limb. Buffon extended this principle to all vertebrates: "There exists . . . a primitive and general design, which may be traced to a great distance, and whose degradations are still slower than those of figure or other external relations: For . . . there is, even among the parts that contribute most to variety in external form, such an amazing resemblance as necessarily conveys the idea of an original plan upon which the whole has been conceived and executed" (III: 400). Buffon asked whether this "does not indicate, that the Supreme Being, in creating animals, employed only one idea, and, at the same time, diversified it in every possible manner, to give men an opportunity of admiring equally the magnificence of the execution and the simplicity of the design?" (III: 402). Such a final cause goes against Buffon's entire approach and was likely disingenuous, especially when one considers that Buffon had been forced by the Sorbonne to include an explicit declaration of faith at the beginning of this very volume. The danger that Buffon perceived (or promoted?) is quite clear:

> If it were proved, that animals and vegetables were really distributed into families, or even that a single species was ever produced by the degeneration of another, that the ass, for instance, was only a degenerated horse, no bounds could be fixed to the powers of Nature: She might, with equal reason, be supposed to have been able, in the course of time, to produce, from a single individual, all the organized bodies in the universe.
>
> But this is by no means a proper representation of Nature. We are assured by the authority of revelation, that all animals have participated equally of the favours of creation; that the two first of each species were formed by the hands of the Almighty; and we ought to believe that they were then nearly what their descendants are at present. (III: 403–404)

Yet, given his later development, it doesn't seem likely that Buffon fully accepted an evolutionary explanation, either.

Maupertuis's *System of Nature* and Buffon's speculations had a profound effect on Diderot, and the following year (1754), he published his *Thoughts on the Interpretation of Nature*. After his imprisonment, Diderot clearly had to be careful about how he expressed his ideas; he could only discuss heretical hypotheses, not adopt them. In this way Diderot manages to give a rather clear statement of the hypothesis of evolution, drawing on both Buffon and Maupertuis (who had written under the pseudonym of Dr. Baumann):

> Nature seems to take pleasure in varying a single mechanism in an infinite variety of ways. She never abandons any one kind of production until she has multiplied the individual examples to produce as many different aspects as possible. If we consider the animal kingdom and perceive that there is not a single quadruped whose functions and parts, above all the internal ones, are not entirely similar to those of another quadruped, would it not be easy to believe that in the beginning there was only one animal, a prototype of all animals, certain of whose organs nature has merely lengthened, shortened, transformed, multiplied or obliterated? . . . But whether we accept this philosophical conjecture, as Dr Baumann does, or reject it as false, as M. de Buffon does, we cannot deny that it must be accepted as a hypothesis essential to the progress of experimental science, to the progress of rational philosophy, and to the discovery and explanation of phenomena which depend on physical constitution. (XII, Mason 1982, 63–64)

One more event of this period is worth mentioning. In the fifth volume of his *Natural History* (1755)—in the article on the pig—Buffon once again denied final causes, in a passage that we will revisit when we come to Cuvier. The curiously reduced lateral digits of the pig are the occasion. Speaking of such anomalies of nature, he states:

> To circumscribe the sphere of Nature is not the proper method of acquiring the knowledge of her. We cannot judge of her, by making her act agreeably to our particular and limited views. We can never enter deeply into the designs of the Author of Nature, by ascribing to him our own ideas. Instead of limiting the powers of Nature, we ought to enlarge and extend them; we should regard nothing as impossible, but believe that every thing which can have existence, does really exist. Ambiguous species, and irregular productions, would not then excite surprise, but appear to be equally necessary as others, in the infinite order of things. They fill the intervals, and constitute the intermediate links of the chain. These beings present to the human intellect curious examples, where Nature, by appearing to act upon an unusual model, makes a greater display of her powers, and affords us an opportunity of recognising singular characters, which indicate that her designs are more general than our contracted views, and that, if she has made nothing in vain, neither are her operations regulated by the designs which we attribute to her.
>
> Does not this singular conformation of the hog merit a few reflections? He appears not to have been constructed upon any original or perfect model; for he is a composition of different animals. Some of his parts, for example, the toes above described, the bones of which are perfectly formed, are evidently of no use to him. Nature, therefore, in the construction of beings, is by no means subjected to the influence of final causes. Why should she not sometimes give redundant parts, when she often denies those which are essential? How many animals are deprived of senses and of members? Why should we imagine, that, in each individual, every part is useful to its neighbour, and necessary to the whole? Is it not enough that they exist together, that they never injure each other, that they can grow and expand without mutual destruction? Every thing which is not so hostile as to destroy, every thing that can subsist in connection with other things, does actually subsist: And, perhaps, in most beings, there are fewer relative, useful, or necessary parts, than those which are indifferent, useless, or redundant. But, as we always wish to make every things refer to a certain end, when parts have no apparent uses, we either suppose that their uses are concealed from us, or invent relations which have no existence, and tend only to throw an obscure veil over the operations of Nature. It is the intention of true philosophy, to instruct us *how* objects exist, and the manner in which Nature acts: But we pervert this intention, by attempting to investigate *why* objects are produced, and the ends proposed by Nature in producing them. (Buffon 1791, III: 503–505)

This view again is a Lucretian vision (with a splash of Descartes) and can be interpreted as an eloquent expression of the conditions for existence. But it is important to note that Buffon's "All that can subsist together does subsist" is ambiguous. It could mean only "Every subsisting thing has parts that can subsist together." This is the principle of the conditions for existence. But it could also mean "All possibilities for existing things are realized," which would make the idea a typical expression of

Leibniz's principle of plenitude (Lovejoy 1936) and thus teleological in a broad sense. The introductory passage in which he states "every thing which can have existence, does really exist" suggests that this latter is closer to his view. Buffon goes on to cite several additional examples of cases where apparently nonfunctional structures exist (reduced digits, the allantois, male nipples). Buffon appears to be perfectly Cartesian here; for the ends of nature, he would substitute "physical relations."

THE LATER ENLIGHTENMENT:
D'HOLBACH AND HUME

The promising beginning given to evolutionary theory by the speculations of Buffon, Maupertuis, and Diderot in the early 1750s did not go any further. The Seven Years War broke out in 1756. Maupertuis died in 1759. Diderot was occupied with the *Encyclopédie*, art criticism, and other endeavors. Buffon continued to publish volumes of the *Natural History*, embracing a limited evolutionism in the treatises on "Animals Common to the Two Continents" (1762) and "Of the Degeneration of Animals" (1766). Meanwhile, the doctrine of the preexistence of germs was revived by Bonnet (who had discovered parthenogenesis in aphids) and supported by Haller; in 1769, Spallanzani's work disproving Needham's spontaneous generation appeared in French translation (Roger 1997, 343).

Against this background, a work appeared in 1770 that was at once the most thoroughgoing and complete system of atheism and materialism since that of Lucretius: Baron d'Holbach's *System of Nature; Or the Laws of the Moral and Physical World* (d'Holbach 2004). This work (published under a pseudonym) openly promoted atheism and materialism, and blamed most of the ills of humanity on religion and superstition. Holbach was a wealthy German who had lived in Paris since 1749. He was a friend of Diderot's and many other *philosophes*, and he had contributed a number of articles to the *Encyclopédie*. Nevertheless, he had been so restrained in his public utterances that he was not widely suspected of authoring this work (to which, in fact, Diderot may have contributed).

For Holbach, matter in motion was all there is. Motion is intrinsic to matter. Unlike Buffon, Maupertuis, and Diderot, he did not attribute intelligence, sensibility, or any such properties to matter. For Holbach, intelligence is always associated with a particular organization of matter: "matter, which is regarded as inert and dead, assumes sensible action, intelligence, and life, when it is combined and organized after particular modes" (d'Holbach 2004, 1: 53). He supported, like Democritus, a complete determinism: all occurs according to the immutable laws of nature; what we call chance is merely relative to our own lack of knowledge. Perhaps because of his acceptance of spontaneous generation, he only argues for an evolutionary hypothesis generally: "With respect to those who may ask why Nature does not produce new beings? we may enquire of them in turn, upon what foundation they suppose this fact?" (1: 64). For him it is unimportant whether man has always existed, though "some reflections seem to favor the supposition, to render more probable the hypothesis, that man is a production formed in the course of time." This view he bases on the law of adaptation; since the conditions on the planet have not always been the same, it is unlikely that the species have been:

> All productions, that they may be able to conserve themselves, or maintain their actual existence, have occasion to co-order themselves with the whole from which they have emanated. Without this they would no longer be in a capacity to subsist: it is this faculty of co-ordering themselves,—this relative adaptation, which is called the ORDER OF THE UNIVERSE: the want of it is called CONFUSION. Those productions which are treated as MONSTROUS, are such as are unable to co-order themselves with the general or particular laws of the beings who surround them, or with the whole in which they find themselves placed: they have had the faculty in their formation to accommodate themselves to these laws; but these very laws are opposed to their perfection: for this reason they are unable to subsist. (1: 62–63)

This clear statement of the principle of the conditions for existence with respect to environmental adaptation is echoed in a consideration of functional adaptation:

> It is said, that animals furnish a convincing proof of the powerful cause of their existence; that the admirable harmony of their parts, the mutual assistance they lend each other, the regularity with which they fulfill their functions, the preservation of these parts, the conservation of such complicated wholes, announce a workman who unites wisdom with power; in short, whole tracts of anatomy and botany have been copied to prove nothing more than that these things exist, for of the power that produced them there cannot remain a doubt. We shall never learn more from these erudite tracts, save that there exists in nature certain elements with an aptitude to attraction; a disposition to unite, suitable to form wholes, to induce combinations capable of producing very striking effects. To be surprised that the brain, the heart, the arteries, the veins, the eyes, the ears of an animal, act as we see them—that the roots of plants attract juices, or that trees produce fruit, is to be surprised that a tree, a plant, or an animal exists at all. These beings would not exist, or would no longer be that which we know they are, if they ceased to act as they do: this is what happens when they die. (2: 105–106)

Not surprisingly, the *System of Nature* was received with horror by the educated reading public, although they did read it. Not only was its thesis repugnant to the sensibilities of the time, but its style was also unpleasantly harsh compared to the works of a Voltaire, a Diderot, a Buffon. Looking back later on his youthful reaction, Goethe described it as "so gloomy, so Cimmerian, so deathlike, that we found it difficult to endure its presence, and shuddered at it as at a spectre" (*Poetry and Truth*, quoted in Willey 1940: 166). D'Holbach's encouragement of atheists and materialists, the culmination of the trend of French free thought in the Enlightenment, was considered by many to have helped unleash the horrors of the French Revolution.

As the last representative of Enlightenment free thinking, I turn to the Scot David Hume (1711–1776). Hume really belongs to an earlier phase of our history; his *Treatise of Human Nature* (1739–1740), written in the years 1734–1737 when he was living in France, draws largely on the empiricism of Locke and Newton, and the skepticism of Bayle. His thought developed in parallel with that of the French free thinkers and apparently there is no evidence for his having been significantly influenced by them, even though in the 1760s Hume came to Paris (as the secretary of the British Embassy) at the invitation of d'Holbach and frequented d'Holbach's salon.

The work that especially concerns us is his *Dialogues Concerning Natural Religion*, which is an extended critique of the argument of design; as noted earlier, philosophers generally consider this to have destroyed the argument. The *Dialogues* were composed in the early 1750s but only published posthumously, in 1779. They are modeled on Cicero's *De natura deorum*; like it, they involve three protagonists, here Demea (a traditional religionist), Cleanthes (the Stoic, or rational theist, corresponding to Balbus), and Philo (the skeptic, corresponding to Cotta, but also using some of Velleius's arguments). Norman Kemp Smith (1947) has argued convincingly that Philo represents Hume's views throughout.

In Part VIII of the work, Philo asks, "What if I should revive the old Epicurean hypothesis? This is commonly, and I believe, justly, esteemed the most absurd system, that has yet been proposed; yet, I know not, whether, with a few alterations, it might not be brought to bear a faint appearance of probability" (Hume 1947, 182). Instead of considering matter infinite, Philo continues, consider it finite but time infinite, and matter continually in movement. If so, we may ask:

> Is there a system, an order, an oeconomy of things, by which matter can preserve that perpetual agitation, which seems essential to it, and yet maintain a constancy in the forms, which it produces? There certainly is such an oeconomy: For this is actually the case with the present world. The continual motion of matter, therefore, in less than infinite transpositions, must produce this oeconomy or order; and by its very nature, that order, when once established, supports itself, for many ages, if not to eternity. But wherever matter is so poised, arranged, and adjusted as to continue in perpetual motion, and yet preserve a constancy in the forms, its situation, must, of necessity, have all the same appearance of art and contrivance which we observe at present. (183)

And Philo here draws the same conclusion as d'Holbach:

> It is in vain, therefore, to insist upon the uses of the parts in animals or vegetables, and their curious adjustment to each other. I would fain know how an animal could subsist, unless its parts were so adjusted. Do we not find, that it immediately perishes whenever this adjustment ceases, and that its matter corrupting tries some new form? It happens, indeed, that the parts of the world are so well adjusted, that some regular form immediately lays claim to this corrupted matter: And if it were not so, could the world subsist? Must it not dissolve as well as the animal, and pass through new

positions and situations; till in a great, but finite succession, it fall at last into the present or some other order? (185)

To this clear statement of the principle of the conditions for existence, Cleanthes replies in a manner that should by now be familiar (see the discussion of Henry More earlier in this chapter), stressing the superfluous and contingent abundance of the universe:

> But according to this hypothesis, whence arise the many conveniences and advantages which men and all animals possess? Two eyes, two ears, are not absolutely necessary for the subsistence of the species. Human race might have been propagated and preserved, without horses, dogs, cows, sheep, and those innumerable fruits and products which serve to our satisfaction and enjoyment. If no camels had been created for the use of man in the sandy deserts of Africa and Arabia, would the world have been dissolved? If no loadstone had been framed to give that wonderful and useful direction to the needle, would human society and the human kind have been immediately extinguished? Though the maxims of nature be in general very frugal, yet instances of this kind are far from being rare; and any one of them is a sufficient proof of design, and of a benevolent design, which gave rise to the order and arrangement of the universe. (185)

Hume's sense for the weakness of the argument couldn't be clearer.

KANT AND THE GERMAN ENLIGHTENMENT

To close out my account of the history of the debate over design, I turn to Germany and to Kant. Kant is a key figure in any account of the history of thought on teleology, but he holds a special place in our history because—as I will argue in the next chapter—Cuvier likely developed his principle of the conditions for existence under Kant's influence. To understand the context in which Kant developed his characteristic teaching on teleology in nature, as expressed in the *Critique of Judgment* (1790), it is necessary to take a step back and briefly look at the development of philosophy in Germany in the early eighteenth century.

The starting point for German philosophy of the time was the rationalist philosophy of Leibniz and its principles of continuity, sufficient reason, and preestablished harmony. Christian Friedrich Wolff (1679–1754), who became a professor of mathematics and natural science at the University of Halle in 1706 under Leibniz's sponsorship, was his most influential follower (Beck 1969). He published a large series of books on philosophy, generally entitled "Rational Thoughts on . . . ," which were adopted as textbooks throughout Germany. Wolff's rationalist teachings put him at odds with the Pietists (traditional religionists within the Lutheran tradition), however, and in 1723 they succeeded in having him banished. Wolff was ordered by Friedrich Wilhelm I to leave Prussia within forty-eight hours or be hanged, for having allegedly taught a determinism that implied that army deserters should not be punished, because they could not help their actions. Wolff settled in Marburg, but his fame grew due to his banishment. In 1740, Frederick the Great ascended to the throne, ushering in a period of free thinking in Germany on the French model (both Maupertuis and La Mettrie took refuge here). One of Frederick's first acts as sovereign was to recall Wolff to Halle.

Wolff's philosophy was dogmatic, long-winded, and scholastic in its presentation. Although there were some differences from Leibniz (Beck 1969), for his generation and after there was a single "Leibniz-Wolffian philosophy." Yet two features of his philosophy are worth mentioning, especially in light of the fact that one of Cuvier's philosophy professors at the Karlsschule was still a die-hard Wolffian in 1784 when Cuvier began his studies there. First, Wolff defined philosophy in general as "the science of the possibles insofar as they can be" (Wolff 1963, sec. 29). He thus focused attention on the relation between possibility and actuality. For Wolff, like Leibniz, possibility was based on lack of logical contradiction. Existence, in contrast, was the "completion of possibility" and thus involved something more than mere possibility. This "something more" was based on the principle of sufficient reason: whatever exists must not only be possible but must have a sufficient reason for existing.* But since the principle of sufficient reason was also based on the lack of contradiction (because it would be contradictory to suppose that God would choose what is worse), then everything that is truly possible, given the constraints imposed by the structure of this best of all possible worlds, exists. This is what Lovejoy (1936) called the principle of plenitude.

Second, Wolff was the first to introduce the term *teleology* into philosophy. For him, teleology is a part of natural science: "A twofold reason can be given for natural things. One reason is to be found in the efficient cause, and the other reason in the final cause. Reasons which are sought in the efficient cause belong to the sciences which we have already defined. Besides these sciences there is still another part of natural philosophy which explains the end of things. There is no name for this discipline, even though it is very important and most useful. It could be called teleology" (Wolff 1963, sec. 85). For Wolff, as a rationalist, truth should be deducible from rational premises. In physics, one experiments to confirm the rational explanation of bodies on the basis of efficient causes (experiments are necessary only because we cannot penetrate to the ultimate reasons for things). Similarly, "teleology is experimental theology insofar as it confirms by a contemplation of the works of nature what is demonstrated about God in natural theology" (sec. 107). In other words, teleology is the argument *from* design, natural theology the argument *to* design. In practice, Wolff's natural theology resulted in a teleological optimism like

---

* Recall that this was how Leibniz had inferred that this is the best of possible worlds: among all logically possible worlds, God would certainly choose the best, because there must be sufficient reason for his choice.

Leibniz's, but more anthropocentric (Beck 1969). This characteristic Wolffian didactic rationalism was the dominant philosophy in Germany in the first half of the eighteenth century and well into the second. It was in the context of the Leibniz-Wolffian philosophy and its critique that Kant developed his philosophy.

Immanuel Kant (1724–1804) was born in Königsberg in East Prussia and lived there his entire life, aside from a short time in a nearby village as a tutor. He entered the university in 1740, the same year that Frederick the Great ascended to the throne, and came under the influence of the philosopher Martin Knutzen. Although Kant was a student in theology, under Knutzen's influence he became interested in the sciences, specifically Newtonian physics, and his earliest writings are in physics, or "natural philosophy," not philosophy as such.

In his *Universal Natural History and Theory of the Heavens, or An Essay on the Constitution and Mechanical Origin of the Whole Universe Treated According to Newton's Principles* (1755), Kant proposed a mechanistic account for the origin and organization of the universe as a whole, and the solar system in particular, based on Newton's law of universal gravitation. He also speculated on the possibility of intelligent life on other planets. It is well to remember that this work appeared soon after Maupertuis's *Essay on Cosmology* (1751) and Buffon's first volumes of the *Natural History* (1749). Kant's theory has obvious similarities both to Descartes' theory of vortices and Laplace's later "nebular hypothesis." It is also similar to the ancient theories of Democritus and Epicurus. However, Kant asserts that "there yet remains an essential difference between the ancient cosmogony and that which I present, so that the very opposite consequences are to be drawn from mine" (Kant 1969, 25). What is this difference?

> The teachers of the mechanical production of the structure of the world referred to, derive all the order which may be perceived in it from mere chance which made the atoms to meet in such a happy concourse that they constituted a well-ordered whole.... All these theorizers pushed this absurdity so far that they even assigned the origin of all animated creatures to this blind concourse, and actually derived reason from the irrational. In my system, on the contrary, I find matter bound to certain necessary laws. Out of its universal dissolution and dissipation I see a beautiful and orderly whole quite naturally developing itself. This does not take place by accident, or of chance; but it is perceived that natural qualities necessarily bring it about. And are we not thereby moved to ask, why matter must just have had laws which aim at order and conformity? Was it possible that many things, each of which has its own nature independent of the others, should determine each other of themselves just in such a way that a well-ordered whole should arise therefrom; and if they do this, is it not an undeniable proof of the community of their origin at first, which must have been a universal Supreme Intelligence, in which the natures of things were devised for common combined purposes? (25–26)

In other words, at this stage Kant is fully in the tradition of the Leibniz-Wolffian philosophy and accepts the argument to design, or what he would later call the physico-theological argument for the existence of God. In doing so, he continues not only the tradition of Leibniz and Wolff but also of Newton (and Maupertuis). However, like Leibniz and Wolff, and especially Maupertuis, but unlike Newton, he insists that all the laws of matter are sufficient to produce the order we see, without the direct intervention of God, and he goes on to promote the argument to design against the "freethinker" who would argue that attributing all order to the laws of nature suggests that there is no need for God. This is again entirely in the tradition of Leibniz and Wolff, and in particular, Leibniz's notion of the "pre-established harmony" of mechanical and teleological laws. Nevertheless, Kant's interest is in the mechanical laws, not in teleological explanations. In this he reminds one more of Maupertuis than of Leibniz or Wolff.

One last point about this early work is worth mentioning. Although Kant is happy to follow Descartes and ascribe the origin of the order of the universe to mechanical laws, to say: "*Give me matter, and I will construct a world out of it!*'" in accordance with those laws, he is not so confident about the ability of mechanism to account for life. "But can we boast of the same progress even regarding the lowest plant or an insect? Are we in a position to say: '*Give me matter, and I will show you how a caterpillar can be produced*'? Are we not arrested here at the first step, from ignorance of the real inner conditions of the object and the complication of the manifold constituents existing in it?" (29). As a scientist, he understands already at this stage the difference between conceiving that a mechanical account based on known laws is possible and being able to provide—or even sketch the outlines of—such an account.

A few years later, in 1763, we find Kant promoting a version of what he calls the "ontological proof" for God's existence in *The Only Possible Argument in Support of a Demonstration of the Existence of God* (Kant 1992). Kant's proof of the existence of God is based on the "internal possibility" of all things. As we have seen, Leibniz and Wolff had asserted that all existence was based on possibility, because God would not deny existence to anything possible, as long as it had sufficient reason for existing (the principle of plenitude). Kant begins by arguing that existence is not a predicate, because the existence of an object must be posited before it can have any predicates whatsoever. Thus, the mere possibility (freedom from logical contradiction) of the concept of an object tells us nothing about the existence of the object; existence does not complete possibility but is rather something else entirely. For Kant, possibility itself presupposes the existence of something. This something is, of course, God, and he goes on to sketch his argument for the existence of God as the necessary ground for the internal possibility of all things.

Once again, however, biology doesn't fit into this Leibnizian-Wolffian-Newtonian mechanical world picture. In examining the problem of biology, or, specifically, of biological development, Kant begins with inorganic forms such as snow crystals, in which their form is explicable on the basis of the laws of nature. To such productions of nature he contrasts plants and animals:

> Nonetheless, nature is rich in another kind of production. And here, when philosophy reflects on the way in which this kind of product comes into existence, it finds itself constrained to abandon the path we have just described. There is manifest in this case great art and a contingent combination of factors which has been made by free choice in accordance with certain intentions. Such art and free choice are the ground of a particular law of nature, which itself belongs to an artificial order of nature. The structure of plants and of animals displays a constitution of this kind; and it is a constitution which cannot be explained by appeal to the universal and necessary laws of nature. . . .
>
> For example: in the light of everything we know, it is utterly unintelligible to us that a tree should be able, in virtue of an internal mechanical constitution, to form and process its sap in such a way that there should arise in the bud or the seed something containing a tree like itself in miniature, or something from which such a tree could develop. The internal forms proposed by *Buffon*, and the elements of organic matter which, in the opinion of *Maupertuis*, join together as their memories dictate and in accordance with the laws of desire and aversion, are either as incomprehensible as the thing itself, or they are entirely arbitrary inventions. But, leaving aside such theories, is one obliged for that reason to develop an alternative theory oneself, which is just as arbitrary, the theory, namely, that since their natural manner of coming to be is unintelligible to us, all these individuals must be of supernatural origin? . . . whether the supernatural generation occurs at the moment of creation, or whether it takes place gradually, at different times, the degree of the supernatural is no greater in the second case than it is in the first. . . . The purpose of these considerations has simply been to show that one must concede to the things of nature a possibility, greater than that which is commonly conceded, of producing their effects in accordance with universal laws. (Kant 1992, 156–157)

One can't help but feel that Kant here is the better scientist than Buffon or Maupertuis. He is willing to concede the possibility that mechanical laws can generate life, but he is unwilling to ascribe such a mechanical generation of life to forces that are "as incomprehensible as the thing itself" or "entirely arbitrary inventions." Nevertheless, he still cannot conceive that the origin of these forms themselves could be due to mechanical causes (i.e., he discounts spontaneous generation); the contingency in their combination is too great.

THE CRITICAL PHILOSOPHY

In 1781, at the age of fifty-seven, Kant entered the stage of history with the appearance of his *Critique of Pure Reason*. The radical "Copernican revolution" of his critical philosophy depended on a reevaluation of the nature of our knowledge of the world (for overviews of this philosophy in its biological context, see

McLaughlin 1990; Grene and Depew 2004; Huneman 2007). Rather than considering all our ideas derived from the senses, like the empiricists, or as based on logical inference from innate truths, like the rationalists, Kant argued that the structure of our mind imposes a structure on our knowledge of the world. He called this point of view "transcendental idealism." We cannot know things-in-themselves, or *noumena*; we can only know things as appearances, or *phenomena*. Starting from a consideration of the nature of causality, he defined twelve innate "categories" through which we structure our knowledge of the world. For Kant, our use of these a priori categories is justified by the consideration that without them, no objective experience would be possible at all. This is not the place to go further into Kant's overall philosophy, although I will have occasion to return to a few relevant aspects later. For us, it will be more useful to turn now directly to consideration of his treatment of teleology in the third critique, the *Critique of Judgment*.

The *Critique of Judgment* appeared in 1790, nine years after the first edition of the *Critique of Pure Reason*. The intervening years had seen Kant publish the *Critique of Practical Reason*, dealing with morality, or human action under the concept of freedom, and the second edition of the *Critique of Pure Reason*, as well as a large number of articles. It is worth noting that this was after Cuvier's school years, although we have good reason to believe that Cuvier read the work soon after its appearance. Kant tells us that the *Critique of Judgment* "concludes his entire critical enterprise," and we thus might view it as a mopping-up operation. This would explain why it deals with two areas that seem unrelated: aesthetics and teleology. For Kant, however—who is a systematic thinker in the tradition of Leibniz and Wolff—these areas are intimately related and form a crucial link in the critical system.

For Kant, "Judgment in general is the ability to think the particular as contained under the universal" (Kant 1987, 179; all page numbers refer to the Akademie edition). He distinguishes two types of judgment, determinative and reflective. "If the universal (the rule, principle, law) is given, then judgment, which subsumes the particular under it, is *determinative*. . . . But if only the particular is given and judgment has to find the universal for it, then this power is merely *reflective*" (179). The foundation of the *Critique of Judgment* lies in Kant's argument that reflective judgment has an a priori principle that lies at the base of all investigations of nature—namely, that of the purposiveness of nature for our understanding. To have any experience at all (to reflectively judge the world in such a way that we can experience it), we must treat the world as if it had been purposely arranged for us to have a coherent experience. Kant gives as examples the maxims "Nature takes the shortest way," that "its great diversity in empirical laws is nonetheless a unity under a few principles" (182), and "that there is in nature a subordination graspable by us of species under genera; that genera in turn approach one another under some common principle" (185). Such maxims are not theoretically justified and thus do not allow us to determine experience in accordance with them; instead, they are guides to investigation, allowing us to reflect on experience by assuming its purposiveness for the understanding. This assumed purposiveness of nature for our understanding is what underlies, and thus connects, the powers of aesthetic and teleological judgment.

In the first section of the *Critique of Teleological Judgment* (the Analytic), Kant argues that it is the obvious contingency of organisms that first leads us to conceive of an objective purposiveness in nature, in addition to mere mechanism. We must assume that the basis for the functional unity of organisms is a purpose, because we can't determine their structure and development on the basis of physical laws. In general, "To say that a thing is possible only as a purpose is to say that the causality that gave rise to it must be sought, not in the mechanism of nature, but in a cause whose ability to act is determined by concepts" (369). Now for Kant there are two types of purposes, purposes of art (*techne*—i.e., human purposes) and natural purposes. If we find a hexagon in the sand, we assume that it could not have been formed by the mechanical laws of nature but must have been formed in accordance with a prior concept of a hexagon. This prior concept is a purpose of *art*:

> If, on the other hand, we cognize something as a natural product and yet are to judge it to be a purpose, and hence a *natural purpose* . . . then we need more. I would say, provisionally, that a thing exists as a natural purpose if it is *both cause and effect of itself* (though of itself in two different senses). For this involves a causality which is such that we cannot connect it with the mere concept of a nature without regarding nature as acting from a purpose; and even then, though we can think this causality, we cannot grasp it. Before we analyze this idea of a natural purpose in full, let me elucidate its meaning by the example of a tree.
>
> In the first place, a tree generates another tree according to a familiar natural law. But the tree it produces is of the same species. Hence with regard to its *species* the tree produces itself: within its species, it is both cause and effect, both generating itself and being generated by itself ceaselessly, thus preserving itself as a species.
>
> Second, a tree also produces itself as an *individual*. It is true that this sort of causation is called merely growth; but this growth must be understood in a sense that distinguishes it completely from any increase in size according to mechanical laws: it must be considered to be equivalent to generation, though called by another name. . . .
>
> Third, part of the tree also produces itself inasmuch as there is a mutual dependence between the preservation of one part and that of the others. . . . The leaves . . . though produced by the tree, also sustain it in turn; for repeated defoliation would kill it, and its growth depends on their effect on the trunk. (370–372)

Kant contrasts the situation of an organism with the favorite Deist example of a watch. Though in both cases we may be able to cognize an object in accordance with a concept of the form acting as a purpose, they are really entirely different, because in one case the purpose (and the means to realize it) is extrinsic to the system and in the other, intrinsic:

> Hence organized beings are the only beings in nature that, even when considered by themselves and apart from any relation to other things, must still be though of as possible only as purposes of nature. It is these beings, therefore, which first give objective reality to the concept of a *purpose* that is a purpose of *nature* rather than a practical one, and which hence give natural science the basis for a teleology, i.e., for judging its objects in terms of a special principle that otherwise we simply would not be justified in introducing into natural science (since we have no a priori insight whatever into the possibility of such a causality). (375–376)

It is obvious why the *Critique of Teleological Judgment* is regarded by McLaughlin (1990) as proposing a philosophy of biology and why Kant's work was foundational for a generation of German biologists to follow (Lenoir 1989; Richards 2002; Grene and Depew 2004; Sloan 2006). Nevertheless, Kant concludes the "Analytic of Teleological Judgment" with a strict reminder of the entirely regulative role of teleology in natural science:

> We must avoid any suspicion, in physics, that we might presume to mix something in with our bases of cognition that does not belong to physics at all—viz., a supernatural cause. That is why, when in teleology we speak of nature as if the purposiveness in it were intentional, we do so in such a way that we attribute this intention to nature, i.e., to matter. This serves to indicate that this term refers here only to a principle of reflective, rather than of determinative, judgment. (It indicates this inasmuch as no one would attribute to lifeless material an intention in the proper sense of the term, and so no misunderstanding can arise). (383)

So in natural science, teleology, the assumption of final causes, is a necessary principle of reflective judgment for dealing with natural purposes (organisms), but this does not mean that it has any objective reality. But how, then, is it possible that the mechanical workings of natural laws can be reconciled with this principle of teleology? This is the question to which Kant turns in the "Dialectic of Teleological Judgment."

In the "Dialectic," Kant formulates the basic problem of teleological judgment as an antinomy of two distinct *maxims* of judgment, both of which appear to be justified:

> The *first maxim* of judgment is this *thesis*: All production of material things and their forms must be judged to be possible in terms of merely mechanical laws.
>
> The *second maxim* is this *antithesis*: Some products of material nature cannot be judged to be possible in terms of merely mechanical laws. (Judging them requites a quite different causal law—viz., that of final causes). (387)

How is the antinomy to be resolved? Kant's relation to the whole history that we have been examining is made clear by his comments in this section on "The Various

Systems Concerning the Purposiveness of Nature." He classifies the possible views as follows: "The systems that deal with the technic of nature, i.e., with nature's power to produce things in terms of the rule of purposes, are of two kinds: one interprets natural purposes *idealistically*, the other *realistically*. The idealistic interpretation maintains that all purposiveness of nature is *unintentional*; the realistic interpretation maintains that some of this purposiveness is *intentional*" (391). Within idealistic interpretations he further distinguishes *casualistic* from *fatalistic* interpretations, the former of which he identifies with Epicurus and Democritus, the latter with Spinoza. For Kant, unlike Hume, the Epicurean hypothesis is not even worthy of consideration. His remarks here echo those of *The Only Possible Argument* some seventeen years before:

> The system that espouses the casualistic interpretation . . . is so manifestly absurd, if taken literally, that we need not let it detain us. (391)
>
> Epicurus . . . completely denies the distinction of a technic of nature from mere mechanism. Instead he adopts blind chance to explain not only nature's technic, i.e., why nature's products harmonize with our concepts of a purpose, but even nature's mechanism, i.e., how the causes of this production are determined to this production according to laws of motion. Hence nothing has been explained, not even the illusion in our teleological judgments, so that the alleged idealism in them has by no means been established. (393)

In other words, Epicurus cannot prove that there is no purpose underlying the apparent purposiveness of nature and, more important, doesn't account for the appearance of purposiveness, which cannot be denied. Kant, unlike Hume, completely fails to see the fundamental point of the "Epicurean hypothesis"—namely, that in the case of an organism, its functional unity, and even its purposive behavior, can be explained as a necessary condition for its existence. At the same time, it is clear why this is so: Kant has the interests of science in mind. For Kant, by denying the philosophical (and religious) significance of the functional unity of organisms, Epicurus (and, by implication, modern mechanists and materialists) provides no basis for cognizing this unity, which is the essence of what it means to be an organism or "natural purpose."

By contrast, for Kant, theism has the advantage that it "is best able to rescue the purposiveness of nature from idealism" (395). Nevertheless, to succeed, theism would have to prove that purposiveness could not possibly be due to mechanism. As he had already shown in the first critique (and before that in *The Only Possible Argument*), this is impossible:

> In fact all we can make out is that the character and limits of our cognitive powers (which give us no insight into the first, inner basis of even this mechanism) force us to give up any attempt to find in matter a principle that implies determinate references of this matter to a purpose, so that we are left with no other way of judging

nature's production of things as natural purposes than in terms of a supreme understanding as cause of the world. That basis, however, holds only for reflective and not for determinative judgment, and is absolutely incapable of justifying any objective assertion. (395)

By recognizing organisms as products of intelligent design, we are only saying that this is how we must regard them. It is entirely possible that there could be a reconciliation between the teleological and mechanical principles or maxims of judgment. This potential reconciliation is based on the possibility of a different sort of understanding than the merely human one (McLaughlin 1990). Kant concludes that

> although the principle of a mechanical derivation of purposive natural products is compatible with the teleological principle, the mechanical one could certainly not make the teleological one dispensable. In other words, when we deal with a thing that we must judge to be a natural purpose (i.e., when we deal with an organized being), though we can try on it all the laws of mechanical production that we know or may yet discover, and though we may indeed hope to make good progress with such mechanical laws, yet we can never account for the possibility of such a product without appealing to a basis for its production that is wholly distinct from the mechanical one, namely a causality through purposes. Indeed, absolutely no human reason . . . can hope to understand, in terms of nothing but mechanical causes, how so much as a mere blade of grass is produced. (409)

This is the objection of the scientist. Kant says we simply don't have the theoretical insight to explain the production of a blade of grass in terms of reductionistic, deterministic mechanical laws.

This seems to be the end of the story, but Kant nevertheless goes on to consider a number of additional issues in an appendix on the "Methodology of Teleological Judgment." I will consider only one, his elaboration on the "necessary subordination of the principle of mechanism to the teleological principle in explaining a thing as a natural purpose." The passage is fundamental for our understanding of Cuvier's project:

> It is commendable to do comparative anatomy and go through the vast creation of organized beings in nature, in order to see if we cannot discover in it something like a system, namely, as regards the principle of their production. We do not have to settle for the mere principle for judging them (it tells us nothing that would give us insight into how they are produced), and do not have to abandon all hope for a claim to *insight into nature* in this area. For there are some facts in this area that offer the mind a ray of hope, however faint, that in their case at least we may be able to accomplish something with the principle of natural mechanism, without which there can be no natural science at all: So many genera of animals share a certain common schema on which not only their bone structure but also the arrangement of their other parts seems to be based; the basic outline is admirably simple but yet was able to produce this great diversity of species, by shortening some parts and lengthening others, by the involution

of some and the evolution of others. Despite all the variety among these forms, they seem to have been produced according to a common archetype, and this analogy among them reinforces our suspicion that they are actually akin, produced by a common original mother. For the different animal genera approach one another gradually: from the genus where the principle of purposes seems to be borne out most, namely, man, all the way to the polyp, and from it even to mosses and lichens and finally to the lowest stage of nature discernible to us, crude matter. From this matter, and its forces governed by mechanical laws (like those it follows in crystal formations), seems to stem all the technic that nature displays in organized beings and that we find so far beyond our grasp that we believe that we have to think a different principle to account for it.

When the archaeologist of nature considers these points, he is free to have that large family of creatures (for that is how we must conceive of them if that thoroughly coherent kinship among them is to have a basis) arise from the traces that remain of nature's most ancient revolutions, and to have it do so according to all the natural mechanism he knows or suspects. He can make mother earth (like a large animal, as it were) emerge from her state of chaos, and make her promptly give birth initially to creatures of a less purposive form, with these then giving birth to others that became better adapted to their place of origin and to their relations to one another, until in the end this womb itself rigidified, ossified, and confined itself to bearing definite species that would no longer degenerate, so that the diversity remained as it had turned out when that fertile formative force ceased to operate. And yet, in giving this account, the archaeologist of nature will have to attribute to this universal mother an organization that purposively aimed at all these creatures, since otherwise it is quite inconceivable how the purposive form is possible that we find in the products of the animal and plant kingdoms. But if he attributes such an organization to her, then he has only put off the basis for his explanation and cannot pretend to have made the production of those two kingdoms independent of the condition of final causes. (418–420)

This is Kant's answer to Lucretius, Buffon, and Maupertuis. Cuvier was, of course, going to be just such an "archaeologist of nature."

. . .

It is ironic that after one hundred years of Enlightenment, Kant arrived at a position not so far removed from that of Bayle, though from a critical, rather than a skeptical, perspective. Kant denies the ability of our reason to cognize final causes determinatively—they are only the way in which our reflective judgment interprets certain phenomena of nature, due to the inherent limitations of human reason. Bayle had likewise emphasized the inability of human reason to arrive at any certain conclusions about metaphysical problems. But where Bayle had made room only for faith, Kant made room both for faith and for science, each within their proper bounds. Cuvier inherited this Kantian legacy.

5

Cuvier and the Principle of the Conditions for Existence

If an animal's teeth are such as they must be, in order for it to nourish itself with flesh, we can be sure without further examination that the whole system of its digestive organs is appropriate for that kind of food; and that its whole skeleton and locomotive organs, and even its sense organs, are arranged in such a way as to make it skillful at pursuing and catching its prey. For these relations are the necessary conditions for the existence of the animal; if things were not so, it would not be able to subsist.

—G. CUVIER, 1798

In traditional accounts of the history of evolutionary theory, Cuvier appears as the villain, the fixist, creationist conservative, who used his immense power and prestige to ridicule the forward-looking—if misguided—evolutionary theories of his rivals Jean-Baptiste Lamarck and Etienne Geoffroy Saint-Hilaire, and thereby held back the development of science for fifty years. And, of course, there is some truth to this perspective. But what is lost in such presentations is any sense of the evidence as it stood at the time, and of who in fact had the more scientific approach to that evidence. As we will see, Cuvier had good reasons for his views, even if the evolutionists eventually—and justifiably—won the day. A child of the Enlightenment, Cuvier brought a critical, rational attitude to the understanding of organisms. His conservatism was the conservatism of any good scientist; and if he rejected explanation by evolution, spontaneous generation, and idealistic principles such as a "chain of being" or "unity of plan," he likewise rejected explanation by divine design. It was this very conservatism, combined with his broad appreciation of the philosophical issues involved, that allowed him to find a solution to the problem of teleology, when so many before him had failed.

BIOGRAPHICAL BACKGROUND

Georges Cuvier (fig. 5.1) was born in 1769 in Montbéliard, now on the Swiss border of France, but at the time a French-speaking town in the Duchy of Württemburg (for a general introduction to Cuvier's life, see Coleman 1964; Outram 1984; Appel 1987; Corsi 1988; Rudwick 1976, 1997, 2005, 2008; Taquet 2006). He came from the Protestant bourgeoisie; his father was a career military officer. He was close to his mother growing up, and by his account it was her encouragement and attention to his study habits that enabled him to excel at school (Flourens 1856, 170), where he received a classical education. In his autobiography, Cuvier tells us that his interest in natural history was initiated when he was twelve or

FIGURE 5.1. Cuvier in 1798, around the time he first articulated the principle of the conditions for existence. Oil painting by Mathieu-Ignace van Brée (1773–1839). © Bibliothèque centrale du Muséum national d'histoire naturelle Paris, 2008. Used by permission.

thirteen, by his discovery of a complete edition of Buffon's *Natural History* in the library of his uncle. He notes that all of his pleasure as a youth consisted of copying the figures and coloring them after the descriptions. Cuvier was slated for the University of Tübingen and a career as a minister, but he failed to gain entry. However, through connections of his godfather, the Comte de Waldner, it was soon arranged that he would attend the Karlsschule in Stuttgart instead.

Cuvier entered the Karlsschule in 1784. At the time it was one of the most advanced universities in Europe; it was modeled on Enlightenment ideals (Uhland 1953). It had been founded by the Grand-Duke Charles-Eugène in 1770, who had been brought up in the court of Frederick the Great and followed his enlightened example; it was awarded university status by the equally enlightened Emperor Joseph II in 1781 (Outram 1984). It was military in style, with uniforms and tight discipline. The students slept in barracks. The course of study consisted of an initial year of philosophy, followed by specialization in law, medicine, administration, military service, or commerce. Cuvier chose to specialize in administration, which was intended to prepare the student for a career in state service. This was an exceptionally broad area: his studies included law, chemistry, mineralogy, zoology and botany, the science of mines, police, commerce, finance, practical and theoretical economics, and geometry (Coleman 1964). At the time he was there, however, there was no official teacher of natural history; Cuvier and several friends (including Christian Pfaff) formed a natural history society to study on their own. The botanist J. G. Kerner took Cuvier under his wings; for his assistance with the collections for Kerner's book on economic botany, Cuvier was given his own copy of the tenth edition of Linnaeus's *Systema Naturae*. Cuvier also received an introduction to comparative anatomy from a fellow student, Carl Friedrich Kielmeyer (1765–1844), who was later a regular instructor at the Karlsschule.

Upon graduating in 1788, Cuvier hoped to obtain the administrative position in the duke's administration his studies had prepared him for, but the position was not forthcoming. Instead, under significant financial pressure, he accepted a position as tutor to the son of the d'Hericys, a Protestant noble family living in Caen, in Normandy. It was during the seven years he was in Normandy that Cuvier became a naturalist. His duties as a tutor were relatively light, and he took advantage of the Norman countryside to botanize, dissect, and otherwise explore the natural realm. These were the years of the French Revolution, and during the Terror, the d'Hericys retreated to their country estate at Fiquainville, near Fécamp on the English Channel. Here Cuvier was first exposed to the diversity of marine life; he took a particular interest in mollusks. His first published works were on the anatomy of invertebrates, including terrestrial isopods, limpets, and insects. Cuvier's zoological and botanical notebooks from this period still exist, and the fascinating recent study of Cuvier's early years by Taquet (2006), relying on both manuscript sources and Cuvier's letters to his school friend Christian Pfaff

(Marchant 1858), demonstrates how Cuvier matured as a largely self-taught naturalist during the years following his departure from the Karlsschule.

The general story of Cuvier's rise to fame is well known, even if the details remain somewhat less clear (Outram 1984; Taquet 2006). From his studies in Normandy, Cuvier had been able to send off several papers to be published in Paris. He was in correspondence with a number of Parisian naturalists, including Etienne Geoffroy Saint-Hilaire, who had recently been appointed to one of the two chairs in zoology at the newly constituted National Museum of Natural History (Lamarck had the other). He also had the recommendation of the Abbé Tessier, a Parisian agronomist who was in hiding in Fécamp, serving as the surgeon to the military regiment there. Finally, he had the promise from the secretary of the Natural History Society, Aubin-Louis Millin, of a small position in the Provisional Commission for the Arts that would provide him with income (Taquet 2006).

Cuvier arrived in Paris in March 1795; within a few days he was reading papers at the Natural History Society and the Philomatic Society. He was nominated to the Provisional Commission for the Arts by late April; the following month, he became a professor of natural history at the Central School of the Panthéon; and in early July, was given the post of "suppléant" or understudy for A.-L. Mertrud in the chair of Anatomy at the National Museum of Natural History. In December he was elected a member of the National Institute, which had replaced the disbanded Royal Academy of Sciences after the end of the Terror. Mertrud died in 1802, and Cuvier was appointed to the chair of comparative anatomy (as he had it renamed) at the Muséum, a position he held for the remainder of his life. This also allowed him to move into a residence at the Muséum, adjacent to which he developed his gallery of comparative anatomy (Sloan 1997).

In subsequent years, Cuvier added many more posts and honors to the ones acquired upon his arrival in Paris (Outram 1984). Most important was his role as the secretary of the First Class (Physical Sciences) of the National Institute,* at first temporary, then permanent (1803). In this role, he was called on to report on developments in a broad realm of scientific endeavors, including chemistry, geology, natural history, and applied arts such as medicine and agriculture. His annual reports to the Institute were widely published (Smith 1993), and his eulogies of deceased members (Cuvier 1861) likewise gave him the chance to encourage what he saw as good science and denigrate that which was bad (Outram 1978). It was for this role, as well as his strictly zoological work, that he earned the title of the "legislator of natural history."

As a member of the staff of the Muséum, Cuvier had full access to the collections of the Buffonian Cabinet du Roi, which were soon enhanced by the seizure of important natural history specimens by the advancing French revolutionary army,

---

* In 1815, after the restoration of the monarchy, the National Institute again became the Royal Academy of Sciences.

and by the specimens collected by Napoleon's expedition to Egypt, which included Geoffroy Saint-Hilaire as naturalist. Cuvier took advantage of his situation and was able to publish a series of important works, beginning with a textbook of zoology (*Tableau élémentaire de l'histoire naturelle des animaux*, 1798), soon followed by his seminal lectures on comparative anatomy, from his course at the Muséum (*Leçons d'anatomie comparée*, 1800–1805). Cuvier's interest in the comparative anatomy and systematic classification of animals continued throughout his life: his *Règne animal* (1817; second edition, 1829–1830) was his crowning achievement in this field, in which he divided animals into four distinct groups, or *embranchements*— the Vertebrates, the Articulates, the Mollusks, and the Radiates (including echinoderms and cnidarians)—based on fundamental differences in body plan. At the time of his death in 1832, Cuvier was deeply involved with a student, Achille Valenciennes, in a work on the anatomy and systematics of fishes, which eventually ran to twenty-two volumes (*Histoire naturelle des poissons*, 1828–1849).

As he prosecuted his studies of comparative anatomy and used the results of his studies to reform the classification of animals, Cuvier also became interested in fossil animals (Rudwick 1976, 1997, 2005, 2008). Together with Geoffroy Saint-Hilaire, officially the mammalogist at the Muséum, he had published a brief examination of the species of living and fossil elephants in 1795, suggesting that not only the mastodon ("the Ohio animal") but also the Siberian mammoth were species distinct from the living forms and presumably extinct. When a natural history collection appropriated from the Netherlands arrived at the Muséum as a conquest of war, it contained the skulls of both African and Indian elephants, and Cuvier was able to confirm his previous suggestions (Cuvier 1796a, translated in Rudwick 1997). At around the same time, a sketch of the mounted skeleton of an extinct ground sloth on display in the royal museum of Madrid arrived at the National Institute. Cuvier's report on the "Paraguay animal" demonstrated that the animal almost certainly represented another extinct species, one that had "affinities" with the living sloths (Cuvier 1796b, translated in Rudwick 1997; see also Rudwick 2005).

Thus stimulated, Cuvier embarked on an ambitious research program on fossil bones (Rudwick 1997, 2005), which enlarged into a project to reconstruct the history of life on Earth, based on the fossils found in the rocks. Together with his colleague Alexandre Brongniart, Cuvier began to study the geology and fossils of the Paris basin. In 1811, they published their *Essai sur la géographie minéralogique des environs de Paris*. In this monograph, developed independently of William Smith's work in England, they established the basic principles of stratigraphy by use of fossils as indicators of relative age, and argued for a succession of freshwater and marine environments to explain the succession of forms seen (Rudwick 1976, 2005).

In parallel with this geological work, Cuvier continued to prosecute his study of the fossil bones of quadrupeds, particularly those from the Paris gypsum formation (late Eocene in modern reckoning). These animals were even more different from living forms. Clearly, there was need for a thorough review of the fossil evidence for extinct species. Cuvier argued in particular that the large quadrupeds were the most useful fossils to focus on, because they were unlikely to represent extant species that were merely unknown to science. He issued an appeal for international collaboration (Cuvier 1800, translated in Rudwick 1997) that was widely distributed, and soon reports on fossil skeletons from all over Europe began to come into the Muséum (Rudwick 1997, 2005; Taquet and Padian 2004). In 1812, Cuvier collected all of the studies he had made of fossils and sketches of fossils that had come his way, together with the geological study of the Paris Basin on which he had collaborated with Brongniart, and published them as a four-volume work, his *Recherches sur les ossemens fossiles de quadrupèdes*. To the individual parts, which had mostly been published previously, Cuvier added an extensive "Preliminary Discourse" that laid out his interpretation of the results of his research; this work was subsequently reprinted separately (as the *Revolutions of the Globe*) and became Cuvier's most popular book.

In his *Ossemens fossiles*, the founding document of vertebrate paleontology, Cuvier presented descriptions of forty-nine fossil quadruped species that were "definitely unknown to naturalists unto now" (quoted in Rudwick 2005, 507) and thus presumably extinct. Martin Rudwick (1976, 1997, 2005) has thoroughly examined the significance of the book; he emphasizes the extent to which Cuvier broke away from the mere description of a "former world" to paint a picture of an ongoing *historical* process of changing depositional environments and organisms, a view that went beyond Smith in the interpretation if not the basic understanding of the relevance of fossils to geology. By 1812, the evidence he had examined convinced him that the sequence of rock formations showed a progressive change in animal life through time. Cuvier laid out the problem as follows:

> What is more important—what indeed comprises the definitive object of all my work and establishes its true relation to the theory of the earth—is to know in which beds each species is found, and whether there are some general laws, relative either to the zoological subdivisions or to the greater or lesser resemblance of the species to those of today.
>
> The laws recognized in this respect are very beautiful and very clear. First, it is certain that oviparous quadrupeds appeared much sooner than the viviparous. The crocodiles of Honfleur and England are below the chalk. The monitors of Thuringia are even more ancient, if—as the school of Werner believes—the coppery shales that conceal them (along with so many kinds of what are believed to be freshwater fish) are among the oldest beds of the Secondary formations. . . . This first appearance of fossil bones thus already seems to show that dry land and freshwater existed before the formation of the chalk; but neither at that epoch nor during the formation of the chalk, nor even for a long time after that, were any bones of land mammals fossilized.
>
> We begin to find bones of marine mammals, namely those of sea cows and seals, in the coarse shelly limestone that covers the chalk in our environs, but there are still no bones of land mammals. Despite the most persistent research, I have been unable to discover any distinct trace of that class, before the formations deposited on the coarse

limestone; but as soon as those formations are reached, the bones of land animals show themselves in great numbers.

Thus, just as it is reasonable to believe that shells and fish did not exist at the epoch of the formation of the primordial rocks, one should also believe that the oviparous quadrupeds began with the fish, from the first times that produced the Secondary formations; but that the land quadrupeds came only a long time later, when the coarse limestones—which already contain most of our genera of shells, though in species different from ours—had been deposited. (Cuvier 1812, translated in Rudwick 1997, 223)

Thus from this Cuvierian project, the first outline of the history of life on Earth emerged.

Cuvier attributed the replacement of one set of organisms by another to a series of "revolutions" or "catastrophes," of varying intensities (Rudwick 1976, 1997, 2005). He followed other continental geologists, particularly Deluc and Dolomieu, in arguing that the evidence suggested a recent revolution, within the last five or six thousand years. This revolution had wiped out the large extinct mammals, such as the mammoth, but was preceded by at least one and probably several additional revolutions. Cuvier argued further that human skeletal remains were known only from after the latest revolution. If we consider that the last ice age retreated from Europe only about ten thousand years ago, leaving piles of gravel, wide valleys, and erratic boulders behind, this view does not seem such an irrational one for Cuvier to have held, though it does rather conveniently allow one to regard the last catastrophe as the Noachian flood. Cuvier was nevertheless at pains to show that not only the Christians, Muslims, and Jews but all the world's peoples had a story of a great catastrophe at some time in the past.

While quite clear on the subject of extinction of species, on the subject of creation of new species Cuvier was less so. On the one hand, he disclaimed any notion of new creations, based on a thought experiment involving Australia:

Moreover, when I maintain that the stony beds contain the bones of several genera, and the superficial beds those of several species, which no longer exist, I do not claim that a new creation was needed to produce the existing species. I only say that they did not exist in the same places, and that they must have come there from elsewhere.

Suppose, for example, that a great irruption of the sea covered the continent of New Holland with a mass of sand or other debris. It would bury there the corpses of kangaroos, wombats, dasyures, bandicoots, flying phalangers, spiny anteaters, and duck-billed platypuses; and it would entirely destroy the species of all those genera, since none of them exist in other countries. If that same revolution left high and dry the many narrow straits that separate New Holland from the continent of Asia, it would open a route for elephants, rhinoceroses, buffalos, horses, camels, tigers, and all the other Asiatic animals, which would come to populate a land where they had previously been unknown. And if a naturalist, after carefully studying all that living nature, were to think of digging into the soil on which it was living, he would find the remains of wholly different beings.

What New Holland would be, in the conjecture we have just made, Europe, Siberia, and a large part of America are in reality. Perhaps one day, when other countries—and New Holland itself—are examined, it will be found that they have all undergone similar revolutions, I would almost say mutual exchanges of animals. (Cuvier 1812, translated in Rudwick 1997, 229–232)

This view seems to deny any overall worldwide progression, in contrast to his earlier statements on the patterns to be seen.

On the other hand, in the concluding passage of the "Preliminary Discourse," we find a rather different vision expressed. Looking forward to the progress of geological science, Cuvier declares:

How good it would be . . . to have the organic productions of nature in their chronological order, as we have the main mineral substances! The science of organization itself would gain from it; the developments of life, the succession of its forms, the precise identification of those that appeared first, the simultaneous birth of certain species, and their gradual destruction, would perhaps tell us as much about the essence of the organism, as all the experiments that we can attempt on living species. And man, to whom has been accorded only an instant on earth, would have the glory of reconstructing the history of the thousands of centuries that preceded his existence, and of the thousands of beings that have not been his contemporaries! (Cuvier 1812, translated in Rudwick 1997, 252)

Interestingly, this concluding passage was deleted in the third (1825) edition of the book, to be replaced by a long section summarizing the results of the studies that were underway to carry out just this program. However, Cuvier's language in the conclusion of this later edition now has a rather different feel; note in particular his invocation of the Creator. He is discussing the animals that lived just prior to the last catastrophe:

These are the principal animals whose remains have been discovered in that mass of earth, of sand and of mud, in that *diluvium*, which everywhere covers our vast plains, fills our caverns, and chokes up the fissures of many of our large rocks. They formed most indubitably the population of the continents at the epoch of the great catastrophe which has destroyed their race, and which prepared the soil on which the animals of the present day subsist. Whatever resemblance certain of the species of the present day offer to them, it cannot be disputed that the total of this population had a totally distinct character, and that the majority of the races which composed it have been annihilated.

It is wonderful, that among all these mammifera, of which at the present day the greater part have a congenerate species in the warm climates, there has not been one quadrumanous animal, not a single bone, or a single tooth of a monkey, not even a bone or a tooth of an extinct species of this animal.

Neither is there any remains of man. All the bones of the human race which have been collected along with those which we have spoken of, have been the result of accident, and besides their number is extremely small, which it certainly would not be if men had then been established in the countries inhabited by these animals. Where then was the human race? Did the last and most perfect work of the Creator exist no where? Did the animals which now accompany him on earth, and of which are no fossil remains to be traced, surround him? Have the lands in which they lived together been swallowed up, when those which they now inhabit, and of which, a great inundation might have destroyed the anterior population, were again left dry? On this head the study of fossils gives us no information, and in this Discourse we must not seek an answer to our question from other sources.

It is certain, that we are at present at least in the midst of a fourth succession of terrestrial animals, and that after the age of reptiles, after that of palæotheria, after that of mammoths, mastodonta and megatheria, the age arrived in which the human species, together with some domestic animals, governs and fertilizes the earth peaceably; and it is only in formations subsequent to this period, in alluvial deposites, in turf-bogs, in the recent concretions, that those bones are found in a fossil state, which all belong to animals known and now existing.

Such are the human skeletons of Guadaloupe, incrusted in a species of travertine with land shells, slate, and fragments of the shells and madrepores of the neighbouring sea; the bones of oxen, deer, roebucks, and beavers of common occurrence in turf-bogs, and all the bones of the human race, and of domestic animals found in the deposites of rivers, in burial grounds, and in fields of battle.

None of these remains belong either to the vast deposite of the last catastrophe, or to those of the ages preceding. (Cuvier 1831, 220–221)

Cuvier's denial of the existence of human fossils was soon to break down (see Rudwick 2008), but he did not live to see this happen.

Toward the end of his career, Cuvier became embroiled in a debate with Geoffroy Saint-Hilaire, his former friend and collaborator and his colleague at the Muséum (Russell 1916; Appel 1987; Asma 1996; Grene and Depew 2004; Le Guyader 2004). After their early collaboration, Geoffroy Saint-Hilaire and Cuvier had grown apart. Geoffroy Saint-Hilaire accompanied Napoleon to Egypt, while Cuvier remained at the Muséum, consolidating his position (Outram 1984; Sloan 1997). On his return, Geoffroy Saint-Hilaire increasingly pursued a project that he termed "philosophical anatomy," involving a focus on homologies (his "analogies") between widely separated forms based on the "principle of connections." Cuvier in general had no problem with establishing homologies within an *embranchement* (though he often disagreed with the homologies Geoffroy Saint-Hilaire proposed). However, he was provoked when two young admirers of Geoffroy Saint-Hilaire, Laurencet and Meyranx, proposed a scheme of homologies between the organs of quadrupeds and that of mollusks (i.e., between two of his *embranchements*) and this was supported by Geoffroy Saint-Hilaire in his report to the Académie (along with some rather nasty comments about Cuvier's views as being representative of a philosophy of "the beginning of the nineteenth century"; translated in Le Guyader 2004, 133). There ensued an exchange of presentations to the Académie, beginning in February 1830 and continuing through the end of March. Cuvier came out ahead in the debate on all substantive points, though in the process, Geoffroy Saint-Hilaire gen-erated a lot of sympathy. This was enhanced by Geoffroy Saint-Hilaire's publication of documents and commentary related to the debate in his *Principes de philosophie zoologique* (1830; translated in Le Guyader 2004), in which he presented himself in the most favorable light possible.

Cuvier died in 1832, unexpectedly, of a progressive paralysis. This condition has been suggested to have been Landry-Guillain-Barré syndrome, an idiopathic, possibly autoimmune neuropathy. Only a few days before, he had been healthy enough to deliver a lecture in his ongoing series on the history of the natural sci-ences to a popular audience at the Collège de France (these were later published in an enlarged, edited form by Magdeleine de Saint-Agy; see Cuvier 1841–1845). He left behind him his unfinished work on fishes, as well as the unfinished update to his *Comparative Anatomy* that was his life's work. He also left behind a number of students, including Jean Pierre Flourens, Henri de Blainville, and Henri Milne-Edwards, who competed for the now-vacant positions at the Muséum and in the Académie.

At his death, Cuvier had an impressive scientific legacy. He had reformed zoological classification, greatly clarifying in particular the structure and relations of the various invertebrate groups. He had established the reality of extinction and the general outlines of the history of life on Earth. His insistence on the primacy of "positive facts" was one of the strands that led to the positivism of Auguste Comte and to the experimentalism of Claude Bernard (discussed later in this chapter). Nevertheless, Cuvier did not leave behind a school of disciples, in the sense of students adhering to his own vision of natural science. In France, Cuvier's former students and colleagues had various and diverging views on the issues most critical to Cuvier: the fixity of species, the history of life on Earth, the number of funda-mental types of animals, and the general approach to be taken to zoological and paleontological studies. In Germany, *Naturphilosophie* continued to exert its influ-ence, and the scientists whom Lenoir (1989) denoted the "teleomechanists" were on the rise, with the fascinating work of Karl von Baer, Heinrich Rathke, Johannes Müller, and others. In England, workers such as Richard Owen adopted an ap-proach that took advantage of some of Cuvier's views, but combined them with ideas from Geoffroy Saint-Hilaire and German zoologists that placed them in a very dif-ferent light (Owen 1992; Sloan 1992), while Charles Lyell (1830–1833) launched his uniformitarian assault on catastrophism that was to be so influential for both Darwin and Wallace.

Cuvier's legacy became one of orthodoxy, an orthodoxy to be pushed beyond. The growing conservatism of his later views, contrasted with the rising tide of romanticism, did indeed make him look "old-fashioned." The worlds of zoology and fossil history Cuvier found when he began his career had been claimed and explored, but not well mapped. He arranged them into well-ordered systems, grounded in the positive facts of comparative anatomy. However his fixist, nonevolutionary interpretation of these systems could not withstand the growing tide of information on development, on comparative anatomy, on geographic distribution, and on the geological succession of forms, which was to lead to the triumph of the evolutionary perspective under the tutelage of Chambers, Darwin, and Wallace.

CUVIER'S PROJECT IN THE CONTEXT
OF ENLIGHTENMENT SCIENCE

James Larson (1994) observed that continental biology in the mid- to late eighteenth century was dominated by three figures he called "the triumvirate": Buffon (1707–1788), whom we have already met; Albrecht von Haller (1708–1777), the Swiss doctor, physiologist, and poet (Roe 1981); and Carl Linnaeus (1707–1778). Although his work is of some interest with respect to Cuvier's intellectual background, I will say no more about von Haller. By contrast, understanding the Linnaean perspective is critical to understanding Cuvier's project in the sciences.

Linnaeus, of course, was the great systematist—his *Species Plantarum* (1753) and *Systema Naturae* (tenth edition, 1758) placed all taxa known to him within an all-encompassing hierarchical system, which still forms the basis for taxonomy today. In the wake of Linnaeus's work, generations of scholars sought to extend his system to ever more forms, while at the same time reforming its basis. This was to be Cuvier's mission.

The recent work of Philippe Taquet (2006) gives us insight into exactly what Cuvier was up to during his school years, in particular the role of G.-F. Parrot and C. F. Kielmeyer, two older students whom Cuvier met in his first year there. In addition to their regular coursework, students could take lectures in other subjects. Cuvier took a course in physics with Parrot and through him met Kielmeyer. The three became friends, and as Cuvier testified in his autobiography, it was Kielmeyer who had the greatest influence: "He had from that time that taste for contemplation and that strength of intellect which make him one of the most profound men of Germany; he taught me to dissect and gave me my first ideas of philosophical natural history. I pursued these studies during hours of recess, and while other students went into town" (Flourens 1856, 173–174, my translation).

Kielmeyer and Parrot left the Karlsschule in the spring of 1786. In June, Cuvier began an herbarium, as well as a botanical journal in which he recorded the plants he identified (by October, he was up to 552 species). In September, he began a zoological journal and collection, largely focused on insects. He had access to Linnaeus's *Species* and *Genera Plantarum*, and *Systema Naturae*, as well as Fabricius's *Systema Entomologiae* and Blumenbach's *Handbuch der Naturgeschichte*. Later that year, Cuvier and some younger friends, all future doctors and natural scientists, formed a natural history society, of which Cuvier was the president and "soul" (see Taquet 2006, 105–108). Thus, while a young student, Cuvier began his training as a naturalist; his focus at first was species identification, and especially the discovery of new species, but already with an eye toward anatomy and classification.

In 1788, at the end of his time in Stuttgart, Cuvier and several of his friends made a week-long exploratory tour of the nearby Swabian Alb, inspired by Benedict de Saussure's recently published account of his conquest of Mont Blanc (1787). Cuvier's account of this tour survives in manuscript and is interesting for our first glimpse of the developing naturalist (Taquet 2006). Written when he was only eighteen, we already find here many of the themes expressed in Cuvier's later work, in particular his focus on well-observed facts and his distrust of system building. He makes observations on the general scenery, geology, and botany of the areas they visited, as well as commerce and other features that would attract a young man trained in administration (he also became enamored of a young lady he met along the way, Louise Glettin). As they passed through Tübingen Cuvier took advantage of the opportunity to buy the work of Storr on the classification of mammals, which influenced his own later classification.

In Normandy, Cuvier settled into his duties as a tutor. He loved the freedom of the lifestyle but missed the intellectual discourse the Karlsschule had provided. Early in his years in Normandy, Cuvier found a model for himself in Aristotle, who for Cuvier was the consummate naturalist, adept both the in the factual details of the subject and in broad generalizations derived from them (Cuvier 1811; see also Coleman 1964; Rudwick 1976; Taquet 2006). He continued to pursue his studies in natural history, including botany, zoology, and mineralogy. His visits to the seashore allowed him to become acquainted with marine forms, as well as occasional exotic specimens (e.g., parrots) arriving by ship. To the extent allowed by his refuge in a provincial part of France during the turbulent revolutionary years, he kept up with the latest developments in natural history, through both his own reading and his correspondence with friends from the Karlsschule. His extensive correspondence with his younger colleague Christian Pfaff (Marchant 1858) gives us some insight into the thoughts of the young naturalist.

In these letters we find Cuvier already exhibiting many of the features for which he would later become famous: an astounding work ethic, a critical view of current developments in natural history, and a high—if not necessarily inflated—opinion of his own worth. In his letters to Pfaff, he enthused over Aristotle, Lavoisier (whose new system of chemistry Cuvier immediately recognized as superior), and René-Just Haüy, the mineralogist who had put the science of mineral classification on firm geometric grounds. In taxonomy, Cuvier took advantage of the work of Linnaeus and Fabricius, but he also began to develop his own ideas about zoological classification from his work dissecting marine forms. A key influence was the work of Antoine-Laurent de Jussieu, professor of botany at the Jardin du Roi and a member of a famous family of botanists. Following his uncle's lead, de Jussieu had reordered the Linnnaean genera after a new arrangement that claimed to be much more "natural." This classification was based on the principle of the subordination of characters, in which the most constant characters were considered of greatest importance for classification, a conception that Cuvier later adopted and greatly extended (see Stevens 1994, especially 422, n. 20). In a 1792 letter to Pfaff, Cuvier promoted the virtues of de Jussieu's work, of which the merit was not necessarily the factual descriptions of species and genera, but rather "the philosophical manner" of viewing the facts (Marchant 1858, 261).

When Cuvier arrived in Paris, he was viewed as part of the up and coming wave of "Linnaeans," in opposition to the "Buffonians" who were trying to hold on to power in the wake of their master's recent (1788) death (Corsi 1988 provides an excellent portrait of the scientific world of Cuvier's early years and his position in it; see also Larson 1994; Gillispie 2004; Rudwick 2005). We have already seen Buffon's denigration of the Linnaeans for excessive focus on nomenclature and artificial groupings, at the expense of real knowledge of the object. As described by Larson (1994), the views of his partisans were that Linnaeus "had introduced order and precision into natural history and had worked with uncommon zeal to extend scientific knowledge of organized beings." By contrast, for the Linnaeans, Buffon "recognized neither order, plan, nor liaison in sublunary nature and replaced method with fantastic and unprovable hypotheses about the autonomy of matter, the generation of living forms, and physical truth" (Larson 1994, 9).

There is no doubt that Cuvier had indeed allied himself with the Linnaeans, as befit his rigorous German training (Rudwick 1976, 2005; Taquet 2006). On Buffon's death, he exclaimed to Pfaff "The naturalists have finally lost their leader; this time, the Comte de Buffon is dead and buried" (Marchant 1858, 49, my translation). Pietro Corsi (1988) notes that Cuvier later "became the fiercest opponent of Buffon's approach to natural history" (4). In the wake of the revolution, the Buffonians, represented by Lamarck (who had been Buffon's protégé) but also by names like Sonnini and Delamétherie, found themselves under attack by the previously suppressed wave of Linnaeans.

Cuvier's position in these early years is clearly painted by Corsi, in discussing the significance of Haüy's mineralogical work:

> It was not hard for those who extolled the method of the new chemistry, the new comparative anatomy, or the technical sophistication of the new taxonomic methodologies to praise Haüy's vast systematic scheme. They saw it as a further proof of the barrenness of the historical method and of the richness of the morphological approach underpinned by the rigor of geometrical procedures. It should nevertheless be stressed once again that the new morphological and ahistorical approach did not come fully into its own until the early 1800s. Between 1795 and 1805 the debate was still open and lively. The advocates of rigor and precision, Cuvier foremost, came under attack from

many of their colleagues and from the cohort of authors who, although not employed in scientific institutions, were nevertheless popular with the educated public. (Corsi 1988, 36)

Thus, while in later years Cuvier may have appeared as a conservative element, in his youth he was the interloper in the field of Buffonians and—along with a number of his colleagues (including Geoffroy Saint-Hilaire, who began his career as a student of Haüy)—fought hard to establish rigorous criteria for scientific natural history based on facts and inference, as opposed to Buffonian "genteel" natural history based on speculative hypotheses about the history of the earth and the nature of life. Cuvier's dismissal of all "systems" had its origin in a rejection of Buffon, and in this light it is much easier to understand his eventual rejection not only of transformism but also of the then frequently related doctrines of spontaneous generation and epigenesis, and of the idealistic morphology of Geoffroy Saint-Hilaire and the *Naturphilosophen*. It is entirely characteristic of Cuvier to find him saying in 1805, "Systems must be swept away.... I rejoice whenever one of them is destroyed by a well-observed fact" (quoted in Corsi 1988, 35). Conveniently, this spirit of overthrowing systems was in keeping with the revolutionary ideals of the day.

THE ENUNCIATION OF THE PRINCIPLE
AND ITS PLACE IN CUVIER'S SYSTEM

As far as I have been able to determine, Cuvier's first public use of the principle of the conditions for existence was in a lecture of October 1798, in the passage quoted as the epigraph to this chapter (Cuvier, 1997 [1798]). This lecture was given the rather lengthy title "Extract from a Memoir on an Animal of Which the Bones Are Found in the Plaster Stone around Paris, and Which Appears No Longer to Exist Today." The context was Cuvier's discussion of his methods for reconstructing the fossil animals from the Paris gypsum, the identical context in which he was to stress the principle some years later, in the "Preliminary Discourse" to his *Ossemens fossiles* (1812).

That same year, Cuvier had published his first book, the *Tableau élémentaire*, a textbook of zoology based on his course in natural history at the Central School of the Panthéon (Smith 1993). Here he notes that in an "organized body" such as a plant or an animal, "all its parts have a reciprocal action upon one another, and concur towards a common goal, which is the maintenance of life" (1798, 5, my translation). This emphasis on the functional wholeness of the organism, and the functional subordination of the parts to the whole, was to be a common theme throughout Cuvier's life. However, while clearly related, this thought is not yet quite the principle of the conditions for existence.

The problem that the zoological world presented to Cuvier at the end of the eighteenth century was the problem of the diversity of biological form, in its broadest aspect. This problem was not just the problem of how best to classify forms—

that is, how to find the most *natural* system of classification (whatever that might mean)—but also how to *interpret* the system achieved. The experience of Haüy with minerals encouraged both Cuvier and Geoffroy Saint-Hilaire to think that something might be achieved in the way of a rational explanation of diversity. But where Geoffroy Saint-Hilaire went looking for empirical—but ultimately idealistic—laws of form, such as the law of unity of plan or composition based on the "analogies" of structure between groups, and the law of *"balancement"* or equilibrium, in which the enlargement of one organ leads to the reduction of another, Cuvier wanted his system to be based on rational laws ultimately grounded in physics and chemistry— and only the latest and best physics and chemistry at that.

When Cuvier came to examine the issue of diversity in a section of the *Tableau élémentaire* on the "natural relations *(rapports)* among organized beings," he noted that the differences among different species of organized bodies, unlike those among species brute matter, consisted not in their material but in their *"organisation"* (1798, 15). Interestingly, at this point Cuvier defined relations among beings largely empirically, based on resemblances; in fact, he defined *rapports naturels* as being any resemblances. However, he went on to posit that the most constant relations are those dealing with the parts of greatest importance in the "organic economy." The parts of little importance are relatively free to vary, independently of each other, but the important parts can't vary without all the other parts feeling the effect; and the more that important parts differ from one species to the next, the more those species differ in their whole organization.

The important relations, then, are the relations that don't vary. Conversely, if we can rationally show the importance of a part, then we can rationally conclude it will have very constant relations. When one takes this scheme and considers it with respect to the Linnaean hierarchy, it suggests that the higher divisions (as of classes) will involve more important parts, while the lesser divisions (as of genera) will involve less important parts. This is the doctrine of the "subordination of characters," which Cuvier derived directly from de Jussieu (Stevens 1994). This natural classification based on the subordination of characters also had the advantage that it most effectively allowed generalizations about the properties of a group, which an artificial system (i.e., one for identification only) would not. Nevertheless, Cuvier was forced to admit that knowledge of the importance of organs was not advanced enough yet to allow a priori ranking and thus, in the composition of his book, he could not stick with the principle of subordination of characters based on first principles but also had to rely on mere comparison among species.

But, although related, the principle of subordination of characters is also not yet the principle of the conditions for existence. Therefore, when Cuvier introduced the latter principle in his 1798 talk, it appears most likely that it was indeed a new principle, one he had derived from his consideration of the rational basis for the reconstruction of fossil forms based on isolated bones. In this talk, Cuvier noted that in his day "comparative anatomy has reached such a point of perfection that, after inspecting a single bone, one can often determine the class, and sometimes even the genus of the animal to which it belonged, above all if that bone belonged to the head or the limbs" (Cuvier 1997 [1798], translated in Rudwick 1997, 36). But how can this be so? Because of the necessary functional relations among the parts: "For these relations are the necessary conditions for the existence of the animal; if things were not so, it would not be able to subsist" (Rudwick 1997, 36, translation slightly modified). Thus, in its initial incarnation, the principle is used as a rational basis for the fact that we are generally able to tell what kind of animal we have from a small sample of its body. Of course, as in his textbook, the functional relations are generally assumed, rather than demonstrated.

By the time the principle reappears in Cuvier's *Leçons d'anatomie comparée* two years later (1800), it plays a central role in his thought. In a section on the "relations (*rapports*) that exist among the variations of diverse systems of organs," he begins by considering the number of possible combinations of organs that would exist if all combinations were possible. "But these combinations, which might appear possible, when one considers them in an abstract manner, do not all exist in nature, because, in the state of life, the organs are not simply brought together, but they act upon one another, and all concur together toward a common goal" (1800–1805, I: 45–46, my translation). There is a decided resemblance between this passage and the remarks in his textbook of two years earlier, but he now goes on: "Accordingly, the modifications of one of these exercise an influence on those of all the others. Those modifications that can't exist together reciprocally exclude each other, while others call each other in, so to speak, and that not only among the organs which are among those with a close relation, but also among those that appear at first glance the furthest apart and most independent" (my translation). After giving some examples of the functional interrelationships of organ systems, he then introduces the principle of the conditions for existence: "It is on this mutual dependence of the functions, and this reciprocal help they lend each other, that the laws that determine the relations of their organs are founded, and which are of a necessity equal to metaphysical or mathematical laws: for it is evident that an appropriate harmony among the organs acting upon one another is a necessary condition for the existence of the being to which they belong, and that if one of these functions were modified in a way incompatible with the modifications of others, that being could not exist" (47, my translation). Thus, the principle of the conditions for existence is conceived of as the principle rationally justifying laws of mutual dependence of form among organ systems. But left open is the question of just how these laws of mutual dependence are to be established.

To show how this principle works out in practice, Cuvier gives some examples of such mutual relations. We see that the mode of respiration depends on the circulation. In animals with a heart and vessels, the blood is always sent out from the heart to the parts and back to the heart. Thus, we see the lungs or gills interposed between the heart and the body. But in the insects, the whole body is bathed in the fluid of the hemocoel, which has no definite circulatory pattern, and thus a centralized lung or gill would not work; instead, tracheae allow the air to penetrate into all parts of the body. Another observed correlation is that animals that breathe air directly into their lungs have the two arterial trunks (pulmonary and aortic) closely approximated and the heart chambers fused, while those that respire by gills either have the two trunks widely separated, as in squids, or have one only, as in other mollusks and in fishes. Although the functional reason for this fact is not clear, we are authorized to believe it to be functionally based because other relations observed to be constant clearly are.

We see a little better the reason for the relations that link the extent and mode of respiration to the general types of movement animals are capable of. Cuvier cites "modern experiments" showing that respiration acts to revive exhausted muscle fibers. He notes that the quantity of respiration decreases from birds, to mammals, to reptiles, and that this correlates well with the general activity level and force of movement of the animals. Moreover, "as it is one of the conditions for the existence of every animal that its needs are proportioned to the faculties which it has to satisfy them, irritability is exhausted all the less easily as respiration is less efficient and less prompt to repair it. This is why it is conserved so well in reptiles, and why their flesh palpitates so long after they have died, while that of warm blooded animals loses this faculty on cooling" (51, my translation). Cuvier also connects the quantity of respiration with digestion: because animals like birds require so much food they have large, efficient stomachs, while reptiles eat little and can fast for many days. But the digestive tract determines the type of food an animal can subsist on; thus, if an animal did not have the means to sense this food, to distinguish it and procure it, it could not subsist. It is at this point that Cuvier again introduces his favorite example: the carnivore, in which all the organs must be suited to the sensing and pursuit of prey, the kill, and the digestion of meat.

For Cuvier, these "laws of coexistence" based on rational deduction from knowledge of the reciprocal influence of the functions and uses of each organ are confirmed by observation, which authorizes us to go in the other direction and from observed constancy of relations deduce the existence of a functional connection, even where one is not evident (he gives the example of the inverse relation in the size of the liver and the lungs, which "makes us think that the liver can, up to a certain point, take the place of" the lungs [57, my translation]).

How do these necessary functional relations relate to the diversity of form, which was the key problem facing Cuvier? As always, he is quite clear:

> However, while always remaining within the limits which the necessary conditions
> for existence prescribed, nature was abandoned to all its fecundity in that which these
> conditions didn't limit, and without ever leaving the small number of combinations
> possible among the essential modifications of important organs, she appears to have

played endlessly with all the accessory parts. It is not required for the latter that a form or disposition be necessary, it even seems often that it need not be useful, to be realized, it need only be possible, that is to say, such that it doesn't destroy the harmony of the whole. (58, my translation)

Recall from chapter 4 that when confronted with the variety of form, Buffon had suggested—in his somewhat loose yet elegant way—that "every thing which is not so hostile as to destroy, every thing that can subsist in connection with other things, does actually subsist." This thought, I noted, could be viewed as a typical example of Leibniz's principle of plenitude or, alternatively, as a statement of the principle of the conditions for existence. In this passage of Cuvier's, we are back again to Buffon's view of the limits on diversity, but Cuvier does not follow Buffon in claiming any sort of universality for his principle that can call forms into being in a Leibnizian manner, according to the principle of plenitude. Instead, it is merely a restriction, a limitation on the forms that can appear. In commenting on this passage, E. S. Russell (1916) noted that "we seize here the relation of the principle of the adaptedness of parts to the problem of the variety of form. The former is in a sense a regulative and conservative principle which lays down limits beyond which variation may not stray. In itself it is not a fountain of change; there must be another cause of change. This thought is of great importance for theories of descent" (39). I couldn't agree more.

I'd like to conclude this section with one more quote, perhaps the most illuminating discussion of the relation Cuvier saw between organismal diversity and the conditions for existence of all. This passage is in his notes for an "essay on zoological analogies," included by William Coleman in an appendix to his biography of Cuvier and clearly written in the context of Cuvier's debate with Geoffroy Saint-Hilaire and the *Naturphilosophen* (Coleman 1964, 189):

> Everything may exist which does not contain within itself the principle of contradiction. There are geometrical contradictions, in terms of 180 degrees contained within 3 angles. There are others which are less obvious and which are found only by long study: the equality of the sum of the distances from the foci of the ellipse to all points on the circumference. Similarly, there are physical contradictions: oxide more than metal in a closed vessel where there is no oxygen to make it. There are also physiological contradictions. A very large bird supports itself: if it is made 4 times longer, its new wing surface will be increased 16 times and its weight 64 times (whose square root is 8). It would need wings 8 times longer. The organism, nourished and growing by intussusception, mobile and hence not tied to the earth, and whose fluids cannot circulate by the action of weight, heat, and evaporation, must contain within itself a reservoir of nutritional material. This is the only material condition of animality.
>
> Among the enormous number of animals made possible by this one condition there are some which are necessarily quite similar and others which are less closely related. All combinations which are not contradictory are possible; in other words, everything which has a "condition of existence," whose parts cooperate in a common action, is possible.

Hence there must be a series of similar creatures which reaches just to the point of this contradiction. They are similar in form, composition, and functions. Differences occur first in things of small importance, in proportions. Later on come the deficiencies and additions. Only much later does a change of plan, of the relationship of the parts, appear.

The constraints imposed by the conditions for existence thus act as a regulative principle to interpret, and thus help explain, the diversity of organic form.

THE PHILOSOPHICAL ORIGINS AND SIGNIFICANCE OF THE PRINCIPLE

Because of his rhetorical emphasis on facts, Cuvier is often seen as the opposite of a theoretician. However, he was a child of the Enlightenment and as such was interested in man's ability to establish rational and general laws of nature; he was no mere Baconian inductivist. At the start, he had Lavoisier's chemistry and Haüy's mineralogy to show him that it was possible to place natural history on rational, even mathematical grounds, and he had de Jussieu's example for how to proceed similarly in biological classification. The problem that presented itself at the time was not just how to classify the animals but how to do so in a way that was "natural," that conformed to intrinsic features of the objects of study, rather than artificial features merely recognized by our mental processes (the Buffonian objection).

Cuvier's philosophical perspective is hard to reconstruct fully, because of the restraint he used in his public utterances. In particular, the degree to which he was influenced by religious considerations is difficult to determine, and no general consensus appears to exist (see Coleman 1964; Outram 1984; Asma 1996; Rudwick 1997, 2005; Cheung 2000; Grene and Depew 2004; Van der Meer 2005). With Cuvier there are always political considerations to keep in mind, and from my perspective, the evidence is still inconclusive as to his true feelings. Nevertheless, an interpretation consistent with the evidence known to me is that Cuvier was indeed religious, at least to the extent of being a deist or perhaps a somewhat enlightened, nondogmatic theist (Outram 1984 provides an excellent discussion of the issue).

Cuvier had grown up in a Lutheran environment, and religion was a key part of the training at the Karlsschule, where there were three chaplains, one for each of the religions of the empire (Flourens 1856, 176). Although in his early years he avoided all religious references in his published work, there is evidence that after Napoleon's rise, he allowed more explicit religious references into his lectures (Corsi 1988; Rudwick 2005). After the restoration of the monarchy in 1815, we increasingly find such references in his published work, as in the invocation of the "Creator" quoted earlier from the 1825 edition of the "Preliminary Discourse."

The most notorious example of such comments came in Cuvier's article on "Nature" in the same year for the *Dictionnaire des sciences naturelles* edited by his brother, Frédéric. Here, after discounting the scale of being, Geoffroy Saint-Hilaire's "unity of composition," the *Naturphilosophen*'s ideas about homologies within organisms, and evolutionary transformation, he proclaimed:

> We, however, conceive nature to be simply a production of omnipotence, regulated by a wisdom, the laws of which can only be discovered by observation; but we think that these laws can only relate to the preservation and harmony of the whole; that, consequently, all must be constituted in a manner that contributes to this preservation and to this harmony, but we do not perceive any necessity for a scale of beings, nor for a unity of composition, and we do not believe even in the possibility of a successive appearance of different forms; for it appears to us that, from the beginning, diversity has been necessary to that harmony, and that preservation, the only ends which our reason can perceive in the arrangement of the world. (Cuvier 1825, 268; translation slightly modified from Lee 1833, 83)

The young Cuvier's enthusiasm for Bernardin de Saint-Pierre's *Études de la nature* (see Marchant 1858), which likewise painted a picture of the harmonious operation of the world as the design of a benevolent creator, though not necessarily the Christian one, suggests that this view was a consistent one throughout his life (cf. Letteney 1999).

Yet however much such considerations may have been in the background, it is clear that they never explicitly entered into any of Cuvier's arguments. In physics, it had been shown since Descartes and Bacon that such considerations were of little use; it was necessary for science to be independent of religion. In considering Cuvier's views, it is important to remember that he was extremely broadly educated, having studied both Latin and Greek in Montbéliard before going to the Karlsschule and pursuing his studies in administration. Moreover, it is important to remember that his first year of study, before choice of a faculty to specialize in, was devoted to philosophy. Cuvier had two professors in this area: Johann Christian Schwab, in charge of courses in logic and metaphysics, and a partisan of Leibniz and especially Wolff; and Johann Friedrich Abel, influenced by the Scotch moralists, French materialists, and Kant (Letteney 1999; Taquet 2006, 82–83). It is thus to be expected that Cuvier had a thorough grounding in philosophy, and we cannot doubt that by the end of his first year at the Karlsschule—if not sooner—he was familiar with most of the history detailed in the preceding two chapters of this book, up to and including the French materialists, Hume, and Kant (save for the *Critique of Judgment*, which had not yet appeared).

Cuvier tells us in his autobiography that he read quite widely and voraciously as a student, though unfortunately we have no list of his readings or record of his thoughts of the time. However, his friend Pfaff did report that during the year they roomed together, Cuvier read the *entire* historical and critical dictionary of Bayle. Pfaff noted that he would frequently fall asleep and wake some hours later to find Cuvier, still awake, sitting like a statue with his Bayle in his hand (Marchant 1858, 17).

Two major possible philosophical sources for the principle of the conditions for existence have been suggested in the secondary literature, Aristotle and Kant. William Coleman (1964), in his classic study, argued in favor of Aristotle, largely on the basis of perceived similarities between Cuvier's principle and Aristotle's views as expressed in his *Parts of Animals* (which I discussed in chap. 3). Coleman states that Cuvier "was either unaware of or rejected the recent criticism brought to the teleological argument by Kant" and further that "Kant's cautious conclusion still influences still influences thinking on the teleological problem, but it had absolutely no influence on Cuvier. Cuvier's theoretical biology was based upon teleological reasoning in its most literal form. He preached in publicly, he adopted it in practice, and he never abandoned it" (42). For Coleman, Cuvier's passage from the *Règne animal* in which he appears to synonymize the "conditions for existence" and "final causes" was sufficient proof of this contention.

In contrast, E. S. Russell, who was deeply sympathetic to Cuvier's point of view and to my mind had a much better understanding of Cuvier's principle, argued precisely the opposite: "Cuvier was indeed a teleologist after the fashion of Kant, and there can be no doubt that he was influenced, at least in the exposition of his ideas, by Kant's *Kritik der Urtheilskraft*, which appeared ten years before the publication of the *Leçons d'Anatomie Comparée*" (Russell 1916, 35). Again, however, no evidence for the connection is given other than the general compatibility of Cuvier's principle of the conditions for existence with Kant's perspective on teleology: "The principle of the adaptedness of parts may be used as an explanatory principle, enabling the naturalist to trace out in detail the interdependence of functions and their organs. When you have discovered how one organ is adapted to another and to the whole, you have gone a certain way towards understanding it. That is using teleology as a regulative principle, in Kant's sense of the word" (34–35).

More recent workers have found various sources for Cuvier's principle. Dorinda Outram (1986) explicitly argued against Coleman's views, stating that "in spite of his admiration for Aristotle, it was not for his teleology that Cuvier singled him out" (345). Like Russell, she points to Kant instead as a source, noting Cuvier's close relationships with Parisian Kantians and what she sees as Kant's conception of life, which views "each function as a *condition* of the existence of the organism, rather than *intended* to contribute toward the existence of the organism" (347).

For Outram, the key to understanding the relevance of Kant's philosophy for Cuvier's project was that for Kant:

> An organic rather than a machine analogy, a model stressing internal rather than externally imposed purposiveness, seemed to provide a better way of understanding the real organization of nature. Through this simple reorientation the teleological version of the argument from design began to disintegrate. By pointing out that living beings are not powered from outside by externally imposed intention, but from inside through an intimate interlocking of internal functions, Kant opened the way to a concept of life deriving from the inner economy of each being, which could be

discussed as a self-perpetuating system independent of first causes. The distinctions necessary to teleological argument, between ends and means and between design and instrument, ceased to raise useful issues. It was in relation to the idea of the inner economy of the organism, rather than to teleology in the natural theological or Aristotelian sense, that Cuvier elaborated his principle of the conditions of existence. (Outram 1986, 348)

A few years after Outram's paper appeared, Stephen Asma examined the issue in his interesting book *Following Form and Function: A Philosophical Archaeology of Life Science* (1996), which covers some of the same ground I have covered here, particularly in its distinction of varieties of teleology. Asma found connections both with Aristotle and with Kant. For Asma, in spite of some significant differences, Aristotle, Kant, and Cuvier are united by support for an "organicist" or "organismic" teleology, a teleology that "takes its start . . . from the irreducible organic principles that make life fundamentally different from the inanimate" (141). But Asma argues that "Cuvier himself did not seem to realize where his real Aristotelian connection lay" (36). Asma's view is most clearly expressed in the following passage:

The functionalism of Cuvier that tacitly refers organisms to a designing Mind is very far from Aristotle. But the internal teleology, which flowed from Cuvier's early defini-tion of life . . . , does bear a very strong connection with Aristotle's biological metaphysics. Granting the two naturalists are working within radically different physico-chemical theoretical systems, we must recognize a common antireductionistic orientation. Cuvier's early concept of life held that simple elements are balanced albeit only temporarily in a harmony of higher physiological functions. Arguably, this internal teleology . . . was inspired by Kant's revitalization of teleological thinking in the *Critique of Judgment* (1790). Yet this antireductionistic holism makes up the bulk of Aristotle's response to his predecessors. (36)

From my perspective, Asma's interpretation of Cuvier's principle of the conditions for existence in a rather Darwinian way, as a principle relating structure to some "external environmental condition of existence," and his thus granting Cuvier only an "incipient form of organicist teleology" (21), is not adequate. His links with Aristotle and Kant also do not go as deep as one might hope. Nevertheless, I sym-pathize with the general tenor of his approach.

By far the best examination of the issue of Cuvier's teleology to date is a doctoral dissertation by Michael Letteney (1999) entitled "Georges Cuvier, Transcendental Naturalist: A Study of Teleological Explanation in Biology." Letteney appears to have been unaware of Asma's work—at any rate, he does not cite it. However, his focus, like Asma's, is on the tension between Coleman, on the one hand, and Russell and Outram, on the other, as to the origin of Cuvier's views on teleology (Aristotle vs. Kant), particularly the principle of the conditions for existence. As the title of his dissertation makes clear, he comes down squarely on the side of Kant (see also Huneman 2006, who likewise argues for a Kantian connection).

Letteney reviews the work of four predecessors in the attempt to understand the origins of Cuvier's principle; three we have talked about (Coleman, Russell, and Outram) and one we haven't (William Whewell, writing in the 1830s; I briefly discuss Whewell's interpretation in the following chapter). Letteney's first major point is that—even ignoring his predecessors' interpretations of Cuvier—their interpretations of Aristotle and Kant differ significantly. In particular, he points out that Outram, in arguing that Cuvier was a Kantian, fails to take account of Kant's distinction between constitutive and regulative principles of judgment. After considering in detail just how Aristotle and Kant differ with respect to their views on teleology, Letteney proceeds to show, by an examination of Cuvier's major writings, that Cuvier's overall perspective on science can indeed quite reasonably be interpreted as Kantian. He pays particular attention to Cuvier's views on the history of science, as expressed in his *Histoire des sciences naturelles* (1841–1845).

Letteney points out that in the *Histoire* Cuvier identifies his principle of the conditions for existence not with the thought of Aristotle or Kant but instead with that of Socrates:

> The sciences owe to Socrates another advantage [in addition to the development of the experimental method]. He introduced the notion of *final causes*, or as we now say, *conditions for existence [des conditions d'existence]*. Socrates recognized that it was in the writings of Anaxagoras that he had imbibed the idea of this principle fecund in useful results. If the universe . . . is as Anaxagoras thought, the work of an intelligent being, all its parts must be in conformity, concurring thus to a common goal. Each organized being must, consequently, be connected with all other beings, forming a link of a vast chain which, from the divinity, descends all the way to the most simple being; further, each being must contain in itself the means of fulfilling the role which is its share. (Cuvier 1841–1845, I: 112; translation slightly modified from Letteney 1999, 108–109)

As Letteney points out, this passage certainly points to Cuvier as indeed an "external" teleologist, one not far from the traditional argument of design, and he is able to demonstrate that Cuvier even appears to credit Anaxagoras himself and Plato's *Timaeus* as predecessors to his approach. Rather strangely, Cuvier never discusses this aspect of Aristotle's philosophy at all (this point had previously been made by Outram). Letteney also notes Cuvier's youthful admiration of Bernardin de Saint-Pierre's *Études de la nature*, concluding that indeed "Cuvier conceives of finality in terms of a creative agent that is *external* to matter" (123). This seems a fair reading of the evidence—this is, after all, the standard deist position, and it is the position that Kant undertakes to critique.

But how does this interpretation of what finality *means* relate to the epistemological problem of how we determine "final causes" or "conditions for existence" in scientific practice? And what is the ontological status of the "conditions for existence" so determined? Letteney points to another passage from the *Histoire* as his epiphany, in his realization that Cuvier was indeed a Kantian, at least in his theory of knowledge or epistemology. The passage relates to Parmenides' separation of the world into two paths, the "way of truth" based on reason and the "way of appearance" based on the senses. The passage reads, "But Parmenides admits that this illusion is subject to fixed laws, in such a way that it is possible to take the variations of this illusion as bases for reasoning, entirely as well as if they were real. The Eleatic school thus would have been able to enter into the method of observation, and extend greatly the domain of the sciences; but, enslaved to vague speculations, they could not follow this road studded with riches" (I: 100–101; translation by Letteney 1999, 386). As Letteney rightly concludes, this passage suggests that Cuvier was indeed an idealist, not a realist. He goes on to show that Cuvier in fact appears to have been transcendental idealist in the Kantian mode—in his phrase, a "transcendental naturalist."

In an interesting discussion of Cuvier's *Dictionnaire* entry on "The soul of animals" *(Ame des bêtes),* Letteney presents Cuvier's position as follows:

> What Kant shows is that while we cannot know the existence of the soul, by the same token we cannot disprove it either, and this because our categories of thought have application only to possible objects of experience, of which the soul is not a member. I think that Cuvier finds this to be a congenial epistemological outlook because it enables him to have his cake and eat it too. By this I mean that by placing the human soul outside the reach of possible knowledge, Cuvier is free to investigate the animal kingdom, man included, *as if* such objects were determined solely by the laws of matter, without thereby threatening the autonomy of morality and religion. In a certain sense, Cuvier wants to *think* like a Humean, but he does not want to *act* like one. Transcendental idealism, I am suggesting, lets him have it both ways, since it provides a principled way of separating science from religion and morality. (402)

This is precisely the point I made with respect to Kant himself at the close of the last chapter.

From here, Letteney moves on to consider the principle of the conditions for existence as Cuvier implements it in practice. His fundamental point is that in spite of his Kantian epistemological stance, Cuvier in practice accords the principle of the conditions for existence a *constitutive* role in the explanation of animal form. He points out that at least in the *Règne animal,* Cuvier indicates that it is only necessary to consult observation *after* "all the laws of general physics and those that result from the conditions for existence are exhausted" (1817, 6, my translation).

Yet as T. H. Huxley noted many years ago, in practice all of Cuvier's "functional correlations" were in fact empirical ones: one can predict that a certain form of tooth, gut, and so forth, will be associated with a carnivorous diet only by observation of living forms, not by a priori functional analysis. Thus, when Cuvier famously exposed the marsupial bones (epipubis) of a fossil opossum in the presence of witnesses, after having predicted their presence, he was merely recognizing that the animal was an opossum by other characters and using the observed empirical correlation to predict the presence of marsupial bones. In other words, although from a Kantian perspective the facts are the empirical correlation, and the functional interpretation is merely a regulative principle by which to understand this perceived relationship, Cuvier believes that when he can "understand" the reason for the correlation as functionally based, it is in fact an a priori law. But the problem for the scientist, of course, is to be able to distinguish between alternative hypotheses for observed correlations; otherwise, one's interpretation is adding nothing to the observed correlation itself.

I think Letteney is also right to locate Cuvier's misuse of his own principle in a constitutive way in his preference for the comparative anatomical, rather than the experimental, approach to understanding the conditions for existence. For Cuvier, the functional integrity of the organism was such a necessary thought that he was unwilling to acknowledge the possibility that one could study functions in *parts* of organisms and attempt to *experimentally* demonstrate the internal conditions for the existence of the organism: "The machines which are the object of our researches cannot be demonstrated without being destroyed; we cannot know what would be the result of the absence of one or several of their cogs, and consequently we cannot know what role each of these cogs plays in the total effect" (Cuvier 1800–1805, I: v, translated by Letteney 1999, 442). But Cuvier claimed that in spite of our inability to experimentally determine the conditions for existence, nature can and has:

> Happily, Nature itself seems to have prepared for us the means of compensating for this impossibility of carrying out certain experiments on living bodies. She presents to us in the different classes of animals almost all the possible combinations of organs; she shows us them united two by two, three by three, and in all possible proportions. There isn't any one, so to speak, of which she hasn't deprived some class or some genus, and it suffices to carefully examine the effects produced by these combinations *(réunions)*, and those that result from these privations, to deduce highly probable conclusions about the nature and function *(usage)* of each organ and each form of organ. (I: vi, translation modified from Letteney 1999, 443)

Cuvier here seems to argue that correlation can allow us to deduce causal relations, something that no modern scientist would generally want to do (though the issue is still with us, in the form of debate over the "comparative method"; see Harvey and Pagel 1991 and chap. 10 in this book). Thus, Letteney appears to be fully justified in arguing that Cuvier's attempt to found a purely observational yet rational science of comparative anatomy was a failure. In fact, it is hard to see how one could hope to have a fully explanatory theory of organic form grounded in observation, especially observation within Cuvier's fixist perspective. However, this interpretation of Cuvier's *practice* does not affect our interpretation of Cuvier's *principle*, which is in no way compromised by the illegitimate uses he made of it.

I will return to this point in a moment, but first I'd like to look at one more view of Cuvier and the principle of the conditions for existence: that offered by Marjorie

Grene and David Depew in their recent *Philosophy of Biology* (2004). Grene and Depew present Cuvier in a broad historical/philosophical context and do not take account of the more recent specialist literature in the area, including only Russell and Coleman in their bibliography. Nevertheless, they are clearly well read in this area and have a good sense for the general meaning of Cuvier's work. It is also of note that they are drawn to translate Cuvier's principle as the "prerequisites of existence," following Corsi (1988). As befits the interest of one of the authors in translating Geoffroy Saint-Hilaire's *Principes de philosophie zoologique* (Le Guyader 2004), their focus is on the Geoffroy Saint-Hilaire/Cuvier debate, and they seem to have some trouble understanding the principle of the conditions for existence (perhaps because their perspective derives from Geoffroy Saint-Hilaire's). In attempting to understand the principle in the context of Aristotle's definition of final cause (for them, Cuvier is a "good Aristotelian"; 313), they ask rhetorically, "But how can conditions be ends? They are what is needed if the end is to be achieved, not the end itself. Perhaps we can say that understanding what is meant by conditions of existence means reasoning in terms of final causes rather than in terms of when-then causes and effects. The end is the very existence, in its proper habitat, of this closely integrated entity, beautifully suited to that habitat and no other. The conditions are the harmonious, well-integrated means to that end" (140). Unfortunately, this is as far as they go; they make no attempt to clarify exactly what they mean by "reasoning in terms of final causes" or to connect this view to any specifics of either Aristotle's or Kant's perspective. This is true in spite of the fact that in this very book, they provide a fair summary of the biological philosophy of both Aristotle and Kant.

One final comment on Grene and Depew is in order. In summarizing the debate, they argue that Cuvier's approach to zoology is "in principle incompatible with . . . transformism," unlike Geoffroy Saint-Hilaire's (2004, 141). While this indeed echoes Cuvier's own view, the present book is an extended attempt to argue precisely the opposite.

To conclude, in reviewing various points of view about the philosophical sources for Cuvier's principle, I have found good reason to agree with the reconstruction offered in the thorough, well-argued examination of Letteney (1999). The picture he paints of Cuvier as a Kantian—in agreement with Russell (1916) and Outram (1986), as well as Sloan (2006) and Huneman (2006)—is consistent both with Cuvier's background and philosophical training at the Karlsschule and with his critical, rationalist approach to science. Letteney is no doubt also right in raising questions about the validity of Cuvier's attempts to use the principle of the conditions for existence in a constitutive manner, attempts that he grounds in Cuvier's refusal to go beyond observation to experiment. In the next section, I will examine how the principle of the conditions for existence in fact came to be used in just such an experimental context, under the influence of the great Claude Bernard. For the present, however, I wish to revisit a few of the issues examined by Letteney.

The first has again to do with Aristotle versus Kant as precursors. One thing that must be borne in mind in considering Cuvier's lack of discussion of Aristotle's ideas on the nature of life, causality, and especially teleology is the Enlightenment context in which he came of age. For most Enlightenment thinkers, like their Renaissance predecessors, Aristotle was metaphysically obsolete, his philosophy irrevocably tainted by its Scholastic interpretation by the Catholic Church. Cuvier, as a liberal Protestant, had little patience with the Catholic Church. For this reason alone, it would have been hard for him to publicly embrace Aristotle's philosophical views, however much he admired his work in natural history.

Nevertheless, as I have argued in chapter 2, there is a connection to be made between Cuvier's principle and Aristotle's views, a connection that revolves around the concept of hypothetical necessity. Here I again find myself and Letteney (1999) in agreement. Letteney likewise notes the passage from *Parts of Animals* 640a33–640b4 quoted earlier, in which Aristotle says, "The fittest mode, then, of treatment is to say, a man has such and such parts, because the essence of man is such and such, and because they are *necessary conditions of his existence*" (Ogle's translation, my emphasis). As Letteney concludes, whether or not there was a direct connection, "Aristotle's conception of hypothetical necessity is very similar to the account of teleology implied by Cuvier's principle of the conditions of existence" (203).

A second point has to do with Kant. Letteney notes that Lee (1833), in her biography of Cuvier, reports that Cuvier lectured on Kant in his course on the history of the natural sciences at the Collège de France. Letteney is unwilling to credit the account of Kant's philosophy in the published version of Cuvier's lectures, however, because the volume that contains it was claimed to be written exclusively by Magdeleine de Saint-Agy (see Smith 1993 for discussion). Nevertheless, if one compares the two accounts, they are in fact quite similar. Lee reports that when Cuvier resumed his lecture series in December 1831, he first "gave a clear and eloquent résumé of the philosophy of Kant, of Fichte, and of Schelling" (121). This description corresponds quite well to the relevant passage from the *Histoire des sciences naturelles*, from a section entitled "The Philosophy of Nature in Germany and France." The passage reads as follows:

> But at that time rival philosophies [to those of Newton and Locke], emanating from Leibniz's principles, began to appear. These were developed and arrived at a mathematical form in the writings and in the teaching of Wolff. They reigned principally in Germany, up to the last third of the eighteenth century.
>
> But Kant, inspired by Hume, then gave metaphysics a new impulse, in publishing in 1781 his *Critique of Pure Reason*, where he examines the question of the scope of human reason, and the things which it can know with certainty. Among other propositions, he declares this: that we can have no certain notion of things in themselves. He expresses ideas on physics from which it would follow that nature is composed of elements of an identical nature. Fundamentally, his physics is dynamic

physics, which from the time of Descartes, is opposed to atomistic physics. He admits a part of the ideas later developed by Fichte, who professed the pure idealism and principles of the Eleatic school. Fichte tried to establish that only the *self* exists for us, that it *poses itself (pose)*, according to the bizarre expression of his students, in opposition to the *non-self*, that is to the external world which is its creation.

This purely idealistic philosophy, born in Germany, the country where mathematics were most cultivated, had a certain vogue there which still exists today. But up to Schelling, it had never been so directly applied to the natural sciences that I should allow it to enter the history of those sciences. (Cuvier 1841–1845, V: 313–314, my translation)

Cuvier goes on to trace the development of *Naturphilosophie*, beginning with Goethe, and then considering Kielmeyer, Schelling, and Oken. The similarity of the quoted passage to Lee's description argues that these are indeed Cuvier's words. But in any case, the context makes it clear why Cuvier might not have been as anxious to highlight his Kantian leanings in his later years: because the *Naturphilosophie* he so vigorously opposed was an indirect outgrowth of Kantian transcendental idealism.

One more passage from the *Histoire* is critical to our story. As just discussed, Letteney (1999) notes that Cuvier claims Socrates as the originator of the principle of the conditions for existence, at least in the form of "final causes." Yet this is not the only invocation of the principle of the conditions for existence in the section of the *Histoire* dealing with the ancients; the principle also occurs in Cuvier's discussion of Lucretius. In fact, Cuvier presented the Lucretian theory in precisely these terms: "Many of the aggregations formed by the coming together of atoms had only an ephemeral duration, because they did not combine the conditions for existence *(conditions d'existence)* necessary *(indispensables)* to the maintenance of life. On the other hand, living bodies that possessed all of these conditions, including the faculty of reproduction, were the source of the species which exist today" (1841–1845, I: 220, my translation).

The fact that Cuvier could discuss the views of both Socrates and Lucretius in terms of the principle of the conditions for existence helps us understand its unique attractions for him. Cuvier clearly sought to develop his science based on rational principles. Yet the problem of teleology was one that had exercised both philosophy and the life sciences for over two thousand years and formed the battleground for the debate between the mentalist teleologists, like Socrates, and the materialistic atomists, like Epicurus and his follower Lucretius. The principle of the conditions for existence was the meeting ground for these two very different views of nature. On the one hand, for the deist or theist, although God had created organisms, in so doing he must have been constrained by the necessary conditions for existence—that is, the functional harmony among parts of organisms and between organisms and their environment that formed part of the larger harmony of the created world. On the other hand, even if the world had simply appeared one day as the outcome of the random concourse of atoms, these same conditions for existence must have been fulfilled. Thus, unlike natural theological "final causes," identified with the intentions of God, the principle of the conditions for existence should be equally acceptable to the theist and to the atheist. What more could one ask of a rational principle on which to ground one's functional explanations?

THE INFLUENCE OF THE PRINCIPLE
IN FRANCE AND GERMANY

In the following chapter, I will take up the question of Cuvier's influence in Britain, in particular the way in which his views were perceived in the intellectual environment that provided the background to Darwin's revolution. For now, I would like to look at a few of the paths that radiated from Cuvier's thought on the Continent. This discussion is by no means intended to be a thorough review but merely to point out that in spite of his failure to found a school of disciples organized around his principles, Cuvier's legacy included more than his factual work in zoology, his classification, and his views on the history of the Earth and of life. In particular, Cuvier's Kantian principle of the conditions for existence influenced German "teleomechanism," Comte's positive philosophy, and, most importantly, Claude Bernard's foundation of experimental biology.

Cuvier and German "Teleomechanism"

In his seminal study, Timothy Lenoir (1989) argued that a coherent approach to biology developed in early nineteenth-century Germany, an approach that he called "teleomechanism." Lenoir argued in particular that this approach or program was distinct from the program of *Naturphilosophie*, with which it was contemporaneous and had often been confused, as well as from that of materialism proper. His teleomechanists were a diverse lot, ranging from Johann Friedrich Blumenbach, C. F. Kielmeyer, and Alexander von Humboldt, to Karl von Baer, H. Rathke, and J. Müller, and again to Hermann Lotze, Theodor Schwann, Carl Bergmann, and Rudolf Leuckart. *Naturphilosophen* like Friedrich Schelling and Lorenz Oken were excluded (though Kielmeyer has also been considered by many, including Cuvier himself, as a *Naturphilosoph*; see Bach 1994).

This thesis has since come under vigorous attack, with many commentators questioning both the general thesis and the specifics used to support it (e.g., Caneva 1990; Nyhart 1995; Richards 2002). In particular, it does not seem that the existence of a true school of teleomechanism distinct from *Naturphilosophie* can be sustained. Nevertheless, the term is perhaps still useful to refer to a group of German zoologists of the early nineteenth century who felt that mechanism alone could not explain organisms. Instead, explaining organisms required reference to something like a "living principle," a "vital force," or a "formative force," whatever the ontological status of such a principle or force might be.

In Lenoir's view, both teleomechanism and *Naturphilosophie* were outgrowths of Kant's insistence that the proper understanding of biology must somehow relate the mechanical and teleological principles, because we cannot understand organisms purely mechanically (see the previous chapter). And if the teleomechanists did not in fact form a distinct school, Lenoir was certainly right to call attention to the centrality of Kant's views on the nature of the organism for many of the German biologists of the late eighteenth and early nineteenth centuries (Sloan 1992, 2006; Richards 2002; Huneman 2006, 2007).

For us, the key question is how Cuvier fits into the German picture. With his German fluency gained at the Karlsschule, his connections with Kielmeyer, and his position at the Paris Muséum, Cuvier was uniquely placed to interact with the German scientists. Lenoir sees him as a seminal figure in the further development of the teleomechanist tradition in the early 1800s:

> Though not himself a member of the research tradition discussed here, Georges Cuvier was responsible for deepening and extending the teleomechanist research program of the vital materialists. In the 1790s all roads led to Göttingen, where the young men in Blumenbach's inner circle were envisioning a comprehensive approach to organic nature. Between 1800 and 1815, however, particularly after the German states had largely become satellites of France and many German universities were closed, those roads led to Paris, which for German zoologists meant they led to Georges Cuvier. During these years, when Alexander von Humboldt made Paris his home base for exploring the world, a small German colony sprang up around Cuvier. These young Germans had several features common in their backgrounds. They were either students of Blumenbach and Reil, or, thorough contact with their students, were about to become enthusiastic converts to the teleomechanist program. A second feature they all shared in common was their opposition to *romantische Naturphilosophie*, feelings which were certainly intensified by contact with Cuvier, who was even less tolerant of it than he was of the views of his rivals, Lamarck and Geoffroy Saint-Hilaire. (Lenoir 1989, 54–55)

The connection that Lenoir points out between Cuvier and the rising generation of German zoologists was not merely one of convenience; Cuvier's principle of the conditions for existence, with its Kantian emphasis on the functional wholeness of the organism in its environment, was quite congenial to the viewpoint of the young Germans.

Nevertheless, Lenoir also points to some significant differences between them. In particular, for the Germans

> the hierarchy of different levels of organization present in each animal was not only the manifestation of the laws of functional organization; it was the trace of a historical lineage of *materially* connected forms, the transformation of the animal type within

the limits set by the physical conditions prevailing within an epoch. Both the historical and materialist dimensions are completely missing from Cuvier's approach. . . . It was the importance of this historical dimension for their view of animal organization that led the vital materialists, in sharp contrast to Cuvier, to include comparative embryology as an essential part of their program. (Lenoir 1989, 64–65)

For the further explication of the views of these German zoologists, I refer the reader to Lenoir's book, which—in spite of its admitted faults—still provides fascinating reading on the history of this phase of teleological thought. For a partial corrective, see Caneva (1990) and Richards (2002).

Before leaving them, I can't resist pointing out one further connection between Cuvier and the Germans. Johannes Müller (1801–1858) was an embryologist, comparative anatomist, and above all the preeminent physiologist of his generation; his *Handbuch der Physiologie des Menschen* (1834–1840) was the founding work of modern physiology. His students included the likes of Theodor Schwann, Hermann von Helmholtz, and Emil du Bois-Reymond. As is well known, Müller was a vitalist; for Lenoir, "Müller's theory of animal organization was based squarely on Kant's conception of teleomechanism" (148). Interestingly, however, the passage Lenoir cites from the *Handbuch* to justify this ascription reads as follows: "Kant says that the cause of the manner of existence of each part of a living body in contained in the whole, while in inert matter the cause is contained in each part itself" (Müller 1844, 19; translated by Lenoir 1989, 148). This passage is a direct quote from Cuvier's introduction to the *Leçons*, a connection confirmed by the fact that in the original German, the wording is almost identical to that used in the 1809 translation of Cuvier's work (Cuvier 1809, 5). It thus appears that Cuvier played a role in the Kantian education of a generation of German doctors, physiologists, and zoologists. Presumably through Müller, Cuvier's version of Kant was influential even into the twentieth century: in his masterful work *On Growth and Form*, D'Arcy Wentworth Thompson quotes the same passage, attributing it to Kant himself (Thompson 1942, 1020).

Comte, Positivism, and the Conditions for Existence

Cuvier's relation to the positivism of Auguste Comte (1798–1857) is complex. Cuvier's religious views, even if we interpret them as being deistic rather than theistic, contrast with the explicit atheism of Comte. Nevertheless, Cuvier's consistent emphasis on the primacy of "positive facts" over metaphysical speculation has led many to link them, and as Pickering (1993, 294–295) shows, Comte's views on epistemology have close links with those of Kant. In fact, Cuvier was himself an early supporter of Comte's work (Pickering 1993; McClellan 2001), and Comte followed the course on general and comparative physiology offered by Henri de Blainville, Cuvier's protégé, in the years from 1829 to 1832. Although de Blainville by that time had become one of Cuvier's greatest detractors (see de Blainville 1890), this was apparently based more on personal animosity than any sharp separation in their views on teleological explanation (Appel 1980). Chris McClellan (2001; see also Gillispie 2004) has recently provided a fascinating examination of the precise nature of the relationship between the Cuvier and Comte, an examination centered on Cuvier's principle of the conditions for existence.

The best-known tenet of Comte's philosophy is his law of the three stages of knowledge: the theological, the metaphysical, and the positive or scientific. As detailed by McClellan, Comte found a special role for biology in his system of the sciences, because it was in attempting to understand the phenomena of life that the greatest temptation toward the theological and metaphysical interpretation occurred. According to Comte, however, biology had now entered into its positive stage, and this because "attacking in its turn and its own manner the elementary dogma of final causes, biology has gradually transformed it into the fundamental principle of the conditions of existence, of which the development and systematization belongs, without any doubt, to biology, although in itself it is essentially applicable to all orders of natural phenomena" (quoted in McClellan 2001, 16). Comte's appreciation of the principle of the conditions for existence is quite in keeping with my interpretation here. Even more interesting is the fact that Comte did not limit the application of this principle to biology (as Cuvier had done, for all practical purposes) but instead extended it to encompass astronomical phenomena. One can deduce a priori certain features of the solar system, he says, "by the sole reflection that since we exist, it follows with all necessity that the system of which we make a part is disposed in a manner to permit this existence, which would be incompatible with a total absence of stability in the principal elements of our world" (quoted in McClellan 2001, 18). McClellan here makes the obvious connection with the so-called weak anthropic principle in cosmology, which I have discussed in chapter 2.

McClellan points out that in addition to Cuvier's effect on Comte's formulation of his positive philosophy, an even more critical legacy of the Cuvier-Comte relationship lies in the experimental work of Claude Bernard (1813–1878), the founder of modern physiology. As already discussed, Cuvier's holist emphasis on the functional unity of the organism made him skeptical that attempts to understand living organisms by taking them apart (vivisection) could ever succeed in general. Nevertheless, he was supportive of experimental work by François Magendie (1783–1855), who was Bernard's mentor, and his own student Jean Pierre Flourens (1794–1867). As McClellan (2001) perceptively notes, "the foundations of Bernard's philosophy of experimental physiology find their origins in the work of Cuvier, mediated through the direct influence of Comte's natural philosophy" (27). It is this relationship I would like to examine in the following section (see also Caponi 2004).

Claude Bernard and the Experimental Determination
of the Conditions for Existence

Claude Bernard is best remembered for his concept of the constancy of the internal environment or *milieu intérieur,* a direct predecessor to Walter Cannon's concept of homeostasis. In his experimental work, he conducted fundamental researches on the functions of the pancreas and liver, and elucidated the effect of curare on the neuromuscular junction. Bernard's precepts for experimentation were laid out in his classic *Introduction to the Study of Experimental Medicine* (1865; translation 1927). In this work, which might be considered the founding document of experimental biology, Bernard refers to Cuvier admiringly, quoting him as saying, "The observer listens to nature; the experimenter questions and forces her to reveal herself" (1927, 6).

However, Bernard goes on to expressly question Cuvier's conclusions about the inability to experimentally investigate the conditions for life:

> All the phenomena of a living body are in such reciprocal harmony one with another that is seems impossible to separate any part without at once disturbing the whole organism. . . . Many physicians and speculative physiologists, with certain anatomists and naturalists, employ these various arguments to attack experimentation on living beings. They assume a vital force in opposition to physico-chemical forces, dominating all the phenomena of life, subjecting them to entirely separate laws, and making the organism an organized whole which the experimenter may not touch without destroying the quality of life itself. They even go so far as to say that inorganic bodies and living bodies differ radically from this point of view, so that experimentation is applicable to the former and not to the latter. Cuvier, who shares this opinion and thinks that physiology should be a science of observation and of deductive anatomy, expresses himself thus: "All parts of living bodies are interrelated; they can act only is no far as they act all together; trying to separate one from the whole means transferring it to the realm of dead substances; it means entirely changing its essence." (59–60)

Nevertheless, Bernard's debt to Cuvier is clear from the general tone and even the language of his approach. In a section entitled "The Phenomena of Living Beings Must Be Considered as a Harmonious Whole," Bernard again explicitly invokes Cuvier:

> Physiologists and physicians must never forget that a living being is an organism with its own individuality. Since physicists and chemists cannot take their stand outside the universe, they study bodies and phenomena in themselves and separately without necessarily having to connect them with nature as a whole. But physiologists, finding themselves, on the contrary, outside the animal organism which they see as a whole, must take account of the harmony of this whole, even while trying to get inside, so as to understand the mechanism of its every part. The result is that physicists and chemists can reject all idea of final causes for the facts that they observe; while physiologists are inclined to acknowledge an harmonious and

pre-established unity in an organized body, all of whose partial actions are interdependent and mutually generative. We really must learn, then, that if we break up a living organism by isolating its different parts, it is only for the sake of ease in experimental analysis, and by no means in order to conceive them separately. Indeed when we wish to ascribe to physiological quality its value and true significance, we must always refer it to this whole, and draw our final conclusion only in relation to its effects in the whole. It is doubtless because he felt this necessary interdependence among all parts of an organism, that Cuvier said that experimentation was not applicable to living beings, since it separated organized parts which should remain united. (88–89)

Even more interesting from the present perspective is Bernard's definition of experimentation in terms of the conditions for the existence of vital phenomena, which clearly draws upon Cuvier's concept of the conditions for the existence of the organism: "We must acknowledge as an experimental axiom that in living beings as well as in inorganic bodies the conditions for the existence *(conditions d'existence)* of every phenomenon are absolutely determined. That is to say, in other terms, that when once the conditions for a phenomenon are known and fulfilled, the phenomenon must always and necessarily be reproduced at the will of the experimenter. Negation of this proposition would be nothing less than negation of science itself" (67–68, translation slightly modified)*. It is from this standpoint that Bernard denies the importance of the debate over vitalism, and in particular over the nature of the "vital force":

It matters little whether or not we admit that this force differs essentially from the forces presiding over manifestations of the phenomena of inorganic bodies, the vital phenomena which it governs must still be determinable; for the force would otherwise be blind and lawless, and that is impossible. The conclusion is that the phenomena of life have their special law because there is rigorous determinism in the various circumstances which constitute conditions for their existence or give rise to their manifestations; and that is the same thing. Now in the phenomena of living bodies as in those of inorganic bodies, it is only through experimentation, as I have already often repeated, that we can attain knowledge of the conditions which govern these phenomena and so enable us to master them. (68, translation slightly modified)

This sentiment is one that would surely be congenial to Cuvier. I cannot but think that had he lived to know of Bernard's work, he would have been an enthusiastic supporter.

But although in his great work we find Bernard constantly speaking of the conditions for the existence *(conditions d'existence)* of vital *phenomena*, and we

---

* Note that Bernard here confuses hypothetical with absolute necessity (i.e., necessary with sufficient conditions for physiological processes to occur). His "determinism" is thus not determinism in the same sense that physical laws determine their phenomena. Nevertheless, this distinction has little effect on his physiological *practice*, which is in fact always concerned with conditional relations.

cannot doubt that he is aware of the source of this phrase in Cuvier, he does not speak of the conditions for the existence of *life* itself, which was Cuvier's preeminent concern. In the hands of Bernard, Cuvier's *conditions d'existence* became transformed into the conditions of the physiological experiment itself; thus, Bernard pays close attention to the effect of varying external and internal conditions on organic phenomena, considering such external conditions as temperature, pressure, and air; and such internal conditions as fasting state, age, and species. It is this shift in meaning, and the associated blurring of absolute and hypothetical necessity, that made possible the rise of experimental physiology. At the same time, however, physiology became disconnected from the existence of the organism as a whole in its environment; it became the study of varying metabolic states, with the conditions for the existence of life only important to the extent they were necessary to prevent the death of the organ or tissue or cell under study. Yet there is no necessary conflict here. Bernard insisted on rigorous methods for the experimental study of the conditions for the existence of phenomena, and life itself can certainly be the phenomenon one chooses to study.

. . .

In this chapter, I have followed Cuvier's path to developing his principle of the conditions for existence *(conditions d'existence)*, its meaning, and its subsequent fate on the Continent. I have argued that the key to the principle is that it recognizes life as a phenomenon that we cannot express as the determinate result of natural law but instead must take as given. Having done so, we can then investigate the internal and external conditions under which it obtains. As a conditional teleological principle, the principle of the conditions for existence resides precisely on the common ground shared by the teleologist and the materialist. It was likely Cuvier's broad philosophical training that allowed him to find this common ground, amidst a vast field of possible alternative views that lean to one side or the other.

However, we have also seen that Cuvier's *application* of the principle was flawed by his insistence in using it in a *constitutive* manner to construct general laws *determining* form, when the facts on which his laws were based were purely observational. As Kant insisted, such observational laws can only be used in a regulative manner; they aid us in interpreting experience but cannot determine it. The temptation for Cuvier must simply have been too great; with his vast knowledge of organic form, he could see the functional necessity for certain observed relations (as in the tooth, claw, and digestive tract of the carnivore); hence, he believed himself to have rationally determined them.

In reforming Cuvier's approach, Claude Bernard pointed the way toward an experimental determination of the conditions for existence of vital phenomena, an approach that has led to the whole of modern physiology, down to the latest discoveries on the molecular physiology of the cell. Yet it was Cuvier himself whom Bernard quotes as saying, "The observer listens to nature; the experimenter questions and forces her to reveal herself." In fact, if we return to Cuvier's examples of his laws of correlation in the *Leçons*, we see that many of them have not withstood the test of time. Those that have include his generalizations with respect to respiration: it is indeed true that mammals and birds respire at a higher rate, that their muscles have a higher metabolism and lose irritability after death much more quickly, than do those of reptiles. But these relations were in fact determined by experiment, not deduction; it was only by careful experimentation that Joseph Priestley was able to show that the part of air necessary for animal life is oxygen, and Albrecht von Haller was able to show that the irritability of muscle fibers lasts longer in some animals than others. Cuvier's missteps remind us that, at best, functional inferences based on the constant conjunction of phenomena can serve as *functional hypotheses*, which must then be tested by experimental means.

Nevertheless, Cuvier's achievement was considerable. Not only did he plant the seed of a critical, rational attitude in the life sciences, which eventually blossomed into Bernard's experimental program, but he also managed to reform zoological classification by overturning the scale of being and to provide the first outlines of the history of the Earth and of life. In his 1973 book on *The Order of Things*, Michel Foucault famously insisted on the novelty of Cuvier's vision. It was Cuvier who "introduced a radical discontinuity into the Classical scale of beings; and by that very fact he gave rise to such notions as biological incompatibility, relations with external elements, and conditions for existence; he also caused the emergence of a certain energy, necessary to maintain life, and a certain threat, which imposes upon it the sanction of death; here, we find gathered together several of the conditions that make possible something like the idea of evolution" (275, translation slightly modified).

It is tempting to speculate about what the history of biology might have been had Cuvier seen fit to devise his own evolutionary theory, one that united his observations on fossils with his zoological classification and the principle of the conditions for existence. But such was not to be. Instead, it was left for an English naturalist by the name of Charles Darwin to attempt such a synthesis, and it is to him I now turn.

6

Darwin, Natural Theology, and the Principle of Natural Selection

One may say that there is a force like a hundred thousand wedges trying to force every kind of adapted structure into the gaps in the oeconomy of Nature, or rather forming gaps by thrusting out weaker ones. The final cause of all this wedging must be to sort out proper structure and adapt it to change.

—CHARLES DARWIN, 1838, NOTEBOOK D

Although teleological explanations based on a deterministic extrinsic teleology had long since been discredited on the Continent, at least in science, this mode of explanation persisted well into the nineteenth century in Britain. In fact, the argument to design underwent a revival in early nineteenth-century Britain, largely as a conservative response to the perceived dangers inherent in Enlightenment thought, as exemplified in the horrors of the French Revolution (Brooke 1989; Grene and Depew 2004, 184–191). This theme was found not only in such overtly religious writers as William Paley (1809) but also in scientific works, most notably in the Bridgewater Treatises (1833–1836), commissioned by the Earl of Bridgewater to demonstrate the "Power, Wisdom, and Goodness of God, as manifested in the Creation."

In this later British natural theology (e.g., Paley 1809; Crombie 1829; Crabbe 1840), much like its predecessors back to Henry More and John Ray, each part of an organism is a "contrivance," exquisitely adapted to the function it serves, and this could only be the result of design; moreover, each species is perfectly adapted to its place in nature. This view was quite clearly expressed by William Whewell (1859) in a discussion of the relation between morphology and teleology on the eve of the publication of the *Origin*: "The arm and hand of man are made for taking and holding, the wing of the sparrow is made for flying; and each is adapted to its end with subtle and manifest contrivance. There is plainly Design" (643). As the epigraph to this chapter makes clear, this revival of the argument of design forms a critical intellectual background to Darwin's theorizing. It is thus here I begin.

ADAPTEDNESS AND EXISTENCE
IN BRITISH NATURAL THEOLOGY

To understand Darwin's path to natural selection, we must examine more closely the implicit assumption underlying the use of the concept of "adaptedness" in natural theology: the assumption that adaptedness is a contingent gift of the Creator. In other words—as we have previously seen in Henry More and Robert Boyle—in the argument to design, a conceptual distinction is made between the *existence* of a species, which is taken for granted, and the *adaptedness* of a species, which is treated as something superadded to this existence, by the grace of God ("superabundant provision," in Boyle's words). Without this conceptual distinction, the argument to design would make no sense whatsoever. If the "adaptedness" of individuals or of a species is no more than their satisfaction of the necessary conditions for their existence, no design on the part of a creator is needed. Moreover, it is this *surplus* above what is strictly necessary that particularly justifies the extension from the existence to the wisdom and goodness of the creator. For the argument to design, it is not enough to argue that the world works; one must argue that it might have worked worse.

For example, we see this argument explicitly used by Paley, in his well-known *Natural Theology* (1809). Against the Epicurean/evolutionary hypothesis that "the eye, the animal to which it belongs, every other animal, every plant, indeed every organised body which we see, are only so many out of the possible varieties and combinations of being, which the lapse of infinite ages has brought into existence; that the present world is the relict of that variety; millions of other bodily forms and other species having perished, being by defect of their constitution incapable of preservation, or of continuance by generation" (63), Paley argued that

> multitudes of conformations, both of vegetables and animals, may be conceived capable of existence and succession, which yet do not exist. Perhaps almost as many forms of plants might have been found in the fields, as figures of plants can be delineated upon paper. A countless variety of animals might have existed, which do not exist. Upon the supposition here stated, we should see unicorns and mermaids, sylphs, and centaurs, the fancies of painters, and the fables of poets, realized by examples. Or, if it be alleged that these may transgress the bounds of possible life and propagation, we might, at least, have nations of human beings without nails upon their fingers, with more or fewer fingers or toes than ten, some with one eye, others with one ear, with one nostril, or without the sense of smelling at all. All these, and a thousand other imaginable varieties, might live and propagate. We may modify any one species many different ways, all consistent with life, and with the actions necessary to preservation, although affording different degrees of conveniency and enjoyment to the animal. (64–65)

Similarly, Alexander Crombie (1829) asked:

> If Nature ever possessed this procreative power, and if animals of various kinds were generated by the mere properties and qualities of matter, how does it happen, that out

of the immense numbers, which would have been fortuitously produced, no species of beings are found with a superfluity, a deficiency, or an incongruity of organs—quadrupeds, for example, with two or more heads, birds with one wing, or with two unequally balanced, animals with three or five legs, human beings without eyes or arms, or unprotected by a skin, and numberless other hideous malformations? Why do we not find animals generating their kind in situations, where their *existence* might have been possible, but ill *adapted* to their nature? If chance had governed, and intelligence had been excluded, is it credible, that out of the innumerable fortuitous formations which must have resulted, multitudes of such monsters would not have been generated? (182–183, my emphasis)

But although the argument to design depends on this conceptual separation between existence and adaptedness, no actual separation is made, because all *existing* species are considered to be perfectly *adapted* to their way of life. Nonadaptedness is merely a hypothetical possibility, with which the actual adaptedness of species can be compared.

Now in the argument to design, adaptedness is primarily a characterization of the relation between organic means and ends, whose similarity to the relation between human means and ends is considered evidence for intentional design by a deity. The head of a hammer is adapted to hitting nails; the wings of birds are adapted to flying. Since we know that the hammer is only adapted to its use by virtue of human design, it must be that the wings of birds are adapted to their use by design as well. Within this context, it is clear that neither the "adaptedness" nor the "adaptations" of an organism explain anything; they merely express our judgment of the ways in which we can understand the structure of an organism as appropriate for its way of life, and thus represent this structure as deriving from an intention similar in kind to our own. Nevertheless, as we have seen, the argument to design can also be worked in reverse. Once one has decided to interpret organic structure as if it were adapted by intentional design for a purpose, the assumed element of design can then be used as a basis on which to *explain* organic structure by its adaptedness for the role it plays. This is the argument *from* design. Just as the head of a hammer is flat *because* it was designed to hit nails, the wing of a bird has its particular structure *because* it was designed to provide lift. The intention explains the structure.

Not only is the structure of the organism explained by the intention of the Creator, however, but the *existence* of the organism is also explained by this intention, because each species is intended to fulfill a certain role in the economy of nature. Not only are the parts of the organism adapted to their role in the organism, but the organism as a whole is adapted to its ecological role. Carnivores exist *because* they were intended to help keep the population of herbivores down. Since it is possible to explain the adaptedness of the organism for its ecological role by the adaptedness of the parts for their functions—having teeth adapted to cutting flesh is one of the ways in which carnivores are adapted for their ecological role as predators—and the organism is designed for just this role, then all structures of the organism are capable of explanation as being designed for the ecological role that it plays.

While I have tried here to explore some of the paths of inference possible within the conceptual framework of the design argument, what is most important to realize, and most difficult to convey, is that this framework was part of a worldview in which the entire universe was seen as a single great system, in which everything was adapted to everything else. Thus, Paley marveled at how the length of the year was perfectly adapted to the growing cycle of plants, and Newton's laws were considered to be laws instituted by the Creator to ensure that the planets stayed in their orbits (now that it had been shown that they would). "If in looking at the universe, we follow the widest analogies of which we obtain a view," stated the influential Whewell (1857, 634–635), "we see, however dimly, reason to believe that all of its laws are adapted to each other, and intended to work together for the benefit of its organic population, and for the general welfare of its rational tenants." Because at the base of this view there always lay the unitary intention of the Creator, this was a universe that was truly unified: it was just as logical to argue that the length of the year is adapted to plants' cycles as to argue that plants' cycles are adapted to the length of the year. This is why the design argument could be worked both ways, both *to* and *from* design. In the first case, adaptedness was considered evidence for design; in the second, design was considered an explanation for adaptedness.

THE CONDITIONS FOR EXISTENCE
MEET NATURAL THEOLOGY

As we have seen, Cuvier was usually quite careful to avoid any overtly religious references, at least in his earlier writings (not so his lectures; see Corsi 1988; Rudwick 1997), and he was certainly never one to explain the structure of organisms by the intentions of God. Nevertheless, on their passage across the Channel, his views were assimilated into traditional British natural theology (Brooke 1989). As many authors have noted (e.g., Eiseley 1958; Rudwick 1976, 1997; Appel 1987), this assimilation was greatly facilitated by the injection of religious references into English translations of Cuvier's works. The struggle of British scientists to incorporate Cuvier's principle of the conditions for existence into the natural theological worldview is illuminating, because it shows how little prepared they were to appreciate what Cuvier had really meant.

As I have discussed in chapter 5, the principle of the conditions for existence, as defined by Cuvier, involves only a conditional teleology. The natural theological use of the concept of adaptedness, in contrast, involves a rather cross deterministic teleology, since the adaptedness of a structure for its role in the life of an organism is considered to be evidence that it was designed for just this role. In the natural theological worldview, the adaptedness of the parts to their functions and of the organism to its way of life does not just allow organisms to live; it was *designed* to allow organisms to live. Coming from an environment in which this worldview predominated, several reactions to Cuvier's principle were possible.

Those British scientists who supported the use of final causes in science objected to the principle on two grounds. First, not understanding that this was the necessary concomitant of a conditional teleological principle, they objected to it because of its self-evidence. Second, they objected that it did not sufficiently incorporate the notion of God's purpose. Thus, William Swainson (1834) reduced the principle to the statement "that every animal is constructed according to the functions it is destined by nature to perform," thereby injecting the notion of purpose into the principle, and then argued that it "is a truth apparent to the most superficial observer" (85). Moreover, he lamented that in perusing the works of Cuvier and his French colleagues, "we look in vain for that pure spirit of religious belief which breathes in the writings of the gentle Ray, or those bursts of lofty praise and enthusiastic admiration of Nature's God which break forth from the great Linnaeus" (88).

A similar view was expressed by Whewell in his *History of the Inductive Sciences*, which Darwin read soon after its appearance in 1837. Whewell, a historian and philosopher as well as a scientist, supported the use of final causes in physiology, although he rejected their use in other fields of science. Like Swainson, he also rephrased the principle of the conditions for existence as being the principle of "*a purpose in organization*; the structure being considered as having the function for its end" (1859, 483). Moreover, he followed traditional natural theology in connecting these functions directly to the purposes of God, since to "the real philosopher . . . it will appear natural and reasonable . . . that after venturing into the region of life and feeling and will, we are led to believe the Fountain of life and will not to be itself unintelligent and dead, but to be a living Mind, a Power which aims as well as acts" (495).

In contrast, a scientist like W. B. Carpenter (1854), who was opposed to the use of final causes in physiology (though he did not fully reject the natural theological worldview), was not prone to mistake the meaning of the principle. He recognized that "it is nothing more than the expression, in an altered form, of the fact, that as the life of an organised being consists in the performance of a series of actions, which are dependent upon one another, and are all directed to the same end, whatever seriously interferes with any of these actions must be incompatible with the maintenance of its existence" (132). For him, however, this fact was of subsidiary interest—what was really interesting were the fundamental laws of organization, which were manifested in the homological relations between different species, especially as they related to von Baer's principle of divergence from a common archetype.

Thus, both those British scientists who supported the use of final causes and those who rejected them felt Cuvier's principle to be lacking something. For the former, it was the element of purpose, which they felt was necessary to account for adaptedness; for the latter, it was those relations that were not explicable in terms of adaptedness for a particular purpose—namely, those relations that Darwin was to explain by common descent.

GEOLOGY AND THE EXPLANATION
OF LIFE'S HISTORY

By focusing his paleontological studies on large mammals, which it was unlikely still existed undiscovered in some remote region of the globe, Cuvier had proved conclusively that animal species that had formerly existed no longer did—that extinction had occurred. As we have seen, Cuvier explained extinction in terms of a series of "revolutions" of the Earth's surface. By attributing the extinctions to intense revolutions, Cuvier avoided having to pay too much attention to their particular causes. For example, in discussing the changes in marine faunas, he simply stated that "during such changes in the general liquid, it was very difficult for the same animals to continue to live in it. And they did not do so" (Rudwick 1997, 189). Cuvier had also managed to avoid the troublesome issue of new creations by suggesting that the faunal change seen in the fossil record was simply due to the immigration of organisms from elsewhere following the revolution (which for Cuvier was always local, not universal). At the time he first wrote, this idea was a tenable view, as many regions of the world had still not been explored for fossils.

It is beyond the scope of this book to discuss the ways in which Cuvier's model was further developed in the next twenty years (see Rudwick 1976 and 2008 for an excellent overview). Suffice it to say that during this period, a picture of the general history of life emerged that involved a succession of faunas and floras, each more similar to living species. While up into the 1820s there were still respected British geologists who interpreted some geological formations as pertaining to the biblical flood (most notably Buckland 1823), by the 1830s, no respectable "diluvialists" were left in Britain, especially none who would argue for divine intervention to produce such an event. Even Buckland had stated that natural causes for the biblical flood could be found. Thus, when in 1830 the first volume of Charles Lyell's *Principles of Geology* appeared (followed by the second in 1832 and the third in 1833), his contrast of the unscientific catastrophists with his own extremely actualistic uniformitarianism (the hypothesis that the uniform action of presently observable causes at presently observable intensities are sufficient to account for all geological phenomena) was partly a rhetorical device to cover up the legitimate doubts that could be entertained as to the intensity and constancy of operation of past causes. A more significant distinction between Lyell and his geological contemporaries was his insistence that the fossil record showed no pattern of progression; there were changes in the organisms present at any time, to be sure, but no overall directionality to the pattern. To support this point, Lyell had to fly in the face of much evidence that the history of life had indeed been "progressive" in some sense, arguing that this appearance was due to the incompleteness of the fossil record.

In spite of their differences, however, both Lyell and his progressionist contemporaries such as Buckland (1841) and Sedgwick (1850) had similar views on the relationship between the history of the Earth and of life (Gillispie 1951; Rudwick 1976, 2008; Ospovat 1981). Geology showed that the Earth had undergone a series of profound changes, involving the raising of mountain chains, their denudation, the incursion of seas over dry land, and their retreat back to the ocean basins. Due to these changes, the environmental conditions clearly had changed as well. In the basic model, those species that were no longer adapted to the changed environmental conditions had gone extinct, to be replaced by newly created species, adapted to the new environmental conditions.

While Lyell attributed climate changes to a shifting distribution of land and sea that affected weather patterns, the progressionists usually linked this change to the progressive cooling of the Earth from an initially molten state (an idea tracing back to Buffon). Either way, there was a one-to-one relationship between the physical state of the Earth and the species that inhabited it. This emphasis on adaptation to the environment had not been so marked in traditional natural theology, but it became necessary for these geologists who were trying to account for biological by geological changes. Thus, Lyell (1830–1833) suggested that if the climate were to warm up again, "then might those genera of animals return, of which the memorials are preserved in the ancient rocks of our continents. The huge iguanodon might reappear in the woods, and the ichthyosaur in the sea, while the pterodactyle might flit again through umbrageous groves of tree ferns" (1: 123). Bakewell (1833) expressed a similar idea more clearly, if less poetically:

> The different classes and orders of molluscous animals that have left their remains in the lower and the upper strata doubtless possessed each, the peculiar organization that best enabled them to exist and multiply under the peculiar condition of our planet that was cotemporaneous with the epoch of their creation. When this condition was changed, their numbers were diminished, or they disappeared entirely, and were succeeded by different races, with an organization adapted to other modes of existence, and to the new circumstances in which they were placed. (23)

De la Beche (1833), in a discussion of the earliest forms of life, shows how explicitly this scenario was still tied in with the traditional design argument: "Whatever the kind of animal life may have been which first appeared on our planet, we may be certain that it was consistent with the wisdom and design which has always prevailed throughout nature, and that each creature was peculiarly adapted to that situation destined to be occupied by it" (429).

But although this conception of Earth history is in one sense indeed only a simple extension of the traditional view of adaptedness, with a temporal element added, it contains one extremely important novelty. Nonadaptedness, which had formerly been only a *hypothetical* state with which the actual adaptedness of organisms could be compared, became an *actual* state that had befallen numerous species in the past. The phenomenon of extinction thereby solidified a relation that had existed *in potentia* in natural theology all along. This was a *causal* relation between existence and adaptedness, in which extinction (the termination of existence) is attributed to lack of adaptedness, and existence is dependent on adaptedness.

As already discussed, in the traditional natural theological view, the existence of a species is explained by its intended ecological role, while its adaptedness for this role is explained by the intention that so adapted it. Because in this view existence and adaptedness always occur together, it is not too surprising that when faced with the problem of extinction, British geologists would have attributed extinction to lack of adaptedness to changed environmental conditions. After all, it would be quite demeaning to the Creator to suggest that he would let nonadapted organisms continue living, since this would disrupt the harmony that he had so obviously intended for the universe. Yet these geologists did not consider it necessary for there to be divine intervention to extinguish nonadapted species; nonadaptedness could lead to extinction through purely natural relations.

By giving adaptedness a causal role in existence, however, these geologists were in fact relying heavily on the conceptual distinction between adaptedness and existence that the natural theologians had already made. Without this distinction, by attributing the extinction of a species to its failure to remain adapted, they would merely have been stating that its failure to continue existing was the cause of its extinction—a statement that could hardly have been controverted. It is clear that Lyell (1830–1833), in particular, felt that there was more to it than this, and, as we shall see, this very issue was a crucial one for the young Darwin. Yet while these geologists were relying on the conceptual distinction between existence and adaptedness, at the same time they were denying the fundamental premise of this distinction—that a species could exist without being adapted (this being prevented only by God's grace). In the geologists' view, as in that of the earlier natural theologians, all organisms that continue to exist are adapted to their way of life. For them, however, this connection between adaptedness and existence was no longer merely an expression of God's grace; it was also a necessary, causal connection. This geological background forms the crucial context for understanding the development of Charles Darwin's (1809–1882) evolutionary perspective.

DARWIN, EXTINCTION, AND EVOLUTION

Darwin's conversion to evolutionism was probably based primarily on biogeographic considerations (Ospovat 1981; Hodge 1982). However, the geological evidence for extensive past changes in the physical conditions of the Earth's surface, as seen especially in the masterful work of Lyell (1830–1833), led Darwin to consider the maintenance of adaptedness in the face of these changes as the major *reason* for evolution to occur. Dov Ospovat (1980, 1981) has shown how this perspective governed the structure of each of Darwin's successive evolutionary theories, up to and including the 1844 version of the theory of "natural selection." In Darwin's evolutionary perspective, when environmental conditions change, species that are no longer perfectly adapted *can* continue to exist, at least long enough for them to become adapted to the new environmental conditions. In other words, Darwin broke (or at least loosened) the rigid connection between existence and adaptedness that the natural theologians had used as evidence of God's benevolence and that the geologists had relied on to explain extinction (Lamarck had taken a similar step). It is this break, of course, that allowed Darwin to develop his concept of *relative* adaptedness (see Ospovat 1981), which is fundamental to the theory of natural selection. What I will argue here, however, is that by breaking the formerly rigid connection between existence and adaptedness, Darwin in fact eliminated the only empirical basis for the concept of overall adaptedness itself. That he was able to do so without noticing it was certainly due to the conceptual distinction that already existed in the work of Lyell and the natural theologians.

In the natural theological view, the *adaptedness* of both the organism to its way of life and of the parts to their functions were explained by the design, or purpose, that so adapted them. The *existence* of the organism, while in a sense taken for granted, in another was explained by its intended ecological role, or its purpose in nature. When the geologists made the connection between existence and adaptedness causal, thus explaining existence by adaptedness, they did not alter the explanation for the adaptedness of the parts and of the organism as a whole. Moreover, since this adaptedness was a result of God's action, it was necessarily perfect. Buckland (1841), for instance, was quick to deny any absolute organic imperfection in his progressionist discussion of the fossil record: "a Polype, or an Oyster, are as perfectly adapted to their functions at the bottom of the sea, as the wings of the Eagle are perfect, as organs of rapid passage through the air, and the feet of the stag perfect, in regard to their functions of affecting swift locomotion upon the land" (1: 90). At the same time, however, other scientists were beginning to realize that all structures are not so easily interpreted as perfectly adapted to their functions. The existence of rudimentary structures with no apparent function, such as the tooth germs that never pierce the gums in embryonic cetaceans, lent support to the notion, promoted by Geoffroy Saint-Hilaire and the *Naturphilosophen*, that one must also take account of an ideal plan or archetype on which organisms were modeled, which limited the perfection of adaptation (e.g., Roget 1840). This view is what Ospovat (1978, 1981) called "limited perfect adaptation," and it was the predominant one in the years that Darwin was formulating his theory.

Within this general framework, it is only a short step from considering the adaptedness of the parts to be limited by the archetypal form on which they're modeled, to considering the adaptedness of the organism as a whole to its way of life to also be more or less perfect, with the existence of the organism being taken for granted. As I will show, this is in fact the path that led Darwin to make the distinction between adaptedness and existence a real distinction.

But at this point in our history, we must remember what the meaning of adaptedness was in the first place: adaptedness was a characterization of the relation between organic means and ends, whose similarity to the relation between human means and ends was considered evidence for intentional design by a deity. As noted earlier, in their original context, the "adaptedness" or "adaptations" of an organism do not explain anything, they are merely ways of expressing our judgment of the degree to which we can understand the structure of an organism as appropriate for its way of life, and can thus represent this structure as deriving from an intention similar in kind to our own. The geologists, by considering that the existence of an organism or species could be explained by its adaptedness, reversed the true logic of the situation.

In reality, it is the *existence* of an organism or a species that is a fact. The *adaptedness* of the parts or of the organism as a whole constitutes at most only a conditional teleological explanation of this fact, since there is no way for us to *determine* the existence of an organism on the basis of its adaptedness, which is only relative to particular qualities or features of the organism. In other words, it is evident that for an organism to be adapted, it must exist, but it does not follow that for an organism to exist, it must be adapted, *unless* one considers all organisms that exist to be adapted. By separating the adaptedness of an organism or species from its existence, Darwin thus broke the only empirical connection that the idea of adaptedness had ever possessed.

To understand the context in which Darwin made this separation, we must delve a little deeper into Lyell's explanation of the extinction and creation of organisms in his *Principles of Geology* (1830–1833), since it was this explanation, in particular, that Darwin was attempting to replace with a theory of evolution. As just discussed, both progressionist geologists such as Buckland and Sedgwick and the nonprogressionist Lyell explained extinction in a similar way: extinction was due to nonadaptedness to changing environmental conditions. The progressionists were content to leave this conception quite vague and could still incorporate occasional intensification of geological processes (revolutions in Cuvier's sense) into their explanations. For the uniformitarian Lyell, however, the processes of extinction and creation were necessarily gradual (one species at a time), and he attempted to explain extinction, like other geological processes, on the basis of "causes now in operation." As many have noted (e.g., Gillespie 1951; Eiseley 1958; Rudwick 1976; Hodge 1982), the structure of Lyell's system almost cries out for a similarly naturalistic explanation of the "creation" of species. It was Lyell's work that brought extinction, and the nonadaptedness that gives rise to it, within the scope of scientific study.

At the base of Lyell's explanation of extinction was the concept of a *station* (see Rehbock 1983), which in his usage is more or less equivalent to the niche of a species at a particular location: "Stations comprehend all the circumstances, whether relating to the animate or inanimate world, which determine whether a given plant or animal can exist in a given locality, so that if it be shown that stations can be essentially modified by the influence of known causes, it will follow that species, as well as individuals, are mortal" (1830–1833, 2: 130). Of course, Lyell proceeded to show that geological changes must necessarily alter the condition of stations, either directly or indirectly, so that "amidst the vicissitudes of the earth's surface, species cannot be immortal," and "there is no possibility of escaping from this conclusion, without resorting to some hypothesis as violent as that of Lamarck" (169). For Lyell, such a hypothesis was untenable, and the corollary of his theory of extinction was that as new stations were created by geological changes, either they would be filled by immigration of organisms from the surrounding regions, or new species would be "introduced" to fill them (although he gave no indication of what process might be responsible for this introduction).

Darwin, however, soon after his return from the *Beagle* voyage, came to the conclusion that the "introduction" of species indeed ought to be explained by "some hypothesis as violent as that of Lamarck." From this it followed that species must somehow be able to avoid extinction. The terms of the problem setup by Lyell were quite clear: the condition of the station is changed by geological processes, and the species must be able to adapt to this change or go extinct. Thanks to the notebooks left by Darwin, the path of his early speculations can be followed quite closely. The first evidence we have of his views is found in the Red Notebook (RN) of spring 1837 (Herbert 1980; Barrett et al. 1987). It is surprising to find that in these early notes, Darwin was a saltationist (Kohn 1980). A suggestive passage hints that this saltationism was at least partly justified in terms of the problem posed by Lyell: the evolution of species is "not *gradual* change or degeneration. from [change of] circumstances: if one species does change into another it must be per saltum—or species may perish" (RN 130, Herbert 1980, 66).

At around the same time that he was considering a saltationist mode of evolution, however, Darwin came to reject a necessary connection between nonadaptedness and extinction. The alternative he supported at the time (Kohn 1980; Hodge 1982) was apparently Giovanni Brocchi's idea of lineage senescence (which Lyell had argued against at length in volume 2 of the *Principles*). In this scenario, a species dies out not due to some change in the physical conditions resulting in its nonadaptedness, but rather due to an internal limitation of its growth potential, similar to that found in fruit trees propagated by vegetative means only. Darwin's conversion to this view was occasioned by the lack of evidence for climatic changes associated with the simultaneous extinction of the *Macrauchenia* and other large mammals of South America (see Kohn 1980), an extinction for which the causes are still debated today, though direct or indirect human impacts are the most likely explanation (Martin and Klein 1984; Martin 2005). Darwin's thinking at the time is clearly expressed in two passages from the Red Notebook:

> Should urge that extinct Llama [*Macrauchenia*] owed its death not to change of circumstances; reversed argument. knowing it to be a desert.—Tempted to believe animals created for a definite time:—not extinguished by change of circumstances. (RN 129, Herbert 1980, 66)

> Dogs. Cats. Horses. Cattle. Goat. Asses. have all run wild & bred. no doubt with perfect success.—showing non Creation does not bear upon solely adaptation of animals.—extinction in same manner may not depend. (RN 133, Herbert 1980, 67)

The reasons for Darwin's rejection of a necessary relation between adaptedness and existence are even more clearly spelled out in a passage from the first edition of the *Journal of Researches ("Voyage of the Beagle")*, also dating from the spring of 1837:

> We see that whole series of animals, which have been created with peculiar kinds of organization, are confined to certain areas; and we can hardly suppose these structures are only adaptations to peculiarities of climate or country; for otherwise, animals belonging to a distinct type, and introduced by man, would not succeed so admirably, even to the extermination of the aborigines. On such grounds it does not seem a necessary conclusion, that the extinction of species, more than their creation, should exclusively depend on the nature (altered by physical changes) of their country. (Darwin 1839, 212)

There is thus good textual evidence to show that it was the failure of the concept of adaptation that Darwin had inherited from Paley, Lyell, and others to explain all attributes of a species as adaptations to its station that led him to conclude that adaptation is not that "close." From this he concluded, logically enough, that extinction is not necessarily dependent on lack of adaptedness (contra Lyell and the other geologists). With the benefit of 170 years' hindsight, however, it is clear that these conclusions should have forced Darwin to take the further step of rejecting the explanatory value of the concept of adaptedness itself.

The geologists had based their explanation of extinction on what was, for all intents and purposes, a hypothesis (though they didn't regard it as such): the existence of species is dependent on their adaptedness. By rejecting this hypothesis, Darwin broke the connection between existence and adaptedness. Since one cannot reject existence, which is a fact, the only logical conclusion is that the adaptedness of a species, as we perceive it, does not explain its existence. But if adaptedness does not explain existence, it has no explanatory value at all, for there is certainly nothing else that it explains (in fact, as I have noted, adaptedness can never be more than a conditional teleological explanation of existence). Darwin did not come to this conclusion, however, which would certainly have been an extremely difficult one for someone raised on Paley to reach. It is likely that at the time his thinking was still too vague for him to be forced into it. He soon gave up his ideas on lineage senescence as the cause of extinction (Kohn 1980), coming back to the view that extinction is due to nonadaptedness (compare the discussion just quoted with that in the second edition of the *Journal of Researches*, published in 1842).

Nevertheless, Darwin's separation of adaptedness and existence, once made, was never healed. In fact, in his subsequent theorizing, up through the 1844 essay on natural selection, Darwin came to see the variation that adapts organisms to environmental change as *caused* by the imperfect adaptedness consequent on this change, which he supposed to have some disturbing effect on the reproductive system (see Kohn 1980; Ospovat 1981). A weaker version of this view is still found in the *Origin* (e.g., Darwin 1859, 8, 82, 131). The empirical support for this premise derived from the great variability exhibited by domesticated compared with natural productions, variability that we would ascribe today to inbreeding and growth under a less stringent environment (less stringent "selection"). Yet this premise in fact depended critically on Darwin's previous conceptual separation of adaptedness from existence, since without this separation, the species would have gone extinct before it could adapt to the new environmental conditions.

Darwin's evolutionary theory at this point was a straightforward modification of Lyell's scenario for extinction: the environment changes, rendering a species imperfectly adapted, but instead of going extinct, it can evolve (adapt) to fit the new environment (station). In Darwin's early theories, the adaptation to new environmental conditions was assumed to be an automatic response of the generative system; later, natural selection became the adaptive force. In fact, to extend this physical analogy, the *force* of natural selection works to improve a species due to the *potential* difference between a current state of imperfect adaptedness and a future state of perfect (or at least more perfect) adaptedness. It was the separation of adaptedness from existence, with the implicit assumption that a future state of more perfect adaptedness is in some sense the goal toward which the evolutionary process is directed, that introduced an element of teleological determinism into Darwin's evolutionary theory. When one considers that Darwin first conceived of natural selection as a process that had a "final cause"—namely, "to sort out proper structure and adapt it to change" (Notebook D, 135, Barrett et al. 1987, 375–376)— this is perhaps not so surprising.

In short, Darwin began by accepting the basic premise of the argument to design—that adaptedness is a problem that demands explanation—but put the designed mechanism of natural selection in place of direct design. He thus considered selection to have solved the design problem Paley and Lyell had posed for him. Or, as he himself put it many year later in his autobiography: "The old argument of design in nature, as given by Paley, which formerly seemed to me so conclusive, fails, now that the law of natural selection has been discovered" (Barlow 1958, 87).

Darwin's acceptance of the underlying premise of the design argument is spelled out even more clearly in a famous passage from the introduction to the *Origin*, in which he notes that

> in considering the Origin of Species, it is quite conceivable that a naturalist, reflecting on the mutual affinities of organic beings, on their embryological relations, their geographical distribution, geological succession, and other such facts, might come to the conclusion that each species had not been independently created, but had descended, like varieties, from other species. Nevertheless, such a conclusion, even if well founded, would be unsatisfactory, until it could be shown how the innumerable species inhabiting this world have been modified, so as to acquire that perfection of structure and coadaptation which most justly excites our admiration. (1859, 3)

By posing the acquisition of "perfection of structure and coadaptation" as the key problem for evolutionary theory, Darwin implicitly claimed that adaptedness is indeed something more than existence, something that requires a special historical process of creation to account for it.

Thus, while I accept Dov Ospovat's (1981) argument that the view that adaptedness is *always* relative became possible for Darwin only when he overcame his previous assumption that variation is infrequent and consequent on environmental change, it is apparent that the seeds of this conception were already sown in Darwin's earliest theories: his conception of relative adaptedness depended on his prior separation of adaptedness and existence. This connection is especially clear in the following passage from the *Origin*, where he presents his basic paradigm of the "probable course of natural selection":

> Taking the case of a country undergoing some physical change . . . into which new and better adapted forms could not freely enter, we should then have places in the economy of nature which would assuredly be better filled up, if some of the original inhabitants were in some manner modified; for, had the area been open to immigration, these same places would have been seized upon by intruders. In such case, every slight modification, which in the course of ages chanced to arise, and which in any way favoured the individuals of any of the species, by better adapting them to their altered conditions, would tend to be preserved; and natural selection would thus have free scope for the work of improvement. (1859, 81–82)

In the *Origin*, as in his earlier theories, Darwin's conception of evolution fundamentally depended on the concept of open "places in the economy of nature," to which organisms that now exist, in spite of being imperfectly adapted, can eventually become better adapted (see Worster 1994). These "places" clearly correspond to Lyell's "stations." For Lyell, however, the new species was created already adapted to the new station, so that there is no period of nonadaptedness. For Darwin, the species does undergo a period of at least relative nonadaptedness, as it gradually evolves into the new station.

The empirical problems inherent in such an approach have been noted by Richard Lewontin (1978, 1984), although he regarded them as a necessary evil: such a conception presupposes that the "places in the economy of nature" (niches) can be defined independently of the organisms that occupy them. It was by thus considering the adaptedness of organisms to be relative to *hypothetical* changes that might better suit them for their current environment that Darwin introduced teleological determinism into his evolutionary theory, since in practice the "place in the economy of nature" to which natural selection adapts the species can be defined only by the achieved *result* of the evolutionary process. This result must then be seen as the cause of its own realization.

The teleology implicit in this approach was only enhanced by the analogy between natural and artificial selection, because artificial selection does involve a teleological agent, the breeder, who can aim to produce a certain result. This connection is clearly expressed in Darwin's discussion of short-beaked tumbler-pigeons, which often prove incapable of getting out of their egg unless assisted by the pigeon fancier: "Now, if nature had to make the beak of a full grown pigeon very short for the bird's own advantage, the process of modification would be very slow, and there would be simultaneously the most rigorous selection of the young birds within the egg, which had the most powerful and hardest beaks, for all with weak beaks would inevitably perish" (1859, 87). It is difficult to see how nature could *have* to make the beak of a bird short and yet be able to wait around for the slow process of selection. Although in nature, as Darwin notes, the evolutionary path toward short-beakedness can only occur if at all stages at least some of the chicks are able to hatch, the occurrence of such an evolutionary path does not justify one in saying that it *had* to occur. Invoking absolute necessity for a process that we are only justified in claiming as possible is exactly what is characteristic of teleological determinism. The fundamentally teleological character of Darwin's views has been recognized by a number of previous authors; see, for example, Croizat (1962), Lennox (1992, 1993, 1994), and Grene and Depew (2004).

Of course, Darwin's discussion of natural selection does not always depend on this teleology; there are many beautiful passages in the *Origin*, particularly in his chapter on the "Struggle for Existence," which are entirely free of such difficulties. For example, he notes, "To keep up a mixed stock of even such extremely close varieties as the variously coloured sweet-peas, they must be each year harvested separately, and the seed then mixed in due proportion, otherwise the weaker kinds will steadily decrease in numbers and disappear" (1859, 75). But while what I am criticizing is merely an element of Darwin's thought, it is a strong and seductive element.

DARWIN AND THE CONDITIONS FOR EXISTENCE

In this section, I wish to consider two separate questions: First, how did Darwin himself understand Cuvier's principle of the conditions for existence? Second, what role did the concept of the conditions for existence play in Darwin's theory? The first question is not difficult to answer, since Darwin explicitly referred to the principle in a well-known passage concluding chapter 6 of the *Origin**:

> It is generally acknowledged that all organic beings have been formed on two great laws—Unity of Type, and the Conditions of Existence. By unity of type is meant that fundamental agreement in structure, which we see in organic beings of the same class, and which is quite independent of their habits of life. On my theory, unity of type is explained by unity of descent. The expression of conditions of existence, so often insisted on by the illustrious Cuvier, is fully embraced by the principle of natural selection. For natural selection acts by either now adapting the varying parts of each being to its organic and inorganic conditions of life; or by having adapted them during long-past periods of time: the adaptations being aided in some cases by use and disuse, being slightly affected by the direct action of the external conditions of life, and being in all cases subjected to the several laws of growth. Hence, in fact, the law of the Conditions of Existence is the higher law; as it includes, through the inheritance of former adaptations, that of Unity of Type. (1859, 206)

It is apparent from this passage that Darwin, like many other British scientists, misunderstood Cuvier's principle. This was pointed out many years ago by E. S. Russell (1916), an ardent admirer of Cuvier and thus particularly sensitive to this distinction: "It is clear that Darwin took the phrase 'Conditions of Existence' to mean the environmental conditions, and the law of the Conditions of Existence to mean the law of adaptation to environment. But that is not what Cuvier meant by the phrase: he understood by it the principle of the co-ordination of the parts to form the whole, the essential condition for the existence of any organism whatsoever" (239).

While I find Russell's characterization of Cuvier's principle too narrow, because the relation of the organism with the environment is certainly also part of the conditions for its existence, I believe that his interpretation of Darwin's meaning is correct. This is quite easily seen if one tries to replace Darwin's *of* with *for*, since the crucial sentence then has natural selection act by "adapting the varying parts of each being to its organic and inorganic conditions for life." Read this way, the passage is

---

* Darwin's initial reaction to Cuvier's principle is recorded in his Notebook B of 1837 (112, Barrett et al. 1987, 197), in the context of reading Geoffroy Saint-Hilaire's 1830 account of his debate with Cuvier in his *Principes de philosophie zoologique*. A passage similar to that in the *Origin* (and citing Geoffroy Saint-Hilaire as his source) also occurs in Darwin's manuscript for his big book on *Natural Selection*, although here he states only that his principle seems to "accord sufficiently well" with Cuvier's (Stauffer 1975, 383–384).

nonsense, because the parts of each being either allow that being to satisfy the conditions for its existence, or they don't—and if the being changes (through evolution), then so do its conditions for existence.

I will reiterate this point because it is the crux of my argument so far: it makes no sense to speak of the adaptedness of an organism to the conditions for its existence, because an organism either fulfills the conditions for its existence or does not. To speak of "adaptedness" in this context is equivalent to saying, with respect to the equation $x + 2 = 4$, that an x equal to 2 is "adapted" to the equation; an x with a value of 2 simply satisfies the conditions of the equation. While we could marvel at the "beautiful adaptedness" displayed by the number 2, such an observation is outside the realm of science. Moreover, it is not just this passage in which Darwin uses "conditions of life" or "conditions of existence" in this sense; a glance at Barrett et al.'s (1981) *Concordance* shows that *every* use of the terms in the *Origin* has the meaning of "environmental conditions." This is not necessarily problematic, however, for it might be that he just uses some other term for the same concept.

This brings us to the second question I asked above: what role does the *concept* of the conditions for existence play in Darwin's theory? As it turns out,, Darwin had little use for Cuvier's concept. For someone who needed to prove that evolution had occurred, and that he had a plausible mechanism for how it had occurred, the conditions that must be fulfilled to allow organisms to exist at any given time were simply not very interesting. The crucial issue for Darwin was instead the changes that might improve the current adaptedness of a species, or the past changes that might be invoked to explain the current adaptedness of a species.

Nevertheless, the conditions for existence do enter into Darwin's theory in several different ways. First, they may be invoked as a limiting factor in evolution (the "checking" action of natural selection). Thus, in his discussion of the evolution of the giraffe's tail and other "organs of little apparent importance" (1859, 195–196), he notes that "any actually injurious deviations in their structure will always have been checked by natural selection." This is only a casual remark, however, and obviously not the main role of natural selection, which is instead conceived of as a deterministic force or agent that, although "it seems at first incredible," might have "adapted [the tail] for its present purpose by successive slight modifications, each better and better, for so trifling an object as driving away flies." Darwin notes further that if this account is not believable, such organs might instead be explained by historical retention (the absence of a force) or by other possible forces ("secondary causes") that could have produced the tail acting "independently of natural selection." The list of these causes includes the direct effect of climate or food, the law of reversion, the correlations of growth, and sexual selection. That Darwin could conceive of causes that act independently of natural selection shows that the checking action of natural selection is not central to Darwin's concept, since it can't be true that these secondary causes act independently of *this* mode of natural selection.

Another interesting way in which the conditions for existence appear in the *Origin* are as conditions whose satisfaction is an obvious, but unimportant, feature of existing organisms. In this context it becomes especially apparent how Darwin's focus on relative adaptedness led him to discount the significance of the adaptations that currently allow an organism to fulfill its conditions for existence, except as evidence of a past selective process. For example, in discussing the varied development of the flying membrane of squirrels, he remarks, "We cannot doubt that each structure is of use to each kind of squirrel in its own country. . . . But it does not follow from this fact that the structure of the squirrel is the best that it is possible to conceive under all natural conditions" (1859, 180). And in discussing variations in the structure of the wing among birds, he notes that "the structure of each of these birds is good for it, under the conditions of life to which it is exposed, for each has to live by a struggle; but it is not necessarily the best possible under all possible conditions" (1859, 182). To paraphrase these passages, the structure of these animals allows them to satisfy the conditions for their existence, but it might not if some change in the environmental conditions occurred.

While this statement is certainly true, what can one legitimately conclude from it? Only that if the environmental conditions change, either the animals will go extinct, or they won't. If they do manage to survive the change, it is possible that members of the species with characters (and conditions for existence) that had been rare in the population will preferentially survive, so that the overall composition of the population will change.

In the case of the squirrels, for example, it might be that the immigration of a new predator makes it more difficult for those squirrels with a smaller flank membrane to fulfill the conditions for their existence and easier for those with a larger flank membrane to fulfill theirs, so that there is a preferential survival of the latter. If the squirrel population manages to avoid going extinct, there will thus be a change in the composition of the population, so that there are more squirrels with large flank membranes (assuming some heritability of membrane size). Yet these hypothetical possibilities have nothing to do with the adaptedness of the original population of squirrels, which we know for a fact was fulfilling the conditions for its existence (or we would not have seen it).

The place where one would particularly expect to find a reference to the conditions for existence in the *Origin* is in Darwin's discussion of the "struggle for existence," since there is a close and obvious tie between the two: those organisms that survive the struggle for existence are those that are able to fulfill the conditions for their existence. Yet in chapters 3 and 4 of the *Origin*, we find that those organisms that survive are always considered to be those with "advantageous," "profitable," or "favourable" variations. These variations give the organisms the "best chance of surviving and procreating their kind" (81). But here Darwin is reversing the logic of the situation—in most cases we have no way of knowing that the variation gave the organism a better chance of surviving, but we know only that those organisms that did survive (fulfill the conditions for their existence) were disproportionately made up of those that had the variation. As noted earlier, one cannot determine the existence of an organism on the basis of its specializations (variations). Moreover, as a by-product of Darwin's focus on single characters, this scheme almost necessitates the consideration of the organisms as passive entities, acted on by the environment, because a profitable variation must be defined with respect to a given environment (the implicit assumption is that "the environment" is the same for all members of the species).

Thus, in spite of the metaphor of the "struggle for existence," which suggests an image of organisms struggling to fulfill the conditions for their existence, in which those that are able to do so by any means survive, we are left with an image in which those surviving the struggle are those that, under particular environmental conditions, show some particular character that is advantageous in the struggle. The property of the organism carrying a profitable variation that allows it to profit by it—the property I am calling the conditions for its existence—receives no name at all.

To see this more clearly, let us reexamine Darwin's paradigm of the action of natural selection, in which changing "conditions of life" create new "places in the economy of nature" to which natural selection adapts the species. It is useful to try to translate this scenario into the language of the conditions for existence. To do so, one would have to say instead that under changing *environmental conditions*, the members of the species that are able to fulfill the *conditions for their existence* are not the same ones that would have be able to do so under constant environmental conditions. Over time, this results in evolutionary change—that is, genetically based change in phenotype and thus in the conditions for the existence of organisms of that species. Because both the conditions for the existence of the species and the environmental conditions are changing, the species will come to occupy a new place in the economy of nature, defined by the interaction between the altered conditions for existence and the altered environmental conditions. In this rephrasing, it becomes clear what is missing in Darwin's account: the organismal side of the equation. Darwin is thereby left with a curiously one-sided theory of evolution, in which the "adaptedness" of the organism is viewed only as it relates to the environment.*

Finally, it is worth examining more closely Darwin's use of the concept of relative adaptedness as it relates to the success of introduced species, since this is apparently the original context in which he came up with the concept (Ospovat 1981). The relevant passage is the following: "Natural selection tends only to make

---

* For an alternative interpretation of Darwin's relation to Cuvier's conditions for existence, see Grene and Depew (2004, 208–212).

each organic being as perfect as, or slightly more perfect than, the other inhabitants of the same country with which it has to struggle for existence. And we see that this is the degree of perfection attained under nature. The endemic productions of New Zealand, for instance, are perfect one compared with another; but they are now rapidly yielding before the advancing legions of plants and animals introduced from Europe" (1859, 201–202). From one perspective, this passage is not problematic: it is certainly a true description of the situation, and it does seem that there is something about the European plants and animals that has tended to make them successful invaders of New Zealand, while the reverse situation, as Darwin notes (337), has occurred less frequently (although the recent successful invasion of Europe by New Zealand mud snails, flatworms, and pigmy weed provides some good examples).

Disregarding the possibility that the success of the invaders was associated solely with new habitats created by European settlers (e.g., by the effect of grazing sheep), this does seem to be evidence for the European plants and animals being "better adapted" (better able to fulfill the conditions for their existence) than the natives. Nevertheless—and this is the crucial point—this is no reason to think that the natives were in any sense "poorly adapted" (unable to fulfill the conditions for their existence) *before* the arrival of the Europeans; relative adaptedness is a concept that only has meaning within the context where it is observed that one species or organism does well to the direct detriment of another.

In other words, relative adaptedness makes sense if one maintains a direct connection between existence and adaptedness, so that the species or organisms doing well (fulfilling the conditions for their existence) are said to be better adapted than the species or organisms not doing well (not fulfilling the conditions for their existence). But Darwin goes beyond this to conclude that "no country can be named in which all the native inhabitants are now so perfectly adapted to each other and to the physical conditions under which they live, that none of them could anyhow be improved," since "as foreigners have . . . everywhere beaten some of the natives, we may safely conclude that the natives might have been modified with advantage, so as to have better resisted such intruders" (82–83). Darwin's separation of adaptedness from existence leads him to conclude that a species that exists (i.e., is fulfilling the conditions for its existence) is poorly adapted, because it might be "modified with advantage" to meet some *future* contingency. Relative adaptedness thereby becomes a fundamentally teleological concept, because it is relative to hypothetical future states.

To summarize my argument to this point: By separating the concept of adaptedness from its empirical base in existence, and then basing his evolutionary theory on this disembodied adaptedness, Darwin introduced a teleological determinism into the heart of his theory. This teleology is expressed in two related conceptions: (1) that evolution is a process going from a less-adapted to a better-adapted state and

(2) that natural selection is a deterministic force, or agent, that directs the evolutionary process toward this better-adapted state.

WALLACE AND THE CONDITIONS FOR EXISTENCE

In the present context, one might expect that Alfred Russel Wallace's (1858; reprinted with some changes in Wallace 1891) independent proposal of an evolutionary theory based on the "struggle for existence" would be well worth examining, since in this early paper Wallace is perhaps the only "Darwinian" whose views were not molded by the structure of Darwin's theory. Wallace and Darwin came from very different backgrounds. Darwin was the gentleman, who inherited wealth from his father and married into the Wedgwood fortune (Desmond and Moore 1991; Browne 1995, 2002). Wallace was from a working-class background; he began his career as a surveyor and conducted all of his natural history collecting not merely for science but to supply the market for beautiful specimens back home. In later life, almost his sole income came from his writings (a number of recent biographies on Wallace are available—e.g., Raby 2001; see also the extensive Web site maintained by Charles H. Smith).

As Adrian Desmond has shown in his wonderful book on *The Politics of Evolution* (1989), during this period of British history the revival of religion in general and natural theology in particular was most common among the aristocracy, wealthy professionals, and clergy, a class that included not only Buckland and Sedgwick, but also Lyell and Darwin. Evolutionary ideas were common in other circles, including among reforming workingmen's champions such as Robert Chambers, the author of the infamous *Vestiges of the Natural History of Creation* (1844). Wallace was just such a workingman, and he became converted to evolutionism by reading Chambers's work soon after it appeared; this determined him to search for evidence in favor of the hypothesis (Wallace 1908a).

In keeping with his background, prior to formulating his evolutionary theory, Wallace had little patience with natural theological explanations, unlike Darwin, who later recalled his "delight" in reading Paley during his undergraduate days at Cambridge (Barlow 1958, 59). In fact, in his notebook from his time in the Malay Archipelago, we find that prior to his enunciation of his version of what Darwin called natural selection, Wallace was quite opposed to such explanations. For example, in commenting on "supposed proofs of Design" in the description of bats in Charles Knight's *Cyclopedia of Natural History*, Wallace scornfully wrote, "As if an animal could have necessities before it came into existence, or as if having come into existence it could continue to exist unless its structure enabled it to obtain food. If the bat had not wings it would of course do without them & would have no more necessity for them than any other animal" (ca. 1855, cited in Pearson 2002, 6). And in commenting on a passage from the same book in which the eye of the coconut was promoted as a "wise contrivance" for the embryo to escape its hard shell, he notes, "Is not this absurd? To impute to the supreme Being a degree of intelligence only equal to that of the stupidest human beings. What should we think, if as a proof of the superior wisdom of some philosopher, it was pointed out that in building a house he had made a door to it, or in contriving a box had furnished it with a lid! Yet this is the kind and degree of design imputed to the Deity as a proof of his infinite wisdom" (Pearson 2002, 8). Finally, one more passage, responding to Agassiz's invocation of a "harmony in numerical proportions among animals," seems to put Wallace quite clearly in the tradition of Lucretius and Holbach, in his recognition of the principle of the conditions for existence: "What are the normal proportions & harmony spoken of. The proportions have continually varied & are varying. Are the horses in S. America harmonious or not? In the tertiary period there were horses, then none now they are again. Whatever exists must be in harmony or it could exist no longer" (Pearson 2002, 24). Thus, Wallace came to his utilitarianism from a background of thorough skepticism toward natural theology and explanations in terms of design (see Kottler 1985).

Darwin, by contrast, became critical of such explanations only *after* he came to the theory of natural selection, and he never fully rejected them. In his reading notes on John Macculloch's *Proofs and Illustrations of the Attributes of God*, apparently written late in 1838, after his reading of Malthus (Barrett et al. 1987), we find many scathing critiques of Macculloch's design arguments, one of which I quoted as an epigraph to this book. But even here we see Darwin struggling to maintain some aspect of the design view. In his notes on Macculloch's discussion of the adaptations of *Fucus* (seaweed) eggs for adhesion as evidence for design, Darwin commented, "Perhaps they are so." But he goes on: "One thing must be admitted there would not be these plants, if there was not some provision for transportation:—But I do not want to deny laws.— The whole universe is full of adaptations" (Barrett et al. 1987, 632). And a little later we find him expressly questioning his own use of design arguments, in the context of the theory that he would later come to call natural selection: "The Final cause of innumerable eggs is explained by Malthus.—[is it anomaly in me to talk of Final causes: consider this!—" (637). Twenty years on, Darwin had not made much progress in this task. In a letter to his American confidant Asa Gray of 26 November 1860, he confessed, "I am conscious that I am in an utterly hopeless muddle. I cannot think that the world, as we see it, is the result of chance; and yet I cannot look at each separate thing as the result of Design" (Darwin 1887, 2: 146).

There are other differences between Darwin and Wallace as well. In his 1858 paper, Wallace did not propose a name for the process he described, and he explicitly denied the relevance of domestic animals, stating that "no inferences as to varieties in a state of nature can be deduced from the observation of those occurring among domestic animals" (1858, 61). Most importantly, a close examination shows that in his original theory, Wallace had not made the critical separation between existence and adaptedness that Darwin made; for Wallace (as for modern "Darwinians"), the variations that are adapted to the new conditions are already present when the environmental conditions change. In a passage parallel to the previously quoted one of Darwin's, we find Wallace hypothesizing that if one were to

> let some alteration of physical conditions occur in the district—a long period of drought, a destruction of vegetation by locusts, the irruption of some new carnivorous animal seeking "pastures new"—any change in fact tending to render existence more difficult to the species in question, and tasking its utmost powers to avoid complete extermination; it is evident that, of all the individuals composing the species, those forming the least numerous and most feebly organised variety would suffer first, and, were the pressure severe, must soon become extinct. (58)

Even more interesting from our present perspective, however, is that Wallace explicitly framed his discussion of the "struggle for existence" in terms of the conditions for existence: "The life of wild animals is a struggle for existence. The full exertion of all their faculties and all their energies is required to preserve their own existence and provide for that of their infant offspring. The possibility of procuring food during the least favourable seasons, and of escaping the attacks of their most dangerous enemies, are the primary *conditions* which determine the *existence* both of individuals and entire species" (54, my emphasis). And in the conclusion of his paper, Wallace summarized his view of the evolutionary process in a way that is remarkably close to the view I am advocating here:

> We believe we have now shown that there is a tendency in nature to the continued progression of certain classes of *varieties* further and further from the original type— a progression to which there appears no reason to assign any definite limits—and that the same principle which produces this result in a state of nature will also explain why domestic varieties have a tendency to revert to the original type. This progression, by minute steps, in various directions, but always checked and balanced by the *necessary conditions*, subject to which alone *existence* can be preserved, may, it is believed, be followed out so as to agree with all the phenomena presented by organised beings, their extinction and succession in past ages, and all the extraordinary modifications of form, instinct, and habits which they exhibit. (62, my emphasis on *necessary conditions* and *existence* only)

All this approbation is not to suggest that Wallace's theory is unproblematic, for in some passages Wallace expresses views that imply a necessary progression, and these views can be interpreted as containing a teleological element. Perhaps the clearest example of such usage is in the following classic contrast with Lamarck's theory: "Neither did the giraffe acquire its long neck by desiring to reach the foliage of the more lofty shrubs, and constantly stretching its neck for the purpose, but because any varieties which occurred among its antitypes with a longer neck than usual *at once secured a fresh range of pasture over the same ground as their shorter-necked companions, and on the first scarcity of food were thereby enabled to outlive them"* (61).

Nevertheless, it is apparent from the fact that Wallace had not made the crucial separation between adaptedness and existence that Darwin had, and that nevertheless most authors have been unable to distinguish the two theories (see Kottler 1985, for an excellent discussion of the differences that have been proposed; also Gayon 1998; Bulmer 2005), that this separation is by no means an essential part of the theory of natural selection.

The differences between Darwin's theory and that of Wallace are brought out quite nicely by considering the few changes Wallace made to the 1870 reissuing of this early essay, for these changes are clearly the result of Darwin's influence on him. The most obvious difference between their presentations of the theory was Wallace's attitude toward domestic animals, the very foundation of Darwin's metaphor of natural selection. In the 1858 essay, Wallace had stated:

> The great speed but slight endurance of the race horse, the unwieldy strength of the ploughman's team, would both be useless in a state of nature. If turned wild on the pampas, such animals would probably soon become extinct, or under favourable circumstances might each lose those extreme qualities which would never be called into action, and in a few generations would revert to a common type, which must be that in which the various powers and faculties are so proportioned to each other as to be best adapted to procure food and secure safety,—that in which by the full exercise of every part of his organization the animal can alone continue to live. Domestic varieties, when turned wild, *must* return to something near the type of the original wild stock, *or become altogether extinct*. (60)

But in 1870 he added a footnote to the end of this passage that read, "That is, they will vary, and the variations which tend to adapt them to the wild state, and therefore approximate them to wild animals, will be preserved. Those individuals which do not vary sufficiently will perish" (Wallace 1891, 31). The focus on the process of modification, rather than the conditions for existence of variant types, is characteristically Darwinian.

Aside from such footnotes, the major change introduced by Wallace into the republished version was the addition of section headings. One of these reads *"The Abundance or Rarity of a Species dependent upon its more or less perfect Adaptation to the Conditions of Existence"* (1891, 26). In the original paper, Wallace had never used the phrase "conditions of existence" in the sense of environmental conditions, instead referring to "circumstances," "physical conditions," or simply "conditions." In the original passage, moreover, the statements of "Adaptation to the Conditions of Existence" are such expressions as "adapted to obtain a regular supply of food, and to defend themselves against the attacks of their enemies and the vicissitudes of the seasons," "capacity for ensuring the means of preserving life," and "capacity of each for performing the different acts necessary to its safety and existence under all the varying circumstances by which it is surrounded" (1858, 57). All of these expressions refer to one thing: the *ability* to satisfy the conditions for existence. Wallace's use of "conditions of existence" in the revised version to mean environmental conditions appears to be an attempt to reconcile Darwin's use of the term with his own understanding of the conditions for existence.

From later works, it seems that Wallace tended more and more to equate the conditions for existence with "environmental conditions." Even in his *Darwinism* (1889), however, one can find such an un-Darwinian passage as the following: "This 'survival of the fittest' is what Darwin termed 'natural selection.' . . . Its primary effect will, clearly, be to keep each species in the most perfect health and vigour, with every part of its organisation in full harmony with the conditions of its existence" (103).

In this context, it is perhaps easier to understand Wallace's extreme adaptationism—which, among other things, led him to reject the possibility that man had evolved by natural selection. This rejection was due to the many human faculties (mathematical ability, etc.) that are not utilized by primitive peoples and therefore could not have evolved by selection. Darwin's answer to this general problem is to state that these faculties had evolved as a by-product of other changes. But it may have been precisely because he had a much more important role for the conditions for existence in his concept of the struggle for existence that Wallace could not accept, as Darwin did, that traits could evolve free from the influence of natural selection (due to "correlations of growth," "laws of inheritance," etc.). Wallace's opposition to the analogy with artificial selection and indeed the very use of the term *selection*, as well as his objection to Darwin's frequent use of teleological language in the *Origin* (see chap. 1 here, as well as his "Creation by Law" in Wallace 1891), may also be traceable to the same cause.

DARWIN, WALLACE, AND INHERITANCE

I have tried to show that Wallace had fewer issues with teleology in his version of natural selection than Darwin did in his, due to a closer connection to the principle of the conditions for existence. But this put him farther away from something that was central to Darwin's conception, and to modern conceptions as well: inheritance (see Gayon 1998). Darwin had from early on been interested in the problem of reproduction, or "generation" (Hodge 1985), and we have seen that the process of generation, which produced variation, was a key part of his theory. The first chapter of the *Origin*, on "Variation under Domestication," contains a discussion of inheritance, including the famous declaration "Any variation which is not inherited is unimportant for us" (Darwin 1859, 12). He here mentions such phenomena as "sports," "reversions," and what now might be called variable penetrance, which are the result of unknown "laws governing inheritance" (13). By contrast, Wallace never mentioned inheritance in his 1858 paper, so that it really applies not to sexually reproducing organisms but instead only to competing asexual lineages of phenotypically constant varieties (whether these be conceived of as individual variants or subspecies). Even in his relatively late *Darwinism* (1889), Wallace has little to say on the subject of such laws of inheritance—for him, it was enough that there was abundant, minute variation, whatever its source or the laws of its transmission to later generations. His later opposition to the nascent Mendelism can be seen in the same light: in a 1908 paper, he referred to Mendelian phenomena as being "of the very slightest importance" to evolution (Wallace 1908b, 139).

In the end, Wallace's theory was focused on competing lineages that do or don't satisfy their conditions for existence, with the implicit assumption that the control of which organismal lineages (and hence traits) survive depends solely on their relative abilities to do so. In contrast, Darwin's theory had a central role for heredity, so that the differential survival and reproduction of variants ("natural selection") interacts with laws of heredity, laws of variation, and so forth, to produce evolutionary change. This is what Gayon (1998) called the "hypothesis of natural selection"; he likewise notes its absence in Wallace's theory. It was Darwin's formulation of natural selection as a process involving the genetic physiology of populations that led to all of modern genetics, as his "provisional hypothesis of pangenesis" gave rise to de Vries's "pangenes," which became Johannsen's "genes" (Gayon 1998, 447–448). This was a path that Wallace's views did not promote.

But while Darwin's understanding of the role of heredity in evolution was indeed more sophisticated than Wallace's, what I have tried to show here is that Darwin's formulation carried with it natural theological baggage that Wallace's did not. By accepting the basic premise of the argument to design, and putting natural selection in the role of the designer, Darwin created an evolutionary theory in which the environment, acting through natural selection, became the teleological guiding force of evolution.

EVOLUTIONARY CONTROVERSIES
BEFORE THE SYNTHESIS

The period from Darwin's death in 1882 through the 1930s was a time of intense controversy in evolutionary biology, a time in which the sufficiency of natural selection as a mechanism for evolutionary change was widely doubted (Kellogg 1908; Bowler 1983; Provine 1985; Gayon 1998). Before turning to the formalization of Darwin's theory that occurred during the rise of the modern synthesis, it is worth our while to take a brief look at the criticisms then made of natural selection. Of course, some of the criticisms were based on the prevalent lack of understanding of genetics, but others focused on the supposed creative nature and dominance of natural selection and seem potentially more justified.

To understand what was (and is) at stake here, it is helpful to divide the evolutionary views of the times not into the customary camps of neo-Darwinians, neo-Lamarckians, orthogeneticists, and mutationists, but instead into views accepting the design argument and those rejecting it. By doing so, it becomes clear that it was the unresolved status of teleology in Darwin's theory, even more than his lack of understanding of genetics, that led to much of the controversy in the years prior to the modern synthesis. Of course, it was perhaps also the unresolved status of teleology that allowed Darwin's theory to triumph. Darwin's theory could make sense to both theistic and atheistic scientists, because it could be interpreted in a variety of ways (see the fascinating discussions in Gray 1963).

When evolution was new, the fundamental question was "Can evolution happen at all?" This was not only the problem of a mechanism of change but also the problem of dealing with the design argument. Materialists had dealt with it by denying its validity, but this response was improbable for naturalists raised on natural theology and focused on issues of adaptation to geological change, as Darwin was. I have already noted Darwin's well-known passage in the introduction to the *Origin*, in which he states that the systematic, embryological, geographic, and geological evidence might lead one to a belief in evolution, but that "such a conclusion, even if well founded, would be unsatisfactory, until it could be shown how the innumerable species inhabiting this world have been modified, so as to acquire that perfection of structure and coadaptation which most justly excites our admiration" (1859, 3). To the fundamental question of whether evolution can happen at all, teleological views respond by saying, "Yes, but only if the end result, the product of the evolutionary process, is somehow built into the process." A teleological view of evolution thus accepts the basic premise of the argument to design: that the organisms we see around us are too complex and orderly to have been produced by chance.

Within such a generally teleological view of evolution, there are three possible sources of the "end": God (as in the case of the "theistic evolutionists"), the organism (as in some orthogenetic views), and the environment (as in Lamarckism and Darwinism). In the terminology introduced in chapter 2, explanation in terms of God involves external representational teleology (ideas in the mind of God), explanation in terms of the organism involves internal representational teleology (goal-directed vital principles), and explanation in terms of the environment involves teleological determinism (unless we can predict the occurrence of the end from knowledge of the organisms and the environment).

What is paradoxical about such teleological views of the evolutionary process is that while they all accept the premise of the argument to design—that the order of the world is too complex to have arisen by chance—they all nevertheless take the general existence, survival, and adaptedness of the organism for granted. The focus is always on evolutionary change, with the implicit assumption that in the absence of a cause for change, organisms would just go on their merry way unperturbed. But in another sense this is entirely consistent—if one accepts the premise of the design argument, one necessarily sees in organismal survival and complexity something more than mere existence, mere ability to survive, and this is true regardless of whether one's evolutionary teleology is rooted in God, the organism, or the environment. Thus, even in orthogenetic and mutationist views that seemingly reject any supernatural design, attribution of organismal adaptedness to the orthogenetic or mutational process itself, rather than to the limits imposed by the conditions for existence (i.e., natural selection in the broad sense), provides a large hole for any naturalistic theory—one that the Darwinians were quite ready to point out.

By contrast, a nonteleological, nondesign view does not worry about whether evolution can occur. Instead, it simply observes that evolution has occurred and therefore accepts that it must have been able to. A nonteleological view begins with the observation that there is a "struggle for existence"—that all organisms exist only by virtue of their ability to satisfy their conditions for existence. From this point of view, then, any evolutionary change must be subject to the conditions for existence. Given such a nonteleological view of the evolutionary process, it remains an open question whether the evolutionary changes that have occurred originated in small variations already existing in populations, small new random variations, large new random variations, effects of changed environments, inherited effects of use and disuse, or even an innate tendency to vary in a particular direction. Thus, Lamarckian inheritance, mutationism, and even orthogenesis as *mechanisms* of evolutionary change are not necessarily teleological.

Given that one accepts a generally nonteleological view of the evolutionary process, the precise nature of the changes involved is a mere technical matter of detail (which is not to say it's not a scientifically important one). The vociferous response of many Darwinians to the proposal of alternative mechanisms of change can't really be justified on technical grounds. What produced such a vociferous response (which continues to the present day) is instead the acceptance by Darwinians of the premise that adaptedness is more than mere existence, and requires a special historical process of creation. Closely coupled with this is the belief that any mechanism of evolution that could produce the harmonious adaptedness we see around us—besides the natural selection of random variations—requires a supernatural cause. The paradox of Darwin's perspective is that he wanted to produce a naturalistic theory of evolution, but began by accepting the premise of the design argument, which is fundamentally antinaturalistic. Thus, the stage was set for intense controversy.

. . .

The long historical path we have followed in these last four chapters, from Socrates to Darwin, shows the amazing tenacity of the design argument. It endured for more than two thousand years, in spite of the best attempts of skeptics and materialists to debunk it, and of Cuvier to avoid it. In the next part of the book, we will move on to examine the rise of the modern evolutionary synthesis in the early twentieth century, particularly the formulation of mathematical population genetics theory at the hands of R. A. Fisher, Sewall Wright, and J. B. S. Haldane. Surprisingly, even this most rigorous part of evolutionary biology has not been immune to the influence of the argument of design.

PART THREE

Evolution in Mendelian Populations: Teleology Gets Mathematical

7

Existence and the
Mathematics of Selection

The Adaptive Landscape versus
the Fundamental Theorem

Something is very wrong with population geneticists' obsession with the mean fitness of a population.

—J. H. GILLESPIE, 1991, THE BURDEN OF GENETIC LOAD

In considering the fallout arising from Darwin's acceptance of the premise of the argument to design, the key issue is the relation between overall adaptedness and existence. In the previous chapter, I pointed out that Darwin's theory of environmentally induced variability depended on the same separation of these concepts implicit in natural theological views, because in his theory a species experiencing a new environment begins to vary precisely because of its lack of adaptedness to the new environment, and natural selection then acts on this variation to improve overall adaptedness to the new environment. Yet whatever conceptual or verbal issues there may have been with Darwin's perspective, with the discovery of Mendelian inheritance and the rise of mathematical population genetics in the 1920s and 1930s it seems that such issues would have been resolved, because the mathematical fitness values that represent the process of selection must precisely account for all increases or decreases in frequency of genes (at least probabilistically).

What I want to show in this and the following two chapters is that this is not the case—that Darwin's separation between adaptedness and existence entered into modern evolutionary theory at its root, in the population genetics work of Sewall Wright and J. B. S. Haldane (but not R. A. Fisher). This separation persists today and is associated with a continued lack of clarity over such fundamental issues as the relation of natural selection to population growth, the meaning of "genetic load," and the distinction between natural selection and genetic drift. As

I will show, the separation of adaptedness from existence is transferred to the mathematical theory most directly in the form of confusion between absolute and relative fitness.

MENDELISM, SELECTION, AND THE MODERN SYNTHESIS

In trying to understand the integration of Mendelism and Darwinism that occurred during the modern synthesis, it is important to remember that genes play two distinct roles in Mendelian inheritance. On the one hand, they function as carriers of genetic information from one generation to the next, through evolutionary time (transmission genetics). On the other hand, they determine the phenotypic features expressed at any given time (physiological genetics). In the basic Darwinian scheme, the traits or features of organisms determine the success or failure of these organisms in the struggle for existence, and inheritance results in traits characteristic of successful organisms becoming more prevalent in the population. When one adds Mendelism into the picture, genes become important both as determinants of phenotype, and thus of success in the struggle for existence, and as a way to measure that success (which is ultimately the success of certain genetic lineages). By the 1920s, when Fisher, Haldane, and Wright developed their evolutionary models, such phenomena as dominance, linkage, epistasis, variable penetrance, and pleiotropy were already apparent. Thus, a key issue for each of them had to be how to integrate physiological and transmission genetics—in other words, how to integrate genes as *causes* of phenotype and thus of differential survival and reproduction with genes as carriers of genetic information and thus what we *count* in measuring success. Only Fisher attempted to deal with this issue seriously. It bedevils evolutionary biology to this day.

It would seem that clarity of thought should be easier in mathematics, in that all relationships among variables are spelled out explicitly. Yet if one's thinking is confused, mathematics can just as easily lead one into error. I will argue that Sewall Wright's metaphor of the adaptive landscape, and the associated mathematical measure of relative fitness, is associated with a teleological view of evolution that is disconnected from much of what is actually occurring in the population. By contrast, R. A. Fisher's perspective, as incorporated in his "fundamental theorem of natural selection," does not partake of such problems with teleology, although it has often been misinterpreted in such a way.

Before looking at how Darwin's teleology translated into Wright's, we must understand the scale on which the height of adaptive peaks is measured. This is a scale of *fitness*. However, because of much controversy over the proper definition of fitness and the varied associations the term has, I will initially avoid it, instead framing my discussion in terms of the *rate of increase* (Fisher 1930a) of genetic lineages.* My discussion will be based on the simple idea that the success of a genetic lineage in a Mendelian population (i.e., its satisfaction of the conditions for its continued existence) can be measured by its rate of increase. This mathematical foundation will serve to ground the ensuing discussion of how to interpret the *causes* of whatever differential rates of increase occur; in other words, the relation of rate of increase to fitness.

RATES OF INCREASE IN MENDELIAN POPULATIONS

Both Darwin and Wallace got their inspiration from Malthus and his law of population. And as we have seen, the greatest overlap between the two was in their concept of the struggle for existence, derived from Lyell as well as Malthus. In its simplest form, the struggle for existence is seen when the rates of increase of two classes of asexual, unicellular organisms are distinct (Gause 1964). With Mendelism, the way was finally clear to construct a mathematical theory of such rates of increase for sexual organisms, treating the genetic rather than the cellular lineages as the elementary particles, and then to relate this theory to the evolution of characters in these populations.

We will be concerned solely with what one might call "evolutionarily closed populations" (box 7.1). In such a population, which is examined over a specified time period, the simplest measure of success, or satisfaction of the conditions for continued existence, is simply the number of descendants of an initial member of the population in the final population (including the founder if still alive). This

---

BOX 7.1. EVOLUTIONARILY CLOSED POPULATIONS

An evolutionarily closed population consists of reproducing entities (organisms, genes, etc.), defined such that

- the population is examined over a given time period, with initial population size N_0 and final population size N_{final};
- all ancestors of members of the final population were members of the initial population (no immigration); and
- each of the i members of the initial population has a defined number of descendants in the final population, N_i, such that $\sum N_i = N_{final}$.

---

* I will use the variable R for absolute rate of increase and r for relative rate of increase. This should not be confused with the use of R for relatedness (e.g., Shuster and Wade 2003) or the use of r for the logarithmic "intrinsic rate of natural increase" in ecology. R in my usage is more common in ecology than in population genetics, where it is denoted the *net reproductive rate* (e.g., Begon et al. 2006, 106); it is often symbolized by λ.

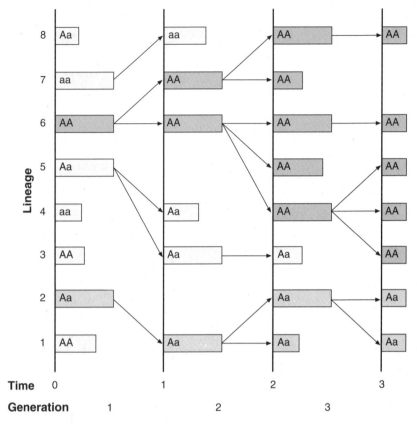

FIGURE 7.1. Schematic representation of a clonal population of diploid organisms over three discrete generations. The genotype at a single locus is shown. For calculations of representative rates of increase for this population, see table 7.1.

measure has the advantage that it expresses an objectively defined feature of the population, which deterministically connects the initial state of the population with its final state (see Arnold and Fristrup 1982). Over multiple generations neither organisms nor genes continue to exist, rather only organismic and genetic *lineages* do. The number of surviving descendants is a measure of the degree to which the lineage founded by each member of the initial population is satisfying its conditions for existence over the time period. Expressed per interval of time, this number *is* the rate of increase of the lineage.

The simplest way to see how one can measure rates of increase is to construct some hypothetical examples of populations. In particular, we want to represent a population of individuals (organisms) carrying genes that are inherited in Mendelian fashion. This is the basic population genetic model. Consider an example (fig. 7.1, table 7.1) of a single genetic locus in an asexual (clonal) population of

TABLE 7.1 Representative Rates of Increase for the Population of Figure 7.1

Variable		Generation				Total/ 3 Gen	Mean/ Gen (G)
		1	2	3	4		
Absolute lineage rate of increase (R_i)	R_2	1.00	2.00	1.00		2.00	1.26
(shown for lineages 2, 5, and 6 only)	R_5	2.00	0.50	0.00		0.00	0.00
	R_6	2.00	2.50	1.00		5.00	1.71
Initial genotype numbers (N_{ab})	N_{AA}	3	2	5	5		
	N_{Aa}	3	3	3	2		
	N_{aa}	2	1	0	0		
Absolute genotypic rate of increase (R_{ab})	R_{AA}	0.67	2.50	1.00		1.67	1.19
	R_{Aa}	1.00	1.00	0.67		0.67	0.87
	R_{aa}	0.50	0.00	—		0.00	0.00
Initial population size (N)		8	6	8	7		
Mean absolute rate of increase ($\bar{R}$)		0.75	1.33	0.88		0.88	0.96
Initial genotype frequencies (f_{ab})	f_{AA}	0.38	0.33	0.63	0.71		
	f_{Aa}	0.38	0.50	0.38	0.29		
	f_{aa}	0.25	0.17	0.00	0.00		
Relative genotypic rate of increase (r_{ab})	r_{AA}	0.89	1.88	1.14		1.90	1.24
	r_{Aa}	1.33	0.75	0.76		0.76	0.91
	r_{aa}	0.67	0.00	—		0.00	0.00
Mean relative rate of increase ($\bar{r}$)		1.00	1.00	1.00		1.00	1.00

diploid, multicellular organisms with discrete generations. In the terminology defined here, we can conceive of this as two coexistent populations, one of organisms, one of genes, with each pair of genetic lineages always mapping to one organismal lineage. This example represents three generations. For organisms, the initial population size is eight, and the final population size is seven; for the genes, they are sixteen and fourteen, respectively. Only two types of events occur in this time period: reproduction and death.

The genetic (allelic) lineages each have the same rate of increase as the corresponding organismal (and genotypic) lineage in this simple clonal case, so there is no need to treat them separately. Applying the simplest measure of rate of increase, number of descendants, to the organismal lineages, we find that in this example only two of these still exist at time 3: lineage 2 has two descendants, and lineage 6 has five. Since we are interested in *rates* of increase, however, we must express these as a rate of increase of two and five per three generations, respectively. All other lineages have a rate of increase of zero (per three generations). The average per-generation rate of increase of each lineage is calculated as the cube root of its rate over three generations (the geometric mean), giving an average for lineage 2 of 1.26 per generation, and for lineage 6 of 1.71 per generation. The geometric mean must

BOX 7.2. ABSOLUTE RATES OF INCREASE IN DISCRETE
GENERATION CLONAL POPULATIONS

For an evolutionarily closed clonal population, the absolute rate of increase of the ith lineage is the number of descendants of the lineage in the final population divided by the total number of generations, or

$$R_i = N_i/m, \qquad (7.2a)$$

where N_i is the number of individuals representing the ith lineage in the final population, and m is the number of generations. The average per generation rate of increase of this lineage is given by the geometric mean:

$$G_{R_i} = \sqrt[m]{N_i}. \qquad (7.2b)$$

Finally, the average rate of increase across lineages over the total period is simply the population growth rate:

$$\overline{R} = \frac{\sum R_i}{N_0} = \frac{N_{final}}{N_0} m^{-1}. \qquad (7.2c)$$

be used because it is the average rate that gives the lineage its actual number of descendants at the end of the period (box 7.2). The average rate of increase across all lineages $(\overline{R})$ is simply the total population growth rate (box 7.2, eq. 7.2c).

This simple situation, however, already brings with it some questions when we try to divide the population into successive subperiods (e.g., generations). We can define the rate of increase of each lineage over a single subperiod as the number of descendants at the end of the period divided by the number at the beginning (box 7.3). With this definition, the rate of increase of a lineage over several subperiods is simply a multiple of its rate of increase over each subperiod. Thus, lineage 6 has a rate of increase of 2.0 (2/1) in the first generation, 2.5 (5/2) in the second generation, and 1.0 (5/5) in the third generation, giving (2/1)(5/2)(5/5) = 5 per three generations (see table 7.1).

The question that arises is what to do with lineages that go extinct in earlier subperiods. For the subperiod in which it goes extinct, the rate of increase of a lineage is clearly zero, but what rate of increase should we assign for later subperiods? This number is undefined. For individual lineages, the value assigned makes no difference to the total calculated by multiplying the values for the remaining subperiods, since one of these is already zero. But the value assigned does matter when we construct averages across the population within each subperiod. One way to deal with this problem is to construct a weighted average, with each lineage

BOX 7.3. TEMPORAL SUBDIVISION OF RATES OF INCREASE IN CLONAL POPULATIONS

For subperiods of the total time period, if $N_i(t)$ is the number of representatives of the ith lineage at time t, and $R_i(t)$ is the rate of increase of the ith lineage from time t to time $t + 1$, then

$$R_i(t) = N_i(t + 1)/N_i(t). \tag{7.3a}$$

Note that this quantity is undefined for the case in which $N_i(t)$ is zero (see discussion in text).

To calculate an average rate of increase *across lineages* from time t to $t + 1$, the arithmetic mean rate of increase for those lineages is used:

$$\overline{R_i}(t) = \frac{\sum N_i(t)R_i(t)/N_i(t)}{N_0} = \frac{\sum R_i(t)}{N_0}. \tag{7.3b}$$

But this is the average rate of increase of the lineages founded by individuals in the initial population, not of the population of individuals that exists at the beginning of that subperiod. Thus, this is not the population growth rate over this subperiod. To find the population growth rate, one must instead construct a weighted average, in which lineages are weighted by the number of representatives of that lineage alive at the beginning of the period:

$$\overline{R}(t) = \frac{\sum N_i(t)R_i(t)}{\sum N_i(t)} = \frac{N(t + 1)}{N(t)}. \tag{7.3c}$$

It is this weighted average for each subperiod that can be multiplied to give the correct total rate of increase over the entire period.

weighted by the number of representatives present at the beginning of each subperiod (box 7.3). Such a weighted average also has the advantage that it gives the correct population growth rate over each subperiod (but note that this is growth of the population of individuals, not lineages; I will return to this point later).

Finally, one might want to construct averages not for the entire population but for some subpopulations within the larger population (e.g., all those sharing some genotype or phenotype). This can easily be done (box 7.4), with the caveat that if one also wants to look at subperiods of the entire period, it is again necessary to weight each lineage by the number of representatives of that lineage alive at the start of the subperiod considered.

All of the derived rates of increase examined so far might be called *absolute rates of increase* since they reflect the increase or decrease in numbers of one or more

BOX 7.4. AVERAGE RATES OF INCREASE IN
CONTEMPORANEOUS CLONAL SUBPOPULATIONS

Rather than measuring the rate of increase of each individual lineage, one might want to measure the average rate of increase of all lineages sharing some characteristic, such as a given genotype at a locus. From time t to $t + 1$, the mean rate of increase for those *lineages* founded by individuals of a particular genotype, ab, is given by

$$\overline{R}_{iab}(t) = \frac{\sum\limits_{i \in ab} R_i(t)}{N_{ab}(0)}, \qquad (7.4a)$$

while that for the *individuals* present at the beginning of each period is

$$\overline{R}_{ab}(t) = \frac{\sum\limits_{i \in ab} N_i(t)R_i(t)}{\sum\limits_{i \in ab} N_i(t)} = \frac{\sum\limits_{i \in ab} N_i(t + 1)}{N_{ab}(t)} = \frac{N_{ab}(t + 1)}{N_{ab}(t)}, \qquad (7.4b)$$

or the total number of ab individuals at the end of the period divided by the total number at the beginning. Once again, it is this weighted average for each subperiod that can be multiplied to give the correct rate of increase over the entire period.

lineages in terms of the absolute number of descendants present at a later time. In some cases, one might be more interested in the *relative rates of increase* of distinct lineages or groups of lineages (box 7.5). Such relative rates of increase can be thought of as the rate of increase or decrease in frequency, rather than number, or alternatively, as the rate of increase compared with that of the population as a whole. Note that the mean relative rate of increase of all lineages ($\overline{r}$) is always 1.0, because the population as a whole neither increases nor decreases in frequency.

For a single population, there is no conflict between relative and absolute perspectives, since for all possible subpopulations relative and absolute rates of increase always have a constant ratio (of $1:\overline{R}$). However, when one considers averages of within-population values across multiple populations, conflicts between the two perspectives do arise (see Karlin and Lieberman 1974; Gillespie 1977). This conflict is not fundamental, however, and I have elsewhere shown that by taking per capita values, rather than averaging across populations, one can overcome it (Reiss 2007).

BOX 7.5. RELATIVE RATES OF INCREASE

For an individual lineage, we can define the relative rate of increase as

$$r_i(t) = \frac{f_i(t + 1)}{f_i(t)} = \frac{R_i(t)}{\overline{R}(t)}, \tag{7.5a}$$

where $r_i(t)$ is the relative rate of increase of lineage i from time t to $t + 1$, $f_i(t)$ is the frequency of members of lineage i at time t ($=N_i(t)/N(t)$), and $\overline{R}(t)$ is the mean rate of increase (population growth rate). Averages across lineages and generations can be calculated similarly to those for absolute rates of increase. Most importantly, the mean relative rate of increase is always unity:

$$\overline{r}(t) = 1.0. \tag{7.5b}$$

Now let's add sex into the picture. The phenomenon of sex is an immensely complicating factor in the consideration of rates of increase. In a sexual population, the rates of increase of organismal lineages are no longer the same as those of contained genetic (allelic) lineages. In fact, it is not even clear what one means by organismal lineages, as the genetic lineages carried by an organism segregate and then mix with others in meiosis and conjugation. Nevertheless, the measures of rate of increase defined earlier can be extended to sexual populations, by focusing on the allelic lineages.

Consider the population of figure 7.2. This represents three generations of a sexual, diploid population with discrete generations; two loci are represented. With reference to the population of alleles (genes), it is clear that the three conditions in box 7.1 are fulfilled: there is a definite population size (number of genes) at each time, there is no immigration, and each of the genes present in the initial population has a definite number of descendants in the final population, which sum to give the final allelic population. Because this allelic population is isomorphic with the asexual population treated earlier (boxes 7.2 to 7.5), the same equations apply.

For various reasons, however, most people are more interested in organisms than alleles. Given the realities of Mendelian inheritance, it is clear that organisms only persist past their death as cellular or genetic lineages. It seems reasonable to define an organismal lineage as being composed of all the genetic (allelic) lineages carried by that organism, and the rate of increase of such an organismal lineage as the average rate of increase of these genetic lineages. This measure is obviously relative to the loci considered (only two are represented in fig. 7.2).

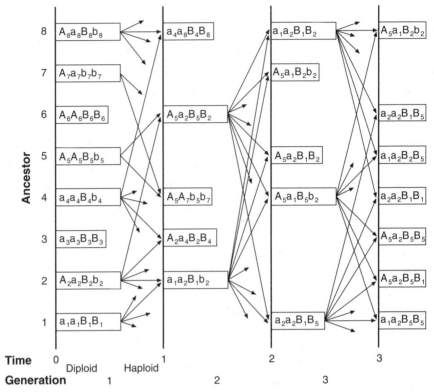

FIGURE 7.2. Schematic representation of a sexual population of diploid organisms over three discrete generations. The genotype at two loci is shown. Alleles are numbered with respect to the ancestor in the initial population ($t = 0$) in which they are found. Note that allelic lineages can be traced directly from the initial to the final population, so that rates of increase can readily be calculated for these lineages.

The "true" rate of increase of an organismal lineage might be defined as that which is averaged over all loci, but the determination of values for all loci is an extremely difficult, perhaps impossible task, and it raises troubling questions about how to define a locus. One possible approach, perhaps not as absurd as it might seem at first, would be to count nucleotide lineages, so that the success of an organism is measured in terms of the total number of nucleotide lineages it contributes to future generations. However, in most cases, it will be both more useful and more practical to determine values with respect to only a few loci (e.g., a gene region or regions) under study. Because of the nature of biparental reproduction, in the first generation, the rate of increase will be the same for all loci (and equal to one-half the number of offspring), but afterward it will of course vary among loci.

Finally, we can examine the situation in a similar diploid population with overlapping generations (fig. 7.3). Since our measures of organismic rates of increase

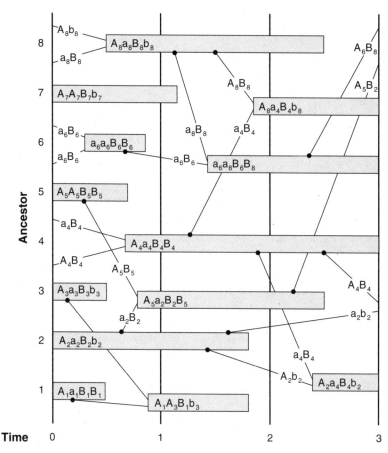

FIGURE 7.3. Schematic representation of an age-structured sexual population of diploid organisms over a period of three time units. The genotype at two loci is shown. Alleles are numbered with respect to the ancestor in the initial population ($t = 0$) in which they are found. Again, note that allelic (genetic) lineages can be traced from the initial to the final population, so that rates of increase can readily be calculated for these lineages.

are defined in terms of allelic (genetic) rates of increase, and allelic lineages are just as continuous when generations overlap as when they are discrete, there is no difficulty using the same equations for continuous time populations, with t now representing astronomical time, rather than generations.

If we want all members of the initial population to be diploids, we must adopt a convention that haploid individuals (e.g., gametes) are only counted as lineages if they are going to pair with other haploids. Of course, with overlapping generations all members of the initial population are not the same age (with the convention suggested, some are even represented by two haploid individuals), and one may feel it rather unreasonable to define an average rate of increase across organisms of all ages in the initial population. In a large enough population, there is no reason why one cannot derive average rates of increase for particular age classes within each genotypic class (cf. Fisher's notion of "reproductive value").

FITNESS IN POPULATION GENETICS

Having defined quantities measuring the rates of increase of various lineages, we are prepared to consider the relation between these quantities and that quantity commonly called "fitness" in population genetics. The volume of literature discussing the true meaning of fitness in evolutionary biology is remarkable. Nevertheless, among population geneticists themselves, there is little confusion about the algebraic meaning of fitness, at least as defined with reference to the discrete generation diploid model. With respect to this model, it is clear that the fitness of an individual or group of individuals corresponds, respectively, to the individual or average rate of increase over a single generation.

Yet this algebraic equivalence does not at all capture the intended meaning of "fitness," which (at least when authors are being careful) is always defined not as the average rate of increase of a group of lineages (of a given genotype) but as their *expected* rate of increase. This allows for the possibility that accidents in any particular population ("genetic drift") may make the average different from the ideal or expected quantity.

The difference between the average and the expected rate of increase gets us back to the fundamental issue in the study of differential lineage survival and reproduction (natural selection)—the distinction between a descriptive and a causal analysis. All of the measures arrived at earlier in this chapter for rate of increase are merely descriptive; they tell us absolutely nothing about the causes of differential increase among various lineages.

In fact, until now I have been dealing with genetics only in terms of gene transmission, not gene expression. Yet in order to say something about the *causes* of differential increase of genetic lineages, it is necessary to treat the rate of increase of each lineage as some function of its *effects* on the external world. The challenge then becomes the definition of such a function.

The simplest assumption would be that each of the alleles at a locus has a constant effect, which completely determines its own expected rate of increase, independent of the circumstances it is in. If we call the total of these circumstances the "environment," then for the allele a, we can write $E(R_a) = f(a, \text{environment}) = f(a) = C$, where C is a constant, and E denotes the expectation, or, roughly, the average across an infinite series of trials. In other words, the expected rate of increase is a function only of the allele itself. Of course such a model is extremely unrealistic for any allele, except perhaps a dominant lethal. In particular, this constant allelic fitness model immediately runs up against a basic fact of Mendelian inheritance in diploid organisms: alleles are always present in pairs, which may interact in their effects (dominance). In other words, the genetic environment of an allele, even at a single locus, is often an important determinant of the effects of that allele.

A slightly more realistic model would be one in which single- (or multiple-) locus genotypes are considered to have constant effects, completely determining their own expected rate of increase: for the genotype ab, $E(R_{ab}) = f(ab, \text{environment}) = C$. It is this model of constant fitness of single-locus genotypes that is at the heart of the standard approach to population genetics (e.g., Hartl and Clark 1997; Hedrick 2000), although many more complicated models have been developed.

It is important to remember that when this model is used, one is treating the rate of increase not only of the genetic lineages at that particular locus but of *all* lineages carried by that organism, as determined (at least probabilistically) by the genotype at that locus. Yet it is obvious that the rate of increase of genetic lineages in organisms of a particular genotypic class depends on more than just that genotype. All of these other factors are lumped together as "environment."

The "environment" thus includes, among other things, the genotype at all loci not considered, the phenotype of the organism from conception to death, the phenotypes of other members of the population, and the characteristics of the physical and biological environment. When one assumes constant fitnesses, one usually implicitly assumes that the genotype at other loci is irrelevant, the phenotype of the organism is constant (since it is determined by the genotype), the phenotypes of other members of the population are irrelevant, and the characteristics of the physical and biological environment are constant.

Let us also assume that mating is random with respect to the locus under consideration, and population size is infinite, so we need not take random effects into account. With these assumptions, we can construct a simple, familiar model of evolutionary change at a single locus due to differences in genotypic fitness, represented by W (box 7.6). Differential fitness at a single locus has only three possible forms under these circumstances: directional selection ($W_{AA} > W_{Aa} > W_{aa}$ or $W_{AA} < W_{Aa} < W_{aa}$), overdominance ($W_{AA} < W_{Aa} > W_{aa}$), or underdominance ($W_{AA} > W_{Aa} < W_{aa}$). At equilibrium, genotype and gene frequency don't change ($p' = p$), so the mean fitness ($\overline{W}$) is a constant (i.e., there is no change in mean fitness). As it turns out, mean fitness is at a maximum for stable equilibrium (overdominance) and a minimum for unstable equilibrium (underdominance), so that the change in mean fitness ($\Delta \overline{W}$) is never negative under this system of evolution. In other words, in this model, as gene frequencies change under natural selection, *mean fitness of the population always increases*. Nevertheless, there is no guarantee that the population will evolve to the highest possible mean fitness, for in the case of underdominance, the homozygote with the lower fitness is nevertheless favored when it is common.

BOX 7.6. NATURAL SELECTION WITH CONSTANT FITNESSES—THE STANDARD MODEL

For the alternative alleles A and a with frequencies p and q $(= 1 - p)$, and assuming random mating, we have the following:

Zygote	Frequency	Fitness
AA	p^2	W_{AA}
Aa	$2pq$	W_{Aa}
aa	q^2	W_{aa}

We thus have the mean fitness:

$$\overline{W} = \sum f_{ab} W_{ab} = p^2 W_{AA} + 2pq W_{Aa} + q^2 W_{aa}. \tag{7.6a}$$

To calculate the frequencies of genotypes in the next generation of zygotes, it is necessary to find the new gene frequencies, p' and q'. We have

$$p' = (p^2 W_{AA} + pq W_{Aa})/\overline{W} = p(p W_{AA} + q W_{Aa})/\overline{W}$$

and

$$q' = (q^2 W_{aa} + pq W_{Aa})/\overline{W} = q(q W_{aa} + p W_{Aa})/\overline{W}. \tag{7.6b}$$

Note that division of the absolute fitnesses by $\overline{W}$, the mean fitness in the population, is required for the new allelic frequencies to sum to unity.

IRONING OUT WRIGHT'S "SURFACE OF SELECTIVE VALUE"

In the single-locus, constant genotypic fitness, random mating, infinite population size model we are considering, mean fitness always increases, and stable equilibria are local peaks of mean fitness in the field of gene frequencies. Thus, one can imagine the population at any time as either sitting on or evolving toward such a peak (fig. 7.4a).

This image of evolution is closely tied to Sewall Wright's (e.g., 1932, 1988) vision of a "surface of selective value" or adaptive landscape on which the population evolves. In fact, the algebraic form of the equations I have been using derives directly from Wright, including the use of W for fitness values (Wright 1937). In most current textbooks this connection is explicit, and an adaptive landscape is defined as a "graph of mean population fitness ($\overline{w}$) against gene frequency" (Ridley 2004, 214) or a

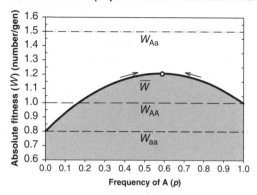

A   Absolute Fitness (Expected Rate of Increase in Number)

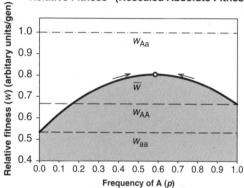

B   "Relative Fitness" (Rescaled Absolute Fitness)

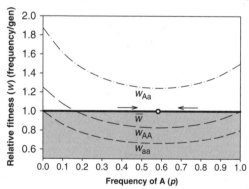

C   Relative Fitness (Expected Rate of Increase in Frequency)

FIGURE 7.4. Comparison of an adaptive landscape for absolute fitness (A) with those for two different representations of relative fitness: rescaled absolute fitness, with the maximum genotypic fitness at 1.0 (B), and standardized relative fitness, with mean fitness at 1.0 (C). The model used is the same throughout: a simple case of overdominance at a single locus in an infinite, random-mating population, with absolute fitnesses $W_{AA} = 1.0$, $W_{Aa} = 1.5$, and $W_{aa} = 0.8$. Note the flattening of the adaptive landscape in C.

"surface of mean fitness for all possible combinations of allelic frequency" (Griffiths et al. 2008, 686). Futuyma (2005) notes that the metaphor is "widely used in evolutionary biology" (287). It is rather disconcerting to find, then, that there are fundamental problems in interpreting Wright's metaphor in these terms.

First, since this metaphor requires that mean fitness increase in time, and mean fitness is just expected population growth rate, population growth rate must continually increase. Yet the image of a constantly increasing population growth rate hardly corresponds to any reasonable view of the normal evolution of a population. The problem obviously lies in our assumption that "fitnesses" of genotypes at a single locus are constant and independent of environment. In any natural situation, one would of course expect absolute fitnesses to at least eventually decrease as the population increases in number, even allowing other aspects of the environment to remain constant.

But if it is obviously absurd to consider absolute fitnesses as constant, it is perhaps not so absurd to think that for some loci, the *ratios* of the fitnesses of the different genotypes might be constant over time. Since we are only dealing with changes in allele frequencies, not numbers, the fitness ratios completely determine the evolution of the population in this model, and there is no need to use absolute fitnesses. If we only care about fitness ratios, we can scale the fitnesses however we please; most commonly (again following Wright), the genotype with the highest fitness is given the fitness value of unity, and all other genotypes are scaled relative to this (fig. 7.4b). These rescaled absolute fitnesses are known as "relative fitnesses," generally represented by a lowercase w. With this rescaling, one can now regard the adaptive landscape as a landscape of mean relative fitness. This is of course the interpretation used in the textbooks quoted here.

Yet now another problem arises in the interpretation of the metaphor: what is the meaning of differences in mean relative fitness among populations? If mean fitness is not (expected) population growth rate, what is it? These questions are particularly important since a major theme of Wright's shifting-balance theory of evolution is that selection alone may cause a population to get stuck on a low adaptive peak, with genetic drift (due to small population size and inbreeding in local populations) necessary for a population to cross the adaptive valley to a neighboring, higher peak. This was stated very clearly by Wright (1932) in his first discussion of the surface of selective value: "The problem of evolution as I see it is that of a mechanism by which the species may continually find its way from lower to higher peaks in such a field" (358–350; this paper is reprinted in Wright 1986, as are all of Wright's I will be discussing).

Unfortunately, it turns out that in terms of the model we are using, the mean relative fitness of a population has no meaning whatsoever. Because we are dividing the absolute fitnesses by an arbitrary value, the mean relative fitness of a population is entirely arbitrary. This point is clearly recognized by most population geneticists and evolutionary biologists, although they do not usually state this as emphatically as I have. Instead, they just caution against trying to compare the mean relative fitnesses of different populations. For example, Griffiths et al. (2008) state, "This maximization of fitness does not necessarily lead to any optimal property for the species as a whole, because fitnesses are defined only relative to one another within a population" (627), and Hartl and Clark (1997) note that "$\overline{w}$ is the average fitness *in* the population, not the average fitness *of* the population" (230).

Then what can these authors mean by employing the metaphor of the adaptive landscape? If the height of a population's adaptive peak means nothing, why be concerned about whether a population is able to reach a higher peak, by genetic drift or otherwise?

I believe that the persistent popularity of Wright's metaphor of the adaptive landscape, in spite of the fact that a critical examination shows it to make no sense (Provine 1986; Gayon 1998), is the result of a persistent confusion between relative and absolute fitness. This confusion comes from the belief that these quantities somehow are different measures of the same thing. Yet if relative fitnesses are to be truly relative, they must be relative to the situation of the population as a whole. Instead, "relative fitnesses," as commonly used, are no more than conveniently rescaled absolute fitnesses. The arbitrary nature of the scaling is reflected in the arbitrariness of $\overline{w}$, which is essentially a dummy variable expressing one's choice of scaling. This can be seen quite clearly in a comparison of figures 7.4A and 7.4B. This problem is in fact the identical one pointed to for Darwin earlier: the existence of entities (here population growth rate) is separated from their adaptedness (here "mean fitness").

By contrast, relative rates of increase, as defined in box 7.5, unlike relative fitnesses under the usual definition, are *not* arbitrary values. Instead, they express the rate of increase or decrease in frequency per generation. Similarly, one could easily define *standardized relative fitnesses* in terms of relative rates of increase, such that $w_{ab} = E(r_{ab})$. Something like this is commonly done in quantitative genetics (Lande and Arnold 1983; Lynch and Walsh 1998). Under this definition, for all periods the mean relative fitness ($\overline{w}$) is a constant fixed at 1.0.

Recognizing that under this model the rate of increase (marginal fitness) of the A allele (w_A) is just the weighted average of the genotypic rates of increase ($w_A = pw_{AA} + qw_{Aa}$), relative fitness is just the per-generation rate of change in frequency of an allele ($p' = pw_A$). Of course, the simpler form of these equations comes at a price: if gene frequencies are changing, the genotypic fitnesses (rates of increase) can no longer be constant; in other words, standardized relative fitnesses are necessarily frequency dependent. In terms of absolute fitnesses, the standardized relative fitness w_{ab} is equal to genotypic absolute fitness divided by mean absolute fitness ($w_{ab} = W_{ab}/\overline{W}$). But since mean absolute fitness (rate of increase) itself increases under natural selection in this model, and the absolute fitness of each genotype is constant, the standardized relative fitness of each genotype must decline.

In fact, with this definition, a population is at a stable equilibrium when the standardized relative fitness of each genotype is at a minimum (fig. 7.4c). This point is thus a joint minimum of relative fitness for all genotypes.

The use of standardized relative fitnesses (fig. 7.4c), while unfamiliar in the context of population genetics, is much less prone to misinterpretation than the use of relative fitnesses that are merely rescaled absolute fitnesses (fig. 7.4b). This is because standardized relative fitnesses have the explicit meaning of expected rate of increase in frequency, rather than being values on an arbitrary scale. A genotype with a standardized relative fitness greater than one is leaving greater than the average number of descendants. Currently, students are told that "selection can also be described as a process that *increases mean fitness*" (Griffiths et al. 2008, 627), which seems to say something real about the effects of selection. Lest they get too confident, however, they are immediately warned that "it is relative (not absolute) fitness that is increased by selection." It is almost unavoidable that such a nebulous concept of relative fitness should acquire some of the associations more proper to absolute fitness.

One obvious problem with the use of such standardized relative fitnesses is that they cannot remain constant while allele frequencies change. However, there is no reason to expect that the value assigned to relative fitnesses should be constant, just because the *ratios* of the fitnesses are constant. No one objects to the fact that the marginal fitnesses (rates of increase) of alleles are not constant with constant genotypic fitnesses. The situation here is somewhat similar. Standardizing the dummy variable $\bar{w}$ at 1.0 simply makes it mathematically impossible for the (standardized) relative fitnesses of genotypes to remain constant as gene frequencies change.

Under the usual definition of "relative fitnesses" as rescaled absolute fitnesses (fig. 7.4b), this necessity for change in their values as frequencies change is concealed by the practice of "normalizing" genotypic frequencies after selection by the mean fitness $\bar{w}$. Unfortunately, the effect of this normalization is often overlooked. Bruce Wallace (1991), in an extremely interesting review of the genetic load controversy (which we will return to in chap. 8), concluded that Haldane's (1957) calculations on the "cost of selection" were entirely erroneous, due to his adoption of the "arbitrary convention that states that the maximum, or optimal, fitness shall be assigned the value of 1.00" (76). He attributes this error to Haldane's "failure to realize that normalization of the population to 100% while maintaining a constant population size resurrected most 'dead' individuals from the category designated 'selective deaths'" (77). If Haldane, Crow, and many other prominent geneticists could be misled by the standard convention into calculating figures for genetic load that have "little to do with fitness as measured by the ratio of daughters to mothers" (Wallace 1991, 135), it is perhaps not unreasonable to think that evolutionary biology would be better served by the use of standardized relative fitnesses.

Wallace (1991) in fact proposes a similar definition himself (though he is thinking of absolute fitness with constant population size): "If the value 1.00 is assigned to $\overline{W}$ rather than to the genotype with the maximum or optimum fitness, the *relative* [i.e., ratios of] Darwinian fitnesses of different genotypes remain unchanged-only the numerical values assigned to them are altered" (134).

What becomes of the adaptive landscape if one uses standardized relative fitnesses rather than traditional relative fitnesses (rescaled absolute fitnesses)? If the adaptive landscape is regarded as the surface of mean relative fitness ($\overline{w}$), then its form becomes quite simple: it is a flat plane at $\overline{w} = 1$ (fig. 7.4c; Wright himself recognized this at various times—e.g., 1949, 376). I do not imagine that such a metaphor of the adaptive landscape would have ever gained much popularity. Alternatively, one could have a landscape measured in terms of the standardized relative fitness of any one of the genotypes: stable points would be adaptive pits; barriers, adaptive ridges (fig. 7.4c). A landscape based on the variance in allelic (marginal) fitness would show both stable and unstable equilibria as pits of height zero.

The question of pits or peaks, however, is not important. What is important is that in none of these alternative cases is there any reason to consider the movement from one adaptive peak (or pit) to another progressive, as suggested by Wright's metaphor of movement "from lower to higher peaks in such a field." There is thus no problem of being stuck on a low peak (or in a shallow pit) for a population to overcome, by genetic drift or otherwise.

But was Wright really so dangerously misled by his metaphor? Is there something else he might have meant other than the interpretation given here, which would allow us to make sense of his conception of the adaptive landscape? To answer this question, we must turn to a short history of the concept. This history is especially important for helping us understand R. A. Fisher's view of fitness, as incorporated in his fundamental theorem of natural selection, and how it differed from the vision expressed by Wright's adaptive landscape.

THE GENESIS OF WRIGHT'S SURFACE

In his excellent scientific biography of Sewall Wright, Will Provine (1986) noted, "Despite its great attractiveness and apparent ease of interpretation, the surface of selective value is one of Wright's most confusing and misunderstood contributions to evolutionary biology" (308). Provine showed that at least three fundamentally irreconcilable concepts of this surface have been used in evolutionary biology, two of them by Wright himself.

The second of these surfaces is the one we have been considering: mean fitness of the population plotted as a function of allele frequencies. A third type of surface was introduced by Simpson (1944), with mean fitness of individuals plotted as a function of their phenotype (this surface has had some recent support; see

Arnold et al. 2001, 2003; McGhee 2007). Provine showed that neither of these corresponds to the concept of the surface first proposed by Wright (1932), in which the mean fitness of genotypes is plotted as a function of those genotypes (see also the detailed discussion in Gayon 1998; Gavrilets 2004).

Wright's 1932 paper was prepared as a nontechnical summary of the results of his mathematical analysis of "Evolution in Mendelian Populations," published the preceding year (see Provine 1986, 283–287). He began by stressing the immense genetic diversity provided by the Mendelian mechanism. Nevertheless, "not all of this field is easily available in an interbreeding population. Suppose that each type gene is manifested in 99 percent of the individuals. . . . The average individual will show the effects of 1 percent of the 1000, or 10 deviations from the type, and . . . only a small proportion will exhibit more than 20 deviations from the type. . . . The population is thus confined to an infinitesimal portion of the field of possible gene combinations, yet this portion includes some 1040 homozygous combinations" (356).

He then introduced the surface of selective value: "If the entire field of possible gene combinations be graded with respect to adaptive value under a particular set of conditions, what would be its nature?" (357). Here adaptive values ("fitnesses") are being assigned to individual multilocus *genotypes* within the entire field of possible genotypes. His figure 1 (here fig. 7.5) illustrates this field for "combinations of from 2 to 5 paired allelomorphs."

In this initial version, each adaptive peak is a cluster of genotypes. The top of the peak (represented by a "+" in both figs. 7.5 and 7.6) represents the favored homozygous "type" genotype, which is surrounded by genotypes that decrease in adaptive value as they get farther and farther away from the favored type (in terms of gene replacements). By positing the existence of adaptive peaks other than the one currently occupied, Wright was arguing that there might be many multilocus homozygous genotypes with high adaptive value that are extremely unlikely to occur in a large, random breeding population. With "many small local races, each breeding largely within itself but occasionally crossbreeding," however, a number of non-"type" alleles could become common in a race, causing one of the other favored multiply homozygous genotypes to occur with high frequency. Thus, Wright concludes that "with many local races, each spreading over a considerable field and moving relatively rapidly in the more general field about the controlling peak, the chances are good that one at least will come under the influence of another peak." He goes on: "If a higher peak, this race will expand in numbers and by crossbreeding with the others will pull the whole species toward the new position. The average adaptiveness of the species thus advances under intergroup selection, an enormously more effective process than intragroup selection" (363). Unfortunately, even in this initial version of the adaptive landscape Wright does not indicate what it might mean for a species to move to a higher

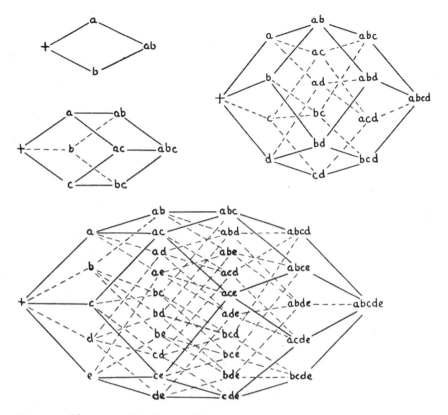

FIGURE 7.5. The gene combinations model underlying the adaptive landscape (Wright 1932, fig. 1). All homozygous combinations possible are illustrated, with each genotype connected to those differing by only a single gene replacement. The legend reads simply, "The combinations of from 2 to 5 paired allelomorphs." Reprinted from Wright (1986), by permission of the University of Chicago Press. ©1986 by The University of Chicago. All rights reserved.

peak and increase in average adaptiveness (though the expansion phase would certainly be tied to a temporary rise in mean rate of increase of the expanding race). As with the standard gene-frequency-based surface discussed here, it is not at all clear on what scale adaptive value is to be measured.

To understand what Wright had in mind, we need to remember that, in Provine's words, "Wright's shifting balance theory of evolution in nature grew directly out of his theory of evolution in domestic breeds" (Wright 1986, 1; see also Gayon 1998). Wright himself was quite aware of this: in his autobiographical 1978 paper on "The Relation of Livestock Breeding to Theories of Evolution," he recalls that "it was apparent . . . from my studies of the breeding history of Shorthorn cattle . . . that their improvement had actually occurred essentially by the shifting

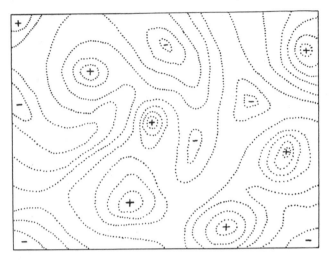

FIGURE 7.6. The first adaptive landscape (Wright 1932, fig. 2). The legend reads, "Diagrammatic representation of the field of gene combinations in two dimensions instead of many thousands. Dotted lines represent contours with respect to adaptiveness." The "+" is used to represent an adaptive peak; the "−," an adaptive valley. Reprinted from Wright (1986), by permission of the University of Chicago Press. ©1986 by The University of Chicago. All rights reserved.

balance process rather than by mere mass selection. There were always many herds at any given time, but only a few were generally perceived as distinctly superior. . . . These herds successively made over the whole breed by being principal sources of sires" (1198–1199).

Wright seems never to have realized that while in the context of domestic breeds it may be reasonable to speak of "improvement" by selection of "superior" herds, where the standard of value is determined by the goals of the breeder, no such a priori standard of overall value exists in nature.

In fact, by focusing on the hypothetical ways in which organisms could evolve toward a future state of improved adaptation, Wright introduced a teleological element into his evolutionary theory, just as Darwin had into his. The underlying pattern of thinking is the same: the current state of the population is conceived of as maladapted with reference to some potential future state (a higher "adaptive peak"), with natural selection then acting as a force to drive the population toward the more adapted state (for Wright, aided by genetic drift in the initial crossing of the valley).

This teleological element was easy for Wright to overlook in 1932, due to the metaphorical way in which he presented his shifting balance theory. Yet as Provine again perceptively noted, "One of Wright's lifelong tasks was to bring his quantitative theory of evolution closer to his qualitative theory" (Wright 1986, 207). He did so by linking his metaphorical version of the surface of selective values to the mathematically defined measure of average fitness, $\overline{w}$. Thus, in a pair of 1935 papers (1935a,1935b), originally written as one (Wright 1986, 207–208), Wright examined the situation in which the quantitative grade of a character on a primary scale depends on the constant effects of alternate alleles at many independent loci, with the optimum grade of the character (in terms of "fitness") being intermediate. He then introduced a secondary scale measuring fitness, based on the squared deviation of the actual value of each individual (S) from the optimum value (O), and constructed a landscape of possible populations on this basis. Wright's basic assumption, that it is "often the case that the best adapted individuals are those nearest the average in every respect" (1935a, 243) was not unreasonable. But Wright went well beyond this assumption, treating the optimum value as a fixed goal toward which the population would evolve (if able to), and with reference to which its average adaptive value could be compared.

Wright took the final step toward the mathematization of his metaphor in 1937, when he first introduced the variable w to refer to the fitness of a genotype. He clearly had in mind not the absolute fitness but the relative fitness of each genotype: "The selective value of zygotes (relative to a certain standard) will be designated w and the mean value for a population, $\overline{w}$" (307). As in the earlier paper, Wright assumed constant fitnesses. However the fitnesses are now assigned to complete multilocus genotypes (as in the 1932 paper), rather than to "the absolute grades of separate elementary characters" (309–310). This approach certainly seems more reasonable, as one can then refer these constant fitnesses of genotypes to a constant external environment. Nonetheless, two major problems with his approach remained (Edwards 2000).

First, to take the partial derivative of the mean fitness ($\overline{w}$) with respect to change in allele frequency, mean fitness must be a function of allele frequency, and thus the distribution of genotypes on which mean fitness depends must also be a function of allele frequency. For this to be true, not only must one assume random mating and independence of loci (no linkage), but one must assume an infinitely large population in which each of these genotypes can occur with its proper frequency, since "there is no reasonable chance that any two individuals have exactly the same genetic constitution in a species of millions of millions of individuals persisting over millions of generations" (Wright 1932, 356). The appearance of increased realism in this new definition is therefore illusory. Nevertheless, because he again made the same simplifying assumptions he had made in the earlier papers (1935a, 1935b), Wright was again able to treat the rate of change in frequency of the allele as determined by the *effect* of this allelic substitution on average fitness—in other words, by the slope of the "surface" of mean relative fitness ($\overline{w}$) along the axis of allele frequency.

This brings up the second and even more serious problem: the significance of the mean relative fitness of the population. As we have seen, the mean relative fitness under the usual definition is mathematically essentially a dummy variable, created by holding relative fitnesses constant when there is genetic change occurring in the population. By contrast, if one instead uses absolute fitnesses, the mean absolute fitness ($\overline{W}$) has the explicit meaning of population growth rate. Under such a definition, however, one must acknowledge the artificiality of Wright's assumptions, which lead to the absurd conclusion that evolution will in general involve a constantly increasing population growth rate. There is simply no other alternative, given the models Wright was using. Thus, in answer to the question posed at the end of the last section, it is clear that Wright was in fact seriously misled by his metaphor. The resulting confusion between absolute and relative fitnesses was bequeathed to us in the form of his equations, which are standard in population genetics texts to this day.

In concluding this 1937 paper, Wright for the first time explicitly treated the surface of selective values as the surface of mean relative fitness ($\overline{w}$). The discussion nevertheless merely reiterates the conclusions reached in the 1935 paper, which in turn derive from the nonmathematical 1932 paper:

> In a large population subdivided into numerous small, partially isolated groups, the combination of directed and random divergences in gene frequencies, associated with intergroup selection, gives a trial and error mechanism under which the system of gene frequencies may pass from lower to higher peak values of $\overline{w}$ and the species may evolve continuously even without secular changes in conditions. . . . The combination of partial isolation of subgroups with intergroup selection seems to provide the most favorable conditions for evolutionary advance. (1937, 319)

This version of the surface of selective values or adaptive landscape as the surface of $\overline{w}$ continued to appear in all of Wright's later works (e.g., Wright 1942, 1949, 1956a, 1969, 1977), although he came to recognize that in any of a number of situations (e.g., frequency dependence of genotypic fitnesses, linkage), the population would not necessarily move toward peak values of mean fitness. In fact, in his later works Wright tried to come up with some other fitness function that would continue to increase, even if mean fitness ($\overline{w}$) was decreasing (see chap. 5 in Wright 1969). In doing so, he appears to be rather desperately struggling to maintain the metaphor for evolution that he found so satisfying (Wright 1977, chap. 13; 1988).

What are we to make of this history? I hope it is now clear that the adaptive landscape, as conceived by Wright, is a fundamentally teleological way of thinking about evolution, based on the confusion between differences in relative fitnesses *within* populations and differences in mean absolute fitness (expected rate of increase) *among* populations. Although one type may have a higher relative fitness than another in a current population, this does not mean that the population as a whole would be better off in any sense if composed solely of this type. However, if one is willing to assume constant absolute fitnesses, then a population composed solely of the type with the highest fitness is better off than any other, in the very real sense that it has the highest population growth rate.

None of these observations is particularly new, but it is clear that they have not been taken to heart by the majority of evolutionary biologists. There appears to be a strong desire among many biologists to believe that evolution, guided by natural selection, is necessarily a process of overall improvement (Ruse 1996). As I have tried to show, the standard Wrightian mathematical model (or paradigm) for evolutionary change, incorporating constant relative fitnesses, only helps support this bias.

FISHER AND THE FUNDAMENTAL THEOREM

By the time he wrote the *Genetical Theory of Natural Selection* (1930a), R. A. Fisher (1890–1962) had been thinking long and hard about just how the reconciliation between Darwinism and Mendelism was to be achieved. It was Fisher, after all, who had provided the theoretical basis for this reconciliation by pointing out that the correlations among relatives observed by the biometricians could be readily explained by Mendelian inheritance (Fisher 1918), and who had developed some of the basics of population genetic theory (including the first recognition of the stability of "heterozygote superiority") in 1922. The key to understanding Fisher's reconciliation of Darwinism and Mendelism, and through this the problem with the standard approach to population genetics, largely derived from Wright (and, secondarily, Haldane), is his fundamental theorem of natural selection (Fisher 1930a).

The fundamental theorem formed the main subject of the second chapter of Fisher's *Genetical Theory of Natural Selection* (1930a). It reads simply, stating merely that "*the rate of increase in fitness of any organism at any time is equal to its genetic variance in fitness at that time*" (35). Unfortunately, its derivation is difficult to follow, and the theorem has been a source of bewilderment for countless evolutionary biologists since. A large number of papers have been generated attempting to prove or disprove the theorem or to explain what Fisher really meant (see Edwards 1994; Lessard 1997; Bennett 1999).

On the surface, the theorem seems to state, at the very least, that under natural selection, fitness is expected to be stable or increase with time, since the genetic variance in fitness cannot be negative. Under this interpretation, the fundamental theorem seems rather similar to Wright's metaphor of the adaptive landscape, in which populations continually move toward peak values of fitness. This is in fact the interpretation that Wright himself gave it (e.g., Wright 1935a, 1988). As it turns out, however, this interpretation is a misinterpretation (one that I, too, was guilty of in an early version of this chapter; see Reiss 1989).

Fisher, with his background in mathematics and theoretical physics, would not have taken the word *theorem* lightly. In my *Webster's* second edition (1941), *theorem* is defined as "2. *Math.* **a** A general statement that has been proved or whose truth has been conjectured. **b** In analysis, a rule or statement of relations as expressed in a formula or by symbols; as, the binomial *theorem*; Taylor's *theorem*." There is good reason to think that Fisher conceived of the fundamental theorem as both a general statement that had been proved and as a statement of relations expressed in a formula. Given the definitions that Fisher used, the fundamental theorem is both a statement of what evolution by natural selection *does* and a definition of what it *is*. Fisher tells us this quite clearly in the introduction to the chapter, where he states that his aim is "to combine certain ideas derivable from a consideration of the rates of death and reproduction of a population of organisms, with the concepts of the factorial scheme of inheritance, so as to state the principle of Natural Selection in the form of a rigorous mathematical theorem, by which the rate of improvement of any species of organisms in relation to its environment is determined by its present condition" (1930a, 22).

Indeed, after pronouncing that "*the rate of increase in fitness of any organism at any time is equal to its genetic variance in fitness at that time*," he explicitly warns that "the rigour of the demonstration requires that the terms employed should be used strictly as defined" (35).

The key terminological distinction he makes is between the "average excess" and "average effect" of a gene substitution, a distinction fundamental to Fisher's reconciliation of Mendelian and quantitative genetics (see also Fisher 1941, 1958a). For a pair of alleles affecting any quantitative measurement, he imagined dividing the population into two parts: one including homozygotes for one allele plus half the heterozygotes, and the other the alternative homozygotes plus half the heterozygotes. He defined the *average excess*, a, associated with the substitution of one allele for another as the difference between the mean value of the measurement in the two groups. In contrast, he defined the *average effect*, α, of the gene substitution as the average change of the measurement that would occur with the substitution of one allele for the other in a random member of the population (i.e., holding all other factors constant, and thus removing any correlations with other factors, as well as nonlinear effects such as dominance). He then defined the (additive) genetic variance due to this factor, showing that it was equal to $pq\alpha\alpha$, where p and q were the frequencies of the alternative alleles.

From this definition, the fundamental theorem directly followed. If the rate of increase of organisms is defined in terms of the "Malthusian parameter" m (the logarithmic rate of increase), "the two groups of individuals bearing alternative genes, and consequently the genes themselves, will necessarily either have equal or unequal rates of increase, and the difference between the appropriate values of m will be represented by a [average excess]; similarly the average effect upon m of the gene substitution will be represented by α. Since m measures fitness to survive by *the objective fact of representation in future generations*, the quantity $pq\alpha\alpha$ will represent the contribution of each factor to the genetic variance in fitness" (34, my emphasis). In this initial definition of fitness, the Malthusian parameter m is defined by Fisher a posteriori, as the *measure* of ability of the two genotypic classes of individuals to survive. But Fisher's formulation leaves one with this question: is evolution by natural selection the difference in "average excess" between these groups, the difference in "average effect," or something else?

The answer is not hard to find. In the previous section of his book, Fisher introduced his formulation of quantitative genetics using stature as the example variate. Here he distinguished between x, the stature of an individual, and X, the expected stature based on summing the average *effects* of all genes influencing stature. This quantity is what is now usually referred to as the "breeding value" of an individual (Falconer 1964). The (additive) genetic variance in stature is then merely the variance of X. If x is treated as rate of increase (m) instead of stature, X would be the *expected rate of increase* of an individual based on the additive effects of all genes influencing rate of increase ("fitness to survive and reproduce"). Evolution by natural selection, then, appears to be defined as the existence of additive genetic variation in expected rate of increase within the population. Moreover, the effect of evolution by natural selection on the mean value of expected rate of increase is quantitatively identical to the additive genetic variance of expected rate of increase. This is the fundamental theorem.

The correct interpretation of Fisher's fundamental theorem was first clearly pointed out by Price (1972a); more recently Ewens (1989), Edwards (1990, 1994, 2002), Frank and Slatkin (1992), Lessard (1997), and Gayon (1998) have contributed some additional perspective. Price (1972a) noted, "The main cause of misunderstanding about the theorem is that everyone has supposed that Fisher was talking about the total change dM/dt [in mean fitness or population growth rate] rather than just the fraction of this due to natural selection" (130). Of course, this is perhaps not so surprising if we recall that Fisher called it the fundamental theorem of *natural selection* and that the first sentence of his book stated quite clearly that "Natural Selection is not Evolution" (Fisher 1930a, vii).

What is surprising about Fisher's view is that "he regarded the natural selection effect on M as being limited to the additive or linear effects of changes in gene frequencies, while everything else—dominance, epistasis, population pressure, climate, and interactions with other species—he regarded as a matter of the environment" (Price 1972a, 130). Thus, the fundamental theorem in effect states that the rate of increase in mean fitness (population growth rate) that *would* occur if all genes always had constant additive effects on fitness (expected rate of increase) is exactly equal to the (additive) genetic variance in fitness. Under this interpretation, there is no reason to think that in the circumstances actually obtaining in any real population, mean fitness (population growth rate) will necessarily increase. Price (1972a) again put this very nicely: "What Fisher's theorem tells us is that natural selection (in his restricted meaning involving only additive effects) at all times acts to increase the fitness of the species to live under the conditions that existed an instant earlier. But since the standard of 'fitness' changes from instant to instant, this constant improving tendency of natural selection does not necessarily get anywhere in terms of increasing 'fitness' as measured by any fixed standard" (131).

Although Price was "disappointed that it does not say more" (139) and felt that there was "a challenge here to find a deeper definition of this elusive concept 'fitness' and to give a deeper and sharper explanation of why it increases" (140), the reason why Fisher's statement is in fact the only possible one should be obvious from the earlier discussion of absolute fitness. Since mean absolute fitness is just (expected) population growth rate, and populations cannot evolve a constantly increasing growth rate, mean fitness cannot be expected to increase. Fisher (1930a) himself was quite clear on this:

> Against the rate of progress in fitness must be set off, if the organism is, properly speaking, highly adapted to its place in nature, deterioration due to undirected changes either in the organism, or in its environment. The former, typified by the pathological mutations observed by geneticists, annul their influence by calling into existence an equivalent amount of genetic variance. The latter, which are due to geological and climatological changes on the one hand, and to changes in the organic environment, including the improvement of enemies and competitors, on the other, may be in effect either greater or less than the improvement due to Natural Selection.
>
> Any net advantage gained by an organism will be conserved in the form of an increase in population, rather than in an increase in the average Malthusian parameter, which is kept by this adjustment always near to zero. (46–47)

Thus, although superficially similar, the fundamental theorem embodies a very different vision of evolution than does Wright's adaptive landscape.

In fact, in a 1941 paper (which well repays reading), Fisher explicitly criticized the formulation used by Wright. He insisted again on the importance of the distinction he had made between the *average excess* in fitness of a gene substitution (*a*), which is merely the difference between the fitness (expected rate of increase) of an allele and that of its alternative, and the *average effect* on fitness of a gene substitution (α), which is the effect that would be produced by randomly substituting the allele for its alternative within the population, and which is thus independent of all other correlated factors. It was by equating these two quantities that Wright (1935b) was able to derive his formula for gene frequency change in the case of an intermediate optimum: "the selective disadvantage must be proportional to the momentary net effect of gene replacement on adaptive value" (258). As Fisher noted, the equation of these quantities also lies at the base of Wright's (1937, 310) more general formula for gene frequency change: "Wright's fundamental formula is merely

$a = \alpha$, a relationship which is certainly not true in general, or approximately true in such a variable as the survival value of different genotypes. . . . The attempt to equate a, measuring the selective intensity in favor of a given gene substitution, to the average effect of that substitution on the mean fitness of the population [is] foredoomed to failure just so soon as the simplifying, but unrealistic, assumption of random mating is abandoned" (1941, 57–58).

More important for us, Fisher also objected to Wright's mode of thinking on philosophical grounds: "In regard to selection theory, objection should be taken to Wright's equation principally because it represents natural selection, which in reality acts upon individuals, as though it were governed by the average condition of the species or inter-breeding group. . . . In Wright's equation the whole evolutionary sequence would appear to be governed by the principle of increasing the 'general good'" (58). In light of this comment, it is of some interest that one of the few significant additions of text to the second edition of the *Genetical Theory* (Fisher 1958a) was a section at the end of chapter 2 criticizing thinking in terms of "the benefit of the species." One might also note a passage in a 1936 paper in which Fisher wryly commented, "It has been proposed, alternatively, that the environment, picturesquely renamed the landscape, governs the course of evolutionary change, much as the field of force determines the trajectory of a comet" (59).

Perhaps the clearest statement of the difference that Fisher saw between his views and those of Wright is contained in a 1956 letter to Motoo Kimura, who was working on a paper discussing the fundamental theorem (published as Kimura 1958):

> In considering the original statement of what I ventured to call "the fundamental theorem of natural selection," I had, of course, considered the relation between such a situation and that in which a potential function existed, for my mathematical education lay in the field of mathematical physics. As you realize, I preferred to develop the theory without this assumption, which of course in another aspect is a restriction. Of course, I do not question that the selective intensities acting instantaneously may well be equivalent to those derivable from such a function, but I think it should be emphasized that both changes in time, that is in the environmental *milieu* and in the gene ratios themselves, that is the heritable constitution of the organism, will change this virtual function in a way that cannot be specified in terms of the quantities used in formulating the fundamental theorem.
>
> Of course I realize that Sewall Wright has often argued as though such a potential function must exist, or as though all systems of forces were conservative, and in such systems, the idea of the mean fitness of the population has, I presume, a meaning more absolute or permanent than the mean value of the Malthusian parameter actually in being. (quoted in Bennett 1983, 229)

An excellent discussion of the differences between Fisher and Wright on this issue can be found in a series of papers by A. W. F. Edwards (1992, 1994, 2000, 2002; see also Gayon 1998). He stresses the mathematical problems associated with

Wright's adaptive landscape, particularly the interpretation of evolution as guided by a potential function, $\bar{w}$. He does not note, as I have emphasized here, that such a view of the evolutionary process is fundamentally teleological—because the potential derives from the difference between the current state of the population and a future improved state.

WHAT IS SELECTED?

Because Fisher's formulation of population genetics considers the dual role of Mendelian genes more explicitly than the standard (Wrightian) approach, it is worth examining his conception of natural selection more fully. As we have just seen, for Fisher fitness is a quantitative property of individual organisms of a specific genotype, and he elsewhere rather consistently defines it as their (expected) proportionate contribution to the ancestry of remote future generations. It thus directly connects genes as *causes* of differential phenotypic success with genes as *measures* of differential success. However, when considering individual allelic types (genes), Fisher generally speaks of the average excess, a (not the average effect, α) as the "selective advantage" (1930a, 92) or "selective intensity in favour of a given gene substitution" (1941, 58), and it appears that this selective advantage or intensity must thus be a *resultant* of the effect of the individual gene substitution (α), combined with any correlations with other genes due to nonrandom mating or other causes. In his 1941 paper criticizing Wright, he notes that the average effect and average excess may even be of opposite sign when more than one factor is involved (as must always be the case with fitness). One might say that it is through his concept of average effect that Fisher brings the integrated, functioning organism into his genetical theory of evolution by natural selection.

Nevertheless, in spite of this clear theoretical distinction, which was obviously quite important to him, in practice Fisher was not always so clear on the distinction between average excess and average effect.* For example, when speaking of his theory of dominance modification, he states simply that "all modifications which tend to increase representation in future generations, however indirectly they may seem to act, and with whatever difficulty their action can be recognized, are *ipso facto*, naturally selected" (1930a, 66), but this suggests that the average effect (the "tendency to increase representation in future generations"), not average excess, controls the evolutionary fate of the modifiers.

A particularly good example of his apparent conflation of average excess and average effect comes in the conclusion of his discussion of metrical characters,

---

* To be accurate, I should distinguish between the average excess or effect of a *gene*, which is defined in relation to the mean value of the population, and the average excess or average effect of a *gene substitution*, which is the difference in average excess or average effect between two genes (Falconer 1964). However, this distinction is not critical for the following discussion.

where he appears to switch back and forth between the two; in each case, I have tried to indicate whether he appears to be speaking of average excess (*a*) or average effect (*α*):

> It will now be clear in what way we should imagine the average value of the measurements to be modified by selection whenever such modification happens to be advantageous. If the optimum value is increased all genes, the *effect* [*α*] of which in contrast to their existing allelomorphs is a metrical increase, will be immediately, or at least rapidly, *increased in frequency* [*a*]. In the case of those factors, the *effects of which upon survival* [*α*] can be completely expressed in terms of their *effect* [*α*] upon the measurement in question, the effects of such a change in frequency will be in some cases permanent and in others temporary. Some genes previously *opposed by selection* [*a*?] will be shifted to frequencies at which they are *favoured* [*a*?], and these may *increase to such an extent* [*a*] during the period of selection that when this dies away they may still be *favoured* [*a*?]. Their subsequent *progress* [*a*] will thus tend to *increase the value of the measurement* [*α*] even after it has attained the new optimum. The same applies to mutant genes the frequency of which is increased past the point of maximum *counterselection* [*a*?], into a region in which *selection* [*a*?], while still opposing mutation, is insufficient to check their *increase* [*a*]. In other cases the *increase or decrease in frequency* [*a*] produced by temporary *selection* [*a*?], is itself only temporary, and these, when the new optimum is attained, will *tend to revert* [*a*] to their previous frequencies. (1930a, 111, my emphasis)

The tension between the two definitions of fitness is clear, as is the reason for it: Fisher wants to talk about the phenotypic effects of genes and connect this to their evolutionary fate, without having to consider genes at all other loci, linkage, mating system, and so forth. The control of evolution by a rather teleologically defined optimum is also characteristic (see my discussion of his concept of adaptation in the next section).

However, if not always clear (at least to me) in his discussion of advantage and disadvantage of individual genes, Fisher is quite clear on the general situation one is faced with in trying to quantify natural selection: one must know the genotype of all members of the population at all loci that affect fitness, as well as their expected rates of increase at all ages. Moreover, even if these data could be obtained for an actual existing population, this would not be sufficient—the theoretical definition requires reference to an even larger population, as is apparent from his comments following his definition of average excess:

> This definition will appear the more appropriate if, as is necessary for precision, the population used to determine its value comprises, not merely the whole of a species in any one generation attaining maturity, but is conceived to contain all the genetic combinations possible, with frequencies appropriate to their actual probabilities of occurrence and survival, whatever these may be, and if the average is based upon the statures [or rates of increase, etc.] attained by all these genotypes in all possible

environmental circumstances, with frequencies appropriate to the actual probabilities of encountering these circumstances. (1930a, 30–31)

It is in the context of this precise, but practically unattainable, definition of what natural selection *is* that we should understand his looseness at times in distinguishing average excess versus average effect, and genotypic versus additive genetic variance (see also Fisher 1930a, 44).

There remains one issue to consider: what is the relation of natural selection of traits or characters to Fisher's definition of natural selection in terms of organisms and genes? On the one hand, as we have seen in his discussion of metrical characters, when it comes to traits, Fisher generally is quite happy to talk in Darwinian terms of "advantage" and "disadvantage" and assume that the genetics will follow. On the other hand, he is rather cautious in his willingness to attribute any particular character to natural selection for that character. Thus, we find the following passage in his discussion of sexual selection:

> It has been pointed out in Chapter II [on the fundamental theorem] that a detailed knowledge of the action of Natural Selection would require an accurate evaluation of the rates of death and reproduction of the species at all ages, and of the effects of all the possible genetic substitutions upon these rates. The distinction between one kind of selection and another would seem to require information in one respect infinitely more detailed, for we should require to know not the gross rates of death and reproduction only, but the nature and frequency of all the bionomic situations in which these events occur. . . . In pointing out the immense complexity of the problem of discriminating to which possible means of selection a known evolutionary change is to be ascribed, or of allotting to several different means their share in producing the effect, I should not like to be taken to be throwing doubt on the value of such distinctions as can be made among the different bionomic situations in which selection can be effected. The morphological phenomena may be so striking, the life-history and instinct may have been so fully studied in the native habitat, that a mind fully stored with all the analogies within its field of study may be led to perceive that one explanation only, out of those which are offered, carries with it a convincing weight of evidence. Every case must, I conceive, be so studied and judged upon by persons acquainted with the details of the case, and even so in the vast majority of cases the evidence will be too scanty to be decisive. It would accord ill with the scope of this book (and with the pretensions of its author) to attempt such a decision in any particular case. (1930a, 132)

FISHER'S GEOMETRICAL MODEL OF ADAPTEDNESS

In spite of such cautions, however, in his evolutionary thought Fisher is rather Darwinian throughout. He focuses on adaptive evolution and explicitly ties the origin of "complex adaptations" to past natural selection (1930a, 156; cf. Fisher 1936). Examples given in his book are of course grounded in the genetical theory developed and include standard adaptations of morphology, behavior, and so forth (with the caveats stated earlier), as well as less common examples such as the recessiveness of mutations (due to the past selection of modifiers of dominance), the relative constancy of meristic characters, the splitting of species due to divergent selection, the tendency to avoid hybridization, the evolution of male ornament due to the "runaway process" of sexual selection, the sex ratio, mimicry, and the decay of human civilizations. He generally excludes adaptations of groups, rather than individuals, with the possible exception of sexual reproduction itself. In all of this there is amazing brilliance of application, but no clarification of the problems inherent in the ascription of all adaptedness to past natural selection.

In the end, Fisher is a big fan of Darwin for the precise reason I pointed out previously: Fisher accepts the premise of the argument to design—that adaptedness requires explanation (see his comment in a letter to Nora Barlow of 1948, praising Paley and promoting his influence on Darwin; Bennett 1983, 178). For him, only natural selection can account for adaptedness by "'known', or independently demonstrable, causes" rather than "hypothetical properties of living matter" (1930a, vii). The analogy of design is clear, as in the following paean to natural selection with its talk of ends and desires:

> In addition to showing adaptation, mimetic resemblances manifest a further characteristic of Natural Selection, in the variety of methods by which the same end is attained. This characteristic, analogous to opportunism in human devices, seems to deserve more attention than it has generally received. For the mechanisms of living bodies seem to be built up far less than might be expected by a human inventor, on simple and effective mechanical, physical, or chemical principles. On the contrary every property of the behaviour of matter, however odd and extraneous it may seem, seems to have been pounced upon as soon as it happened to produce a desirable effect. (1930a, 156)

This passage anticipates by almost forty years François Jacob's (1977) well-known analogy of natural selection as a "tinkerer" and is similarly teleological, though perhaps only in a heuristic way.

Where teleology enters most prominently into Fisher's view is his conception of the nature of adaptedness, as developed in his famous multidimensional geometrical model of adaptedness. Fisher begins rather well: "an organism is regarded as adapted to a particular situation, or to the totality of situations which constitute its environment, only in so far as we can imagine an assemblage of slightly different situations, or environments, to which the animal would on the whole be less well adapted; and equally only in so far as we can imagine an assemblage of slightly different organic forms, which would be less well adapted to that environment" (1930a, 38). There is nothing objectionable here—if environmental adaptedness means anything, it must indeed mean a local fit between an organism and its environment. However, he then introduces his geometrical model, in which the degree of adaptedness is represented by "the closeness with which a point A approaches a fixed point O." It is this model that Fisher uses to argue that the probability of improvement has a maximum value of 0.5 for small changes and drops off rapidly as the magnitude of the change is increased, from which he concludes that the mutations important in evolution are likely to be those of small effect.

Unfortunately, teleology enters Fisher's formulation because adaptedness is represented as relative to a fixed optimum, which then becomes the goal of selection. Thus, he gives the example of a microscope, asking how likely it is that a change in the position of the lenses (or in the refractive index of the glass, etc.) would improve its functioning. Obviously, a small change in any of these parameters is more likely to improve its functioning than a large change. But in the case of a microscope, the function can indeed always be assessed relative to a particular fixed goal or optimality criterion—for example, the best resolution, with the least loss of light. We are assessing its performance as a microscope, and we know what we want a microscope to do. In evolution, however, there is no fixed goal. A change in the structure of an organism, or in its environment, must be seen as changing the conditions of the problem. The eyespot of a flatworm and the image-forming eye of a vertebrate may both be analyzed as adapted to the lifestyle and environment of the organisms. From this perspective, it is not so clear that large changes will always be detrimental (though of course Fisher is certainly right that *all* changes are more likely to be detrimental than beneficial).

Still, in the end, Fisher's teleology, as expressed in his model of adaptedness, is relatively harmless compared with Wright's, as expressed in his metaphor of the adaptive landscape. This is precisely because he was too good a mathematician to lack clarity on the issue of absolute versus relative fitness. Thus, the fundamental theorem remains a valid insight into the nature of evolution by narrow sense natural selection, whatever one might think of its importance, while the adaptive landscape is merely a teleological metaphor for evolution.

THE REEMERGENCE OF THE ADAPTIVE LANDSCAPE

After a period of decline, the metaphor of the adaptive landscape appears to making a comeback. For example, Steve Arnold and coworkers (Arnold, Pfrender, and Jones 2001; Arnold 2003) have argued for the usefulness of Simpson's phenotypic adaptive landscape, and a book-length treatment promoting this view in the context of theoretical morphospaces was recently published by George McGhee (2007). While it is true that the variation in fitness in a population at any given time can be reasonably portrayed as such a landscape, as soon as one tries to picture the evolution of a population on this landscape, one runs into the same problems as with the gene frequency version, in which evolution is viewed as a process leading to higher and higher "adaptedness" under the influence of natural selection (fig. 7.7).

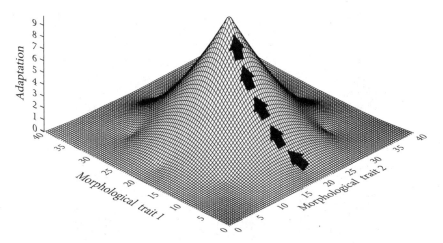

FIGURE 7.7. Phenotypic adaptive landscape (McGhee 2007, fig. 2.4). Note that the vertical scale is "adaptation," which is undefined. Later in the book, it becomes clear that "adaptation" here means the efficiency of performance of some specific function, such as filter feeding. © Cambridge University Press 2007. Reprinted by permission of Cambridge University Press.

Much more common today is the use of the adaptive landscape in the context of protein and nucleic acid evolution, a use dating back to John Maynard Smith (1970) and more recently promoted by Stuart Kauffman and his colleagues (Kauffman and Levin 1987; Kauffman 1993). For any given length protein or nucleic acid, one can define a "sequence space" consisting of all sequences of a given length and assign each a fitness. This view is clearly allied to Wright's original (1932) gene combination version of the adaptive landscape (cf. Gavrilets 2004), though here the combinations are occurring within a single gene. How is fitness measured in such models? Kauffman (1993) begins by defining fitness as "any well-defined property" such as "the capacity of each protein in protein space to catalyze a specific reaction under specified conditions" (37). As with the phenotypic landscape, such a definition is certainly valid for a given population at a given time, where the specific reaction one assumes must be catalyzed is one that is already being catalyzed. To translate this into a view of the evolutionary process, however, one must assume that the need to catalyze a particular reaction is a goal toward which the evolutionary process is directed, a step that Kauffman is not long in taking (1993, chap. 4).

One of the most striking examples of such a teleological view of the evolutionary process in the context of protein evolution comes in a recent paper by Frank Poelwijk and colleagues (2007), entitled "Empirical Fitness Landscapes Reveal Accessible Evolutionary Paths." The authors looked at the possible trajectories between an ancestral form of bacterial β-lactamase and a derived form differing by five mutations, which confers resistance to the antibiotic cefotaxime. They assumed

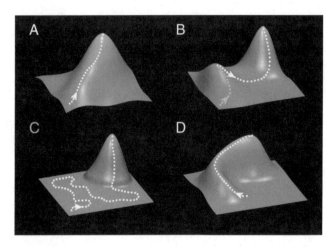

FIGURE 7.8. Mutational adaptive landscape (Poelwijk et al. 2007, fig. 1). The legend reads, "Schematic representations of fitness landscape features. Fitness is shown as a function of sequence: the dotted lines are mutational paths to higher fitness. A, Single smooth peak. All direct paths to the top are increasing in fitness. B, Rugged landscape with multiple peaks. The yellow path has a fitness decrease that drastically lowers its evolutionary probability. Along the blue path selection leads in the wrong direction to an evolutionary trap. C, Neutral landscape. When neutral mutations are essential, evolutionary probabilities are low. D, Detour landscape. The occurrence of paths where mutations are reverted shows that sequence analysis may fail to show mutations that are essential to the evolutionary history." Reprinted by permission from Macmillan Publishers Ltd., *Nature* (*London*) 445:383–386, © 2007.

that this one particular evolved form of β-lactamase is the goal toward which the evolutionary process is necessarily directed when bacteria are exposed to cefotaxime and that natural selection would find this form; their task was merely to show the paths selection could use to do so. The teleology inherent in this assumption is beautifully shown by their schematic representation of the adaptive landscapes involved (fig. 7.8).

In the context of speciation theory, Sergey Gavrilets (2004) has recently produced an elegant synthesis of mathematical models of speciation grounded in the idea of "fitness landscapes." Here he provides an excellent overview of the many versions of the adaptive landscape that have appeared, beginning with Wright (1932). Gavrilets is able to avoid most of the problems I have identified because he insists that the adaptive landscape is only a metaphor, because he is concerned mainly to distinguish "high-fitness" or viable genotypes from "low-fitness" or inviable ones and, most importantly, because his preferred model of evolution on "holey landscapes" does

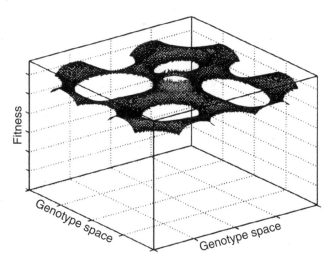

FIGURE 7.9. Holey adaptive landscape (Gavrilets 2003, fig. 5b). Note that the fitnesses of the populations all lie in a narrow band near 1.0, so that evolution on such a landscape does not involve any overall improvement in fitness. Reprinted by permission of Wiley-Blackwell. © Wiley Blackwell, 2003.

not involve the evolution of improved fitness over time. Gavrilets's holey landscapes are essentially current fitness landscapes dragged through evolutionary time to create ridges of high fitness along which a population can evolve (fig. 7.9). As he points out, this is a very different vision than Wright's.

· · ·

It is ironic that the teleological view that evolution by natural selection involves constantly improving adaptedness—implicit in Wright's metaphor of the adaptive landscape—has been associated by many, including Wright himself, with Fisher's fundamental theorem of natural selection. Fisher's fundamental theorem of course says no such thing (Edwards 1994). I would argue that even at the level of metaphor the fundamental theorem succeeds better than the adaptive landscape (cf. Burt 1995). It is readily understandable at the metaphorical level (if difficult at the mathematical) that evolution by narrow sense natural selection acts to increase mean absolute fitness (population growth rate), while other processes (environmental change, increased competition due to increases in population size, change in gene and genotype frequencies, mutation, etc.) act against this tendency, so that any existing population persists by virtue of a balance between all these processes. In the following chapter, we will further examine some of the quantitative aspects of this balance, as they relate to population growth and the concept of genetic load.

8

Population Growth, Genetic Load, and the Limits of Selection

Evolution Happens
—BUMPER STICKER

One of the main contentions of this book is that by separating overall adaptedness from existence, and linking this separation to a teleologically conceived natural selection, Darwin's formulation of his theory has obscured—or at least needlessly hindered understanding of—the real relationships between organisms and their environment. In the previous chapter, I showed how this separation led to the teleological metaphor of the adaptive landscape popularized by Sewall Wright, tied to the confusion between absolute and relative fitness. In this chapter, I turn to another aspect of the relation between absolute and relative fitness: the relation of overall population growth rate to the variance in relative rate of increase in a population (generally known as the opportunity for selection or standardized variance in reproductive success). What I am interested in here is the *natural history* of differential survival and reproduction in Mendelian populations, and how this relates to whatever natural selection is occurring.

I begin by considering the limits to the total amount of natural selection that can occur in a population. Such limits have been recognized at least since Darwin (1859) and have been elaborated upon by Crow (1958) and Robertson (1968), among others. Historically, such limits have been connected with the concept of "genetic load," defined as the depression of mean population fitness due to the presence of individuals of lower than optimum fitness within the population (Crow 1958). I review the history of the genetic load concept. I show that Haldane's figures for the "cost of selection" have little to do with anything really going on in a population. In contrast, I show that Muller's arguments about our "load of mutations" are basically correct. All of this discussion, as it turns out, is most usefully framed in terms of the variance in relative rate of increase (the "opportunity for selection"), which allows one to talk about the limits to selection without invoking any hypotheses about the relation between such selection and the health of the population (as measured by its mean absolute rate of increase). Nevertheless, we will see that the variance in relative rate of increase is indeed intimately related to population growth (and decline).

The concept of genetic load is not merely of historical interest; it is central to much current thinking in conservation biology (as considered in greater detail in chap. 12). I will try to show that the confusion between absolute and relative fitness has greatly increased the difficulty of dealing with the issues surrounding genetic load, even if, ultimately, these issues can only be addressed empirically.

VARIANCE IN RATE OF INCREASE:
THE OPPORTUNITY FOR SELECTION
(AND DRIFT) IN NATURAL POPULATIONS

It is obvious that the total variation in rate of increase (realized fitness) among individuals provides an upper limit to how much selection (differential survival and reproduction of classes of individuals) is possible in a population. If there is no variation in rate of increase, there is clearly no selection going on. The converse does not hold, of course: there can be much variation in rate of increase without any of it being selective among *classes* of individuals. This rather obvious relation was not formulated mathematically until the seminal paper of James Crow (1958), though implicit in Fisher's work. In this paper Crow defined an "Index of Total Selection," I, as the total variance in absolute rate of increase over the mean absolute rate of increase squared, or, equivalently, the total variance in relative rate of increase. I prefer to call this the *standardized variance* in rate of increase.

This quantity was introduced into the study of sexual selection by Michael Wade (1979). After consulting with Crow (and Russ Lande), it was renamed the "opportunity for selection" by Arnold and Wade (1984b), who observed that "fitness variance places an upper bound on the force of selection that can act on any phenotypic character" (713). I might note that it likewise places an upper limit on the "force of selection" that can act on genotypes (at least by differences in organismal survival and reproduction).

In the years immediately following Arnold and Wade's (1984a, 1984b) work, much interest focused on the standardized variance as a metric for comparing natural populations, culminating in the appearance of Tim Clutton-Brock's edited collection of studies on *Reproductive Success* (1988c). The standardized variance (opportunity for selection), I, was used by many of these studies as an estimate of the variability of reproductive success (Clutton-Brock 1988a). However, in reviewing these studies, several of the authors were critical of the validity and/or utility of Arnold and Wade's measure, or at least their development of it. David Brown (1988) suggested an alternative partitioning of the standardized variance, which we will examine in more detail later. Alan Grafen (1988) noted that the standardized variance in reproductive success, while indeed forming an upper limit to selection, "measures something to do with selection in progress, not with adaptive value" (462). On this basis, he argued that the "opportunity for sexual selection" is uninteresting, because it answers the question about "selection in progress" but tells us nothing about what we really want to know—the "adaptive value," which depends on *past* selection.

Clutton-Brock himself (1988b) used I as an index to compare the variation in breeding success among studies in the book, noting that for females, I was lowest for zygotes produced, intermediate for surviving young, and highest for recruits to the adult breeding population, but that no such pattern was evident for males. In contrast, the ratio of male to female standardized variance (I_m/I_f) did show a clear dependence on the sex ratio of breeding adults, rising with increasing number of females per male. He concluded, however, like Grafen, that this relationship is not of interest, but in his case because it is *not* directly related to current selection on phenotype: "sex differences in morphology or behavior should instead be associated with differences in the extent to which particular behavioral or morphological traits affect fitness in the two sexes" (476).

While this book was in press, an even more damaging assessment of the utility of the opportunity for selection metric appeared (Downhower, Blumer, and Brown 1987). Among their other criticisms, Jerry Downhower and his colleagues (1987) noted that if one assumes a random (Poisson) distribution of rate of increase, which has a mean equal to its variance, the standardized variance in rate of increase should be the inverse of the mean rate of increase (population growth rate) (box 8.1). Moreover, they showed that a dependence of standardized variance on mean rate of increase does occur in data for a wide variety of animal species, when each is compared across the same part of the life cycle, although this standardized variance is typically at higher than Poisson levels. They considered this a problem. They granted the index any validity only for comparing males and females in the same population, and thus experiencing the same population growth rate, and even then they noted that it was necessary to take the sex ratio into account (so that the mean number of offspring of males and females is the same).

These criticisms certainly had their effect, although in the subsequent years, some work has continued to include the opportunity for selection as a measure of the total amount of selection that can occur, particularly in comparisons between the sexes in the context of sexual selection theory. The use of this index was vigorously defended against the criticisms of Downhower et al. (1987) by Shuster and Wade (2003), who pointed out that there is nothing surprising in the opportunity for selection being sensitive to population growth rate, because the basic definition of opportunity for selection has population growth rate in

BOX 8.1. VARIANCE IN RATE OF INCREASE
AND THE POISSON ASSUMPTION

If we assume a Poisson distribution of descendant number, with mean number of descendants equal to the variance in descendant number, then the *standardized variance in rate of increase* (I) is

$$I = \text{var}(r) = \text{var}(R)/\overline{R}^2 = (1/\overline{R})\text{var}(R)/\overline{R} = 1/\overline{R}, \qquad (8.1a)$$

where R is the absolute rate of increase (realized fitness) of each lineage, and r is the relative rate of increase. In other words, the standardized variance is the inverse of the population growth rate (Downhower et al. 1987).

In the case of biparental reproduction, if the raw number of offspring is k per individual, but each offspring counts only as ½, then $R = k/2$, and

$$\text{var}(R) = \text{var}(k/2) = (1/4)\text{var}(k). \qquad (8.1b)$$

Yet the standardized variance is not changed, because the correction factor enters both top and bottom of the equation.

the denominator. This cannot therefore be considered a problem with the metric. Nevertheless, the standardized variance or opportunity for selection is not a common metric in evolutionary biology, as evidenced by the fact that it does not appear at all in three popular recent textbooks (Ridley 2004; Futuyma 2005; Freeman and Herron 2007). This omission seems rather strange when one considers that the standardized variance certainly does limit the total amount of selection that can occur, and it is much easier to *measure* than any such selection that may be occurring.

Although the standardized variance is a quantity that is calculable for any natural population in which individual rate of increase (realized fitness) can be estimated, it is one that many studies of natural selection do not report. A common attitude seems to be that expressed by Brodie, Moore, and Janzen (1995): "This parameter is useful as an upper bound on how strong selection could be, but is not especially satisfying because it does not describe any relationship between fitness and a trait" (313). Hersch and Phillips (2004) found that of sixty-three studies included in the survey of natural selection by Kingsolver et al. (2001), only seventeen reported variance (or standard deviation) in rate of increase. Nevertheless, they demonstrated that this would indeed be an important metric to include in such studies, because the power to detect selection depends on the standardized variance (opportunity for selection).

STANDARDIZED VARIANCE
VERSUS POPULATION GROWTH: DATA

Since the publication of the collection of studies on *Reproductive Success* by Clutton-Brock (1988c), and the overview by Downhower et al. (1987), a large number of studies have in fact examined variation in survival and reproductive success among individuals for some or all of the life cycle. What do these figures for variance in survival and reproduction look like? With the caveat that many of these studies do not strictly account for the distribution in offspring production over time, simply summing them over the entire lifetime or some fraction thereof (clearly not appropriate for age-structured populations), an overview from a variety of studies is given in table 8.1 (see also fig. 8.1). In each case, where data from more than one period or situation were available, I have included the two with the highest and lowest population growth rate. I also included the figures from Howard's (1979) classic study of bullfrog reproduction, even though only one year was available, because of the historical role it played in Arnold and Wade's (1984a) work.

One simple mathematical point is worth making at the outset. In examining data on rates of increase, it is important to be clear on whether one is considering uniparental or biparental reproduction. With obligatory biparental reproduction, each offspring should of course be credited as a half offspring for each parent, so that when each individual on average has two offspring, the mean population growth rate will be 1.0. This allows survival rates and reproductive rates to be expressed in the same currency (box 8.1, eq. 8.1b). In the case of intermediates between biparental and uniparental reproduction, common in plants (Vogler and Kalisz 2001), each offspring should be counted either as 1 or $\frac{1}{2}$, depending on its selfed or outcrossed status. Because the opportunity for selection only involves time (its units are time^{-2}), its value will be the same however one counts offspring. Nevertheless, population growth rate is affected, and thus the appropriate correction has been made to all the data presented in table 8.1.

In these studies, absolute rates of increase ranged from a low of 0.035 (one in thirty survive) for seedling survival in the zebrawood, to a high of 13,050 for seeds produced per tree, also in the zebrawood (the only species for which two different life history segments were examined). The range of standardized variance (opportunity for selection) is also almost encompassed by the zebrawood data, with a value of 27.6 for seedling survival but only 0.85 for seeds produced per tree. The general pattern that low mean absolute rate of increase is associated with high standardized variance in rate of increase holds in many other species as well. For example, in song sparrows females of the 1978 cohort had a mean number of offspring surviving to reproductive age of just 0.1, whereas the 1976 cohort had 2.8. The corresponding variances are 11.0 and 0.43. Likewise, in a laboratory study of mating success in *Drosophila*, series A had a mean mating success of 3.52; series C, 0.59.

TABLE 8.1 A Survey of Mean and Variance in Rate of Increase

VERTEBRATES

Species	Reference	Subgroup Examined	Measure of Increase	Period	Year, Sex, or Other Qualifier	Mean Absolute Rate of Increase ($\bar{R}$)	Variance in Absolute Rate of Increase (var(R))	Standardized Variance (I = var(r) = var(R)/$\bar{R}^2$)
Homo sapiens, human (Basque)	Alfonso-Sanchez et al. 2004	30-year "cohort" of females (at birth)	Total number of offspring (live births) (*½)	Lifetime	1860–1889 1920–1949	1.01 1.70	2.47 2.53	2.42 0.88
Ovis aries, Soay sheep	Coltman et al. 1999	Cohort of males (at birth)	Total number of offspring (live births) (*½?)	Lifetime (birth to death) = 7 years	1988 1989	0.25 3.25	0.66 12.15	10.54 1.15
Ovis aries, Soay sheep	Coltman et al. 1999	Cohort of males (at birth)	Survival	First winter	1988 1989	0.05 0.88	0.05 0.11	19.84 0.14
Mirounga angustirostris, northern elephant seal	Le Boeuf and Reiter 1988	Cohort (of weaned pups)	*Males* Estimated number of weaned offspring pups (*½) *Females* Number of weaned pups (*½)	Lifetime (weaning to death) = 14 years	*Males* 1965 1966 *Females* 1973 1974	4.98 0.21 0.59 0.85	173.41 1.16 0.61 0.82	7.00 26.00 7.01 4.58

Species	Reference	Cohort	Measure	Duration	Year			
Tamiasciurus hudsonicus, red squirrel	McAdam and Boutin 2003	Cohort (at birth)	Survival	200 days (one winter)	1993 1999	0.65 0.03	0.23 0.03	0.54 32.90
Agelaius phoeniceus, red-winged blackbird	Weatherhead and Boag 1997	Adult males with territories	Number of fledglings sired ($\times\frac{1}{2}$)	One breeding season	1988 1991	1.75 2.71	3.50 2.74	1.14 0.37
Tetrao tetrix, black grouse	Kruijt and de Vos 1988	Adult males with territories	Number of copulations	One breeding season	1969 1980	10.00 0.30	72.00 0.90	0.72 10.00
Melospiza melospiza, song sparrow	Smith 1988	Cohort	Number of offspring surviving to breeding age of birds surviving to breeding age ($\times\frac{1}{2}$)	Lifetime (1 year old to death) = 5 years	*Females* 1976 1978 *Males* 1976 1978	1.40 0.05 0.85 0.30	0.84 0.03 0.68 0.74	0.43 11.00 0.94 8.22
Rana catesbeiana, bullfrog	Howard 1979 Arnold and Wade 1984b	All males (adult)	Number of hatchlings sired ($\times\frac{1}{2}$)	One breeding season	1976	5,582	72,830,000	2.34

(continued)

TABLE 8.1 (continued)

Species	Reference	Subgroup Examined	Measure of Increase	Period	Year, Sex, or Other Qualifier	Mean Absolute Rate of Increase ($\bar{R}$)	Variance in Absolute Rate of Increase (var(R))	Standardized Variance ($I = \text{var}(r) = \text{var}(R)/\bar{R}^2$)
			INSECTS					
Pollistes annularis, paper wasp	Queller and Strassman 1988	Female foundresses of nests (cohort)	Reproductive females produced (*½*)	One breeding season	1977 1978	1.18 2.91	8.93 40.73	6.41 4.81
Drosophila melanogaster, fruit fly	Joshi et al. 1999	Adult males (in laboratory populations)	Copulations > 5 min (in experimental tests) Series A—added at 0 hour Series C—added at 8 hours	Time of intro to 14 hours (averaged over five replicates for each series)	Series A Series C	3.52 0.59	1.82 0.57	0.16 1.64
			MARINE INVERTEBRATES					
Macoma balthica, Baltic tellin	Honkoop and van der Meer 1997	Adults (cored)	Number of eggs spawned (*½*)	One season	Site B 1995 1996	7,977 16,848	42,928,000 67,938,000	0.67 0.24

Species	Reference	Stage	Measure	Time	Category			
Datura stramonium, Jimson weed	Stone and Motten 2002	Mature plants (annuals)	Seeds produced (selfed)	One season	FS3 pop. IR4 pop.	20,200 1,100	337,360,000 4,500,000	0.83 3.72
Microberlinia bisulcata, African zebrawood tree	Green and Newberry 2002	Mature trees	Seeds produced (*½)	One season	1995	13,050	145,200,000	0.85
Microberlinia bisulcata, African zebrawood tree	Green and Newberry 2002	Seeds (cohort)	Survival to 31-month seedling stage	31 months	1995	0.04	0.03	27.59
Linanthus parryae, desert phlox	Schemske and Bierzychudek 2001	Adults	Seed set (*½)	One season	*Blue* 1991 1993 *White* 1991 1993	130.06 1.39 151.21 0.56	13,914 35 16,884 9	0.82 18.15 0.74 29.08
Ceanothus cuneatus, buck brush	Deveny and Fox 2006	Adults	Seed production (*½)	One season	Browsed Unbrowsed	661 37,568	613,089 1,405,100,000	1.40 1.00

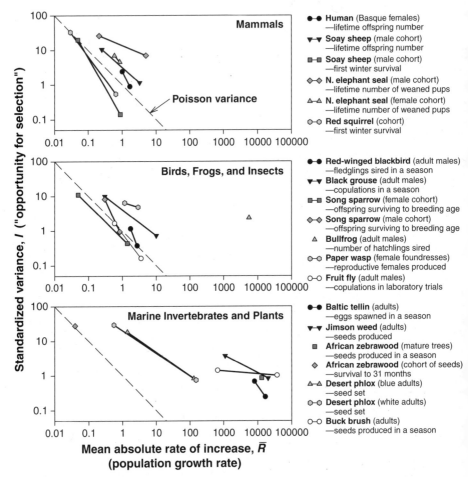

FIGURE 8.1. Trends in the standardized variance of rate of increase (I) with respect to mean rate of increase ($\bar{R}$) or population growth rate, for the species and life history periods listed in table 8.1. Note the general negative relation between these two quantities within populations, and the tendency for less variation in standardized variance at higher population growth rates. The Poisson expected variance is shown by the dashed line.

The corresponding variances are 0.16 and 1.64. In fact, of seventeen comparisons between similar measures within the same species, *all* of them showed a similar relation (fig. 8.1). This empirical relation supports the mathematical connection between the mean absolute rate of increase (population growth rate) and the standardized variance in rate of increase (realized fitness variance) noted by Downhower et al. (1987). I would now like to look at this mathematical connection more closely.

STANDARDIZED VARIANCE VERSUS POPULATION
GROWTH: MATHEMATICAL CONSIDERATIONS

As noted by Downhower et al. (1987), if one assumes a Poisson distribution of rate of increase in a population, the standardized variance in rate of increase should be inversely proportional to the population growth rate. To further examine the question of the relation between population growth and standardized variance, it is helpful to begin with simple survival, rather than survival and reproduction. This removes any issues about the precise form of the distribution of rates of increase, because only two values are possible: survival (1) and death (0). Let us assume that we are examining a population over a given time period, and x_0 individuals die, x_1 individuals survive (box 8.2). For this simple situation, we find that the standardized variance is simply equal to one less than the reciprocal of survival rate (box 8.2, eq. 8.2c). This number is zero at complete survival and increases rapidly as survival rate declines (fig. 8.2). Moreover, the standardized variance is simply the number dying/number living (box 8.2, eq. 8.2d), as first shown by Crow (1958). If 1 percent of a population survives, then ninety-nine die for every one that lives, and the standardized variance is 99 (cf. Crow 1958; Barrowclough and Rockwell 1993; Burt 1995). If only one dies for every one that lives (i.e., half survive), the standardized variance is only unity. Thus, considered only with respect to survival, the empirical trend for low mean rates of increase to be associated with high standardized variance in rate of increase is indeed a mathematical necessity.

One further relation brings home the apparent naturalness of the standardized variance as a measure of overall differential survival and reproduction. In the case of pure survival (or equal reproduction among all survivors, which comes to the same thing), the *relative* rate of increase of the survivors itself increases as mean survival decreases, because they are having to balance out a greater and greater number of deaths for the mean to remain unity. As the relative rate of increase of the survivors grows, so does the standardized variance. In fact, the relation is linear (box 8.2, eq. 8.2e).

How do these simple relations change if we consider differential reproduction as well? I will look only at asexual populations, because the situation is basically the same (if more complicated) in sexual systems (see chap. 7; for a good overview of the application of the opportunity for selection to sexual systems, see Shuster and Wade 2003). As we have just seen, for a given survival rate, the case where all individuals that survive have the same absolute rate of increase (same number of offspring, etc.) has the same standardized variance as the case of pure survival. If the population is replacing itself, then the part of the population surviving to reproduce does so at a rate inversely proportional to the mean survival rate. If there is variation in reproduction across the population, we must also consider the precise

BOX 8.2. STANDARDIZED VARIANCE IN RATE
OF INCREASE DUE TO SURVIVAL ONLY

Assume that of the total number of individuals existing at the beginning of a period, x_0 die and x_1 survive. Then the mean absolute rate of increase (survival rate) is

$$\overline{R} = \frac{x_1}{x_0 + x_1}, \tag{8.2a}$$

and the variance in absolute rate of increase is

$$\mathrm{var}(R) = \overline{R}(\overline{R} - 1)^2 + (1 - \overline{R})(\overline{R} - 0)^2 = \overline{R} - \overline{R}^2 = \overline{R}(1 - \overline{R}). \tag{8.2b}$$

The standardized variance is the ratio of this number to the mean rate squared, or

$$\mathrm{var}(r) = (\overline{R} - \overline{R}^2)/\overline{R}^2 = (1 - \overline{R})/\overline{R} = 1/\overline{R} - 1. \tag{8.2c}$$

Moreover, in this case of simple survival, we find that if we substitute equation 8.2a back into equation 8.2c,

$$\mathrm{var}(r) = x_0/x_1. \tag{8.2d}$$

Finally, because the mean relative rate of increase of the survivors is the inverse of the overall survival rate

$$\overline{r}_{\mathrm{survivors}} = \frac{x_1 + x_0}{x_1} = 1 + \frac{x_0}{x_1}, \tag{8.2e}$$

we have the simple linear relation:

$$\mathrm{var}(r) = \overline{r}_{\mathrm{survivors}} - 1. \tag{8.2f}$$

distribution of rates of increase, which can only be determined empirically. Nevertheless, we can establish limits to the possible values of the standardized variance based solely on mathematical considerations.

For any given level of population growth (mean absolute rate of increase), it is clear that standardized variance in rate of increase is maximal for the case where the population is reproduced from a single individual (see Hedrick 2005). This provides an upper limit to the level of variance in rate of increase that can occur. If a population of 1,000 is replacing itself, 999 die without leaving offspring, and the remaining one has 1,000 offspring (requiring asexual reproduction), the standardized variance is 999.

BOX 8.3. MATHEMATICAL CONSTRAINTS
ON THE STANDARDIZED VARIANCE

If the maximum number of descendants (i.e., rate of increase) that can occur among individuals in a population is R_{max}, then the *maximum standardized variance* possible (I_{max}) occurs when all individuals that reproduce do so at the maximum level, because this gives the greatest fraction of deaths. This situation is similar to that of pure survival (see eq. 8.2c):

$$I_{max} = var(r) = \frac{R_{max}}{\overline{R}} - 1. \qquad (8.3a)$$

Likewise, for decreasing populations, the *minimum standardized variance* for a given rate of population decrease occurs when all of the individuals who survive reproduce at the minimum rate (R_{min}), because this gives the lowest fraction of deaths. Thus,

$$I_{min} = var(r) = \frac{R_{min}}{\overline{R}} - 1. \qquad (8.3b)$$

Finally, another sort of minimum is set by the total *fraction of deaths* occurring at any level of population growth, because the case of pure survival or equal reproduction again sets a lower limit for the standardized variance. This minimum is determined by equation 8.2d, where x_0 is now the fraction of lineages that die and x_1 the fraction that survive:

$$I_{min} = \frac{x_0}{x_1}. \qquad (8.3c)$$

The situation is exactly the same in a large population if there is a maximum number of descendants possible per individual (box 8.3, eq. 8.3a). For example, consider instead a population of 1,000,000, where 1 in 1,000 survive to reproduce, the maximum number of descendants is 1,000, and all reproducing individuals achieve this maximum (no variation among them). In this case, the standardized variance in survival and reproduction is still 999. If instead individuals reproduce at less than maximum rate (still assuming equal reproduction among reproducers), then the fraction of individuals dying without reproducing must decrease, which can only reduce the variance. For example, if 2 of every 1,000 individuals survive to reproduce, and each has 500 offspring, the standardized variance is now only 499. The maximum standardized variance possible at different population growth rates has been plotted for several values of the maximum possible number of descendants (R_{max}) in figure 8.2.

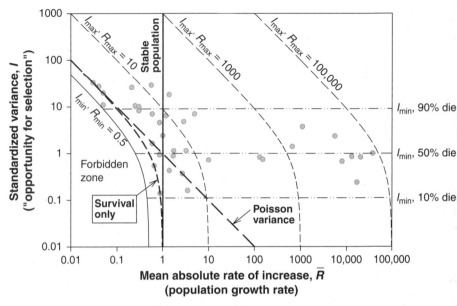

FIGURE 8.2. Mathematical constraints on the standardized variance. At any given population growth rate, the Poisson expected variance is shown by the straight line at $I = 1/\bar{R}$. The curve for pure survival (viability only) converges with Poisson variance for shrinking populations but has a maximum at 1.0 (complete survival). At any given population growth rate, the maximum standardized variance (I_{max}) is determined by the maximum possible number of descendants of an individual (R_{max}). The minimum standardized variance (I_{min}) for shrinking populations is set by the smallest possible number of descendants (R_{min}), defining a "forbidden zone" at the lower left. Another sort of minimum (I_{min}), constant across varying population growth rates, is set by the total fraction of deaths (zero values for rate of increase) across the population. It is graphed here for values of 10, 50, and 90 percent overall survival. Data points from figure 8.1 are plotted in the background to facilitate comparison.

For decreasing populations, a similar argument applies to the minimum possible standardized variance. For example, if one looks over a single generation, it is impossible for an individual who reproduces at all to have less than one offspring, which, when corrected for biparental reproduction, gives a minimum rate of increase (R_{min}) of 0.5. This minimum (box 8.3, eq. 8.3b) is also plotted in figure 8.2, where it defines a "forbidden zone" at the lower left of the graph: values of population growth and standardized variance that cannot occur together.

Another sort of minimum is set by the total fraction of deaths occurring at any level of population growth, because the case of pure survival or equal reproduction sets a lower limit for the standardized variance (box 8.3, eq. 8.3c). For example, in the case described earlier, if one of the two individuals that reproduces has 250 and the other 750 offspring, rather than each having 500, the standardized variance is increased from 499 to 624. This minimum is also plotted in figure 8.2, for survival rates of 90 percent, 50 percent, and 10 percent.

We thus have maximum and minimum values possible for standardized variance based on maximum and minimum rates of increase (descendant number) and the overall survival rate observed over a time period. These values give a fairly tight range for low maximum rates of increase. For example, in a stable population if the maximum number of descendants per individual is 10 and survival rate is 50 percent, the standardized variance necessarily lies in the range from 1 to 9 (fig. 8.2). Conversely, if the maximum number of descendants is 100, then the standardized variance can range between 1 and 99. The value realized depends on the exact distribution of rates of increase.

Thus, both in the case of a Poisson distribution of offspring number and in the case of pure survival, the standardized variance in rate of increase depends on the overall rate of population growth. The mathematical relation that exists between absolute rates of increase and standardized variance in rate of increase seems like it should be highly relevant to the question of how evolution by natural selection occurs. It is of course the standardized variance that is indeed the "opportunity for selection" that sets the upper limit for the rate of evolution by natural selection (in the narrow sense). Because limits and costs to selection form the basis for the concept of genetic load, this is precisely the background we need to examine the question of genetic loads from a somewhat different perspective.

GENETIC LOAD: THE DARK SIDE OF NATURAL SELECTION

We have already had occasion to touch on genetic load in several contexts, but I would now like to delve a little deeper. The background developed in chapter 7 will allow us some insight into the basic mathematical issues, because they fundamentally concern the relationship of absolute and relative rate of increase (realized fitness). For the history of the concept of genetic load, we are fortunate in having Bruce Wallace's (1991) excellent survey. Wallace, a student of Theodosius Dobzhansky's, is not only aware of the historical dimensions of the debate but has himself been a key player.

The concept of genetic load had its origins in 1937, when Haldane published a paper on "The Effect of Variation on Fitness." Considering populations in equilibrium, he noted that "every species observed with sufficient care has been found to include members less fit than the average and whose lack of fitness is heritable." The presence of less fit individuals he considers the result of deleterious genes, which are present either because of recurrent mutation or because they are advantageous in other combinations. The fundamental issue for Haldane was "the effect of such deleterious genes on the fitness of the species" (338).

The "fitness" he considers is absolute rate of increase, assigned to genotypes: "half the mean number of progeny left by an individual of that genotype." Moreover, he is quite clear that "the mean fitness of all members of a species must always be very close to unity, if we average over any length of time" (339). Nevertheless, he notes that "in any species some genotypes have a fitness less than unity, ranging to zero in the case of lethal genes and genes causing complete sterility. So it is clear that the fitness of the standard type containing no deleterious genes must exceed unity. A population composed of such a type would of course increase until, owing to its pressure on the means of subsistence, the fitness was again reduced to unity" (340). Haldane thus assumes at the outset that absolute rate of increase of a particular genotype as determined in a particular population would be the same in a different population, as long as population number was the same. In other words, he allows density-dependent effects on absolute rate of increase, but not frequency-dependent ones.

The main case he considers is that of mutation-selection balance. In typical Haldane fashion, he treats successively the cases of autosomal dominants, sex-linked autosomals, and recessive autosomals. He also considers the situation of heterozygous advantage (overdominance). In all of these cases, he calculates the "loss of fitness" (depression of rate of increase) to the species due to the presence of unfit individuals. He notes in particular that the loss of fitness is much greater for overdominance than for any case of mutation-selection balance. Finally, based on lethal mutation rates, he gives a quantitative estimate of the actual loss of fitness experienced by species due to recurrent mutation as about 4 percent for *Drosophila melanogaster* and 10 percent for *Homo sapiens*. "This may be taken as a rough estimate of the price which the species pays for the variability which is probably a prerequisite for evolution" (348).

Twenty years later, Haldane (1957) returned to the issue in a paper on "The Cost of Natural Selection" (see also Haldane 1954, 1956a, 1956b, 1956c, 1961). He began with the observation that

> breeders find difficulty in selecting simultaneously for all the qualities desired in a stock of animals or plants. . . . In slow breeding animals such as cattle, one cannot cull even half the females, even though only one in a hundred of them combines the various qualities desired.
>
> The situation with respect to natural selection is comparable. . . . In this paper I shall try to make quantitative the fairly obvious statement that natural selection cannot occur with great intensity for a number of characters at once unless they happen to be controlled by the same genes. (511)

These comments lead one to think he is going to discuss the total opportunity for selection available, and the fraction of this utilized by the total selection in any given population. Haldane instead returns to the question of the "loss of fitness" to the species.

Beginning with the example of the peppered moth, he notes that upon the change of environment by soot darkening tree trunks, the *carbonaria* homozygotes survived at perhaps twice the rate of wild-type moths (Kettlewell 1956b). "Now if the change of environment had been so radical that ten other independently inherited characters had been subject to selection of the same intensity as that for colour, only $(\frac{1}{2})^{10}$, or one in 1024, of the original genotype would have survived. The species would presumably have become extinct" (511). Note that here selection intensity refers to absolute fitness (survival), and the assumptions made are that characters are determining survival, the genes for these characters may be (and are) independent, and, finally, absolute fitness is both multiplicative and frequency independent.

Haldane's assumption of frequency-independent absolute rate of increase is justified in a discussion of population regulation (see also Haldane 1956a). He notes that "loss of fitness in genotypes whose frequency is being lowered by natural selection will have different effects on the population according to the stage of the life cycle at which it occurs, and the ecology of the species concerned." He considers the "reproductive capacity" of the species (equated to "natural rate of increase") to be lowered by natural selection. He points out that in negative density-dependent population growth, a "moderate" lowering of reproductive capacity may have no effect on density. For example, he considers it possible that "as the result of larval disease due to overcrowding the density is not appreciably higher in a wood containing mainly *carbonaria* than in a wood containing the original type." This situation is actually a case of frequency-dependent absolute rate of increase, because the absolute rate of increase of the *carbonaria* morph is higher in populations where it is not common. Haldane does not recognize this, however, merely pointing out that in at least some areas, such as the margins of the range of a species, population size is determined by density-independent factors, and here reproductive capacity must be utilized "to the full" (513). It is presumably this situation he is considering, because he never returns to the question of frequency-dependent absolute rates of increase; for the remainder of the paper, he assumes that population size is controlled by invariant absolute rates of increase of genotypes, so that the observed population growth rate is less than it would be if there were no "deleterious" genotypes of less than maximal rate of increase (fitness).

The basic "unit process" Haldane considers is gene substitution at a single locus, so that while his 1937 paper had examined what have subsequently been called *mutational load* and *balance load*, here he is examining *substitutional load*. This is what he means by the "cost" of natural selection.

How is the cost of selection to be measured? He does so by comparing the rate of increase of the whole population with that of the "optimal" type. In particular, following a 1954 treatment, he defines "intensity of selection" as $I = \ln (s_0/S)$, where s_0 is the expected rate of increase of the optimal type; S, that of the deleterious mutant type. As in 1937, he assumes that the absolute rates of increase are frequency independent and thus compares the actual population to one composed only of the fittest, or optimal type. Selective deaths are those that are due to the population containing suboptimal types: "if all genotypes had survived as well as the optimal genotype, s_0 individuals would have survived for every S which did so. That is to say, of the $1 - S$ deaths, $s_0 - S$ were selective" (512).

Haldane envisions the following scenario: "A population is in equilibrium under selection and mutation. One or more genes are rare because their appearance by mutation is balanced by natural selection. A sudden change occurs in the environment. . . . The species is less adapted to the new environment, and its reproductive capacity is lowered. It is gradually improved as the result of natural selection. But meanwhile a number of deaths, or their equivalents in lowered fertility, have occurred" (514). Reproductive capacity is here clearly tied to fitness and adaptedness, but there is no real consideration of how the population manages to survive during this period when it is paying the "cost of selection." Adaptedness is measured with respect to an ideal population that had already been "improved" by natural selection—this ideal serving as the goal of the process. In this respect, Haldane's basic scenario is similar to Darwin's (see chap. 6) and Wright's (see chap. 7).

Haldane proceeds to calculate the selective cost of gene substitutions in various genetic situations. He concludes that, in general, substitution of one gene by another usually involves a number of deaths about ten to twenty times the number in a population. Based on this rather startling figure, he concludes further that this process must of necessity occur rather slowly, suggesting that three hundred generations would be a common period for the substitution of one gene by another, which could not be occurring at many loci simultaneously, with perhaps three hundred thousand generations necessary to generate an interspecific difference. He ends with this caveat: "I am quite aware that my conclusions will probably need drastic revision. But I am convinced that quantitative arguments of the kind here put forward should play a part in all future discussions of evolution" (523).

The third major source of the concept of genetic load, and the one that gave it its name, was a 1950 paper by H. J. Muller entitled "Our Load of Mutations." Muller was one of T. H. Morgan's infamous "fly lab" students. In the late 1920s, he discovered the mutagenic effects of X-rays, work for which he received the Nobel Prize in 1946. In the aftermath of the bombings of Hiroshima and Nagasaki, there was much concern about the effects of radiation on human populations. It was in the context of this concern that Muller's paper (based on a talk given the previous year) was written.

The basic situation he considers is the same one considered by Haldane in 1937 (whom he cites, thought noting that he came to his views independently): mutation-selection balance. He begins with the work of Danforth (1923), on the relation between mutation frequency, persistence, and the equilibrium number in the population, and develops the math from this view. His fundamental equation is thus $f = np$, where f is the equilibrium number, n the number of mutants per generation, and p the persistence in number of generations. Persistence is connected to selection by noting that "if i designates the amount of impairment or 'selective disadvantage,' i.e. the chance . . . of dying out at any given manifestation, we have the relation $i = 1/p$ or $p = 1/i$. On the other hand, the mutant gene's chance of survival, s, in any generation, often called its survival value, is $1 - i$" (Muller 1950, 114–115).

Let us look at this calculation carefully. The mutants are graded with respect to an optimal (nonmutant) genotype with an assumed value of 1.0. Any chance of dying due to a detrimental mutation depresses the "survival value" of the gene (and the individual bearing it), creating "genetic deaths." Implicit in these calculations is again the assumption of frequency independence of rate of increase: a population composed entirely of the nonmutant type would necessarily be doing better (in some way) than one with mutants in it. Whereas Haldane (1937) had focused on absolute rates of increase, however, Muller is really looking at relative rates of increase. Nevertheless, he assumes that these relative rates of increase are relevant to absolute rate of increase—a detrimental gene is detrimental not just relatively but in some absolute sense. Such a view is natural to a geneticist accustomed to working with laboratory mutants. With this general background, he notes that at equilibrium detrimental mutants are eliminated at the same rate they arise, and thus ascribes the total genetic death due to a mutant as equal to the mutation rate.

For present purposes, the most interesting discussion of the paper begins in a section on the "Minimum and Maximum values for Selective Elimination in Man." Here he calculates that the "rate of elimination of individuals" due to selection against detrimental mutations is at least 1 in 5, or 0.2, meaning that, on average, 1 in 5 people die for genetic reasons. Noting that in humans it is unlikely that the population could be maintained if each individual had to have more than three surviving offspring (six per couple), he suggests further that the rate of elimination probably has a maximum of about 1.0, at which point a Poisson distribution of survivorship suggests that about 1 in 3 individuals would be fortunate enough to escape "genetic extinction" (138). The closeness of his minimal estimate for overall mutation rate and his calculation of how much mutation humans could tolerate suggests that we may be close to the point at which we simply can't handle our mutational load, with extinction sure to follow.

In looking at the situation in the population at any given time, he notes, "On the whole, those individuals who died the genetic deaths would simply be that fifth of them who, on the whole, carried a more detrimental assortment of these genes, and they would average only about one more detrimental gene per individual than the others" (142). Calculating that the average human carries eight detrimental mutations in the heterozygous state, each of which is approximately 2.5 percent less viable than the normal type, he concludes further that "the average man must be in one way or another, all told, at least 20% below the par of the fictitious all-normal man" (142). This leads to the suggestion that we are now in a situation in which there is an imbalance between the level of mutations we harbor and our survival:

> Our germ plasm was selected, in our more primitively-living ancestors, for a world without central heating or refrigerators, without labor-saving mechanisms in the home, in industry or in agriculture, without sewers or bathrooms, and without knowledge of contraceptives, asepsis, antibiotics, calories, vitamins, hormones, surgery or psychosomatic treatment. And so now for the first time, with the newly found aid of all these devices and methods, the average American, in spite of his eight or more inborn disabilities, adding up to at least a 20% natural disadvantage, manages to get by for almost the three score and ten years which, surprisingly enough, ancient tradition declared to be the "normal" span. (144)

Given this situation, this imbalance between current mutation rates and survival, what are our evolutionary prospects? Muller points out that if n the number of mutants is constant, and p the persistence is increasing because of modern technology, f the equilibrium number will rise, and thus mutations will tend to accumulate until they reach a point at which genetic deaths again are equal to the mutation rate. The picture he paints of our future is not promising:

> In correspondence with this, the amount of genetically caused impairment suffered by the average individual, even though he has all the techniques of civilization working to mitigate it, must by that time have grown to be as great in the presence of these techniques as it had been in paleolithic times without them. But instead of people's time and energy being mainly spent in the struggle with external enemies of a primitive kind such as famine, climatic difficulties and wild beasts, they would be devoted chiefly to the effort to live carefully, to spare and to prop up their own feeblenesses, to soothe their inner disharmonies and, in general, to doctor themselves as effectively as possible. For everyone would be an invalid, with his own special familial twists. (145–146)

This condition he calls "relaxation of selection," an interesting phrase that we will have occasion to revisit later. In this context, it clearly means that many mutant alleles that in former times would have caused death no longer do, although they do still cause phenotypic degeneration.

Muller goes on to examine the effect of increasing mutation rate by exposure to radiation, noting that the long-term effects are likely to be significant, even if no short-term effects can be measured. His solution to the general problem, and to that occasioned by increased mutation rate, is to recommend a program of negative eugenics, where the individuals with the greatest number of mutant alleles would voluntarily forgo reproduction. Nevertheless, he concludes, "None of us can cast stones, for we are all fellow mutants together" (169).

From the roots established by Haldane and Muller, the theory of genetic load expanded greatly. Dobzhansky (1955) picked up the concept in his influential "Review of Some Fundamental Problems and Concepts in Population Genetics." Another key paper was the same one we have already noted by Crow (1958), in the context of the opportunity for selection. Here he distinguished mutational, segregational (balance), and incompatibility loads, and defined the load of a population as "the extent to which it is impaired by the fact that not all individuals in the population are of the optimum type" (5–6). By 1970, when Bruce Wallace first surveyed the field, he could justifiably claim that "genetic load has become a major preoccupation of population geneticists" (vii). This was due at least partly to the interest maintained by the controversy between Muller and Dobzhansky. Muller had argued that there could not be a significant amount of overdominance or heterozygote advantage in natural populations, because this would lead to excessive genetic loads. Instead, most loci should have the normal or "wild-type" allele at high frequency, with a low frequency of deleterious mutations maintained by mutation-selection balance. Dobzhansky thought that there were many loci at which heterozygous advantage obtained, with much genetic variation thereby maintained in populations. This is the contrast between what Dobzhansky (1955; see also Lewontin 1974) called the "classical" and "balance" theories of genetic variation.

When Lewontin and Hubby (1966) first demonstrated the extensive polymorphism of enzyme loci by electrophoresis in *Drosophila*, the natural tendency was for them to interpret the results as a blow to the "classical" theory and support for the "balance" theory. However, they pointed out the problem entailed for the balance theory by the effect of segregational load, assuming independence of the effects of different loci on rate of increase (fitness):

> If our estimate is correct that one third of all loci are polymorphic, then something like 2,000 loci are being maintained polymorphic by heterosis. If the selection at each locus were reducing population fitness to 95% of maximum, the population's reproductive potential would be only $(.95)^{2000}$ of its maximum or about 10^{-46}. If each homozygote were 98% as fit as the heterozygote, the population's reproductive potential would be cut to 10^{-9}. In either case, the value is unbelievably low. While we cannot assign an exact maximum reproductive value to the most fit multiple heterozygous genotype, it seems quite impossible that only one billionth of the reproductive capacity of a Drosophila population is being realized. (606–607)

This situation is what Lewontin (1974) called the "paradox of variation"—there is simply too much variation to explain by heterozygote superiority under standard assumptions about genetic load. Lewontin and Hubby considered the possibility that this variation was irrelevant to selection and was instead due to "adaptively equivalent isoalleles" (605), but they rejected this explanation on grounds of implausibility. At the dawn of the age of molecular biology, then, it was quite clear that calculations of genetic load could result in some rather strange figures.

As previously noted, Bruce Wallace (1968a, 1968b, 1970, 1991) has been a persistent critic of the theory of genetic load from early on. In 1968, he introduced the concept of "hard" and "soft" selection, stating that his goal was to "show that the apparent ability to assign causes—genetic as well as environmental—to the deaths of zygotes in a population is largely an artifact of static concepts." In particular, "genetic load cannot be used to predict the extinction of the population" (1968a, 88). Wallace's basic point, reiterated in all of his later works, is that variation in relative rate of increase among genotypes within a population has no necessary relation to variation in mean absolute rate of increase between populations composed of these genotypes. He argues this on the basis of density-dependent population regulation, such that under pure "soft" selection, the type of individual has no effect on survival probability: if genetically superior individuals are lacking, genetically inferior ones will take their place. Of course, it is unrealistic to think that no relatively inferior genotypes will be absolutely inferior as well, so he introduces a second concept, that of density-dependent population regulation that varies among genotypes. In this model, a population of an inferior genotype will be smaller than one of a superior genotype, but it will not go extinct. Thus, Wallace does allow for some effect of intrinsic genetic quality ("population fitness") on population size: "genetic changes which lead to an increased equilibrium size of a population can be regarded as changes that increase population fitness." However genetic load, as formally defined by Crow (1958), "is unrelated to population size and, consequently, to the fitness of a population" (1968a, 107).

These criticisms by Wallace and others (e.g., Feller 1967) certainly had their effect, and the concept of genetic load, while standard in population genetics texts, has nevertheless been regarded with some suspicion by many workers. In recent years, however, the concept has been rejuvenated, and we find it playing a significant role in the literature of evolutionary biology, especially evolutionary conservation biology. One of the key steps in this rejuvenation was the appearance of a paper by Michael Lynch and Wilfried Gabriel in 1990, arguing that "mutational meltdown" due to the accumulation of deleterious mutations was likely to lead to the extinction of small populations.

I will come back to the relation between genetic load and population survival in chapter 12, in the context of conservation biology. For now, I would like to return to Crow's (1958) other measure of the intensity of natural selection, the standardized variance, or "index of total selection." Doing so will allow us to examine the quantitative limits of selection and drift in natural populations without worrying about the health of the population itself.

LIMITS TO SELECTION AND THE STANDARDIZED
VARIANCE IN RATE OF INCREASE

As we have seen, both Haldane (1937, 1957) and Muller (1950) were concerned about the limits to natural selection's ability to deal with mutations, both favorable and unfavorable. Thus, the main case considered by Haldane (1937) was mutation-selection balance. In his quantitative calculation that *Drosophila melanogaster* suffered a 4 percent drop in fitness (mean rate of increase) due to mutation, he assumed that the total "loss of fitness" from nonlethal deleterious genes was twice that of deleterious lethals. The deleterious lethal rate per generation he estimated at 0.0125, or 1.25 percent, which when tripled gave the "about 4 percent" figure. If we consider this from the perspective of standardized variance, we see that Haldane is looking at a situation where 96 of every 100 survive and 4 die for genetic reasons; this is equivalent to a standardized variance of $4/96 = 0.042$. How does this compare to total standardized variance? Given that most of these lethals act at the egg stage and that a *Drosophila* female can easily lay 1,000 eggs, of which on average only 1 survives, there is a *minimum* standardized variance of 999 from egg to adult. Of this total variance, then, about $0.04/999 = 0.00004$, or 0.004 percent, of the minimum total variance is used by this selection.

In his 1957 paper on the cost of selection, Haldane was concerned to calculate the cost of replacement of one allele by another. As we saw, he calculated that the replacement of one gene by another usually involved a number of "selective deaths" equal to thirty times the population size in any generation. If we consider the standardized variance in rate of increase involved in the replacement, from beginning to end of the process, we see that it is approximately equal to the population size, assuming that a single mutant allele spreads through the entire population. Thus, for a population of a million, the variance is about a million (actually 999,999). This sounds like a lot, though it could obviously be achieved in a single generation by death of all but the mutant type. But let us look at the process of replacement. Haldane considered the usual intensity of selection, given by the natural log of the ratio of the maximum possible rate of increase to the mean rate of increase, to be on the order of 0.1. Thus, the ratio is equal to $e^{0.1}$ or about 1.1. If the best genotype is 10 percent more "fit" than the average one, and the overall population is a million, in the first generation the standardized variance required to produce this difference is about 10^{-10}; or if we consider an increase from one to two mutant individuals (rather than to 1.1), it is 10^{-6}. In the middle of the replacement, when the mutant and ancestral forms are equally common, the standardized variance involved reaches a peak at 0.025. Finally, in the last generation, when the last nonmutant is lost from the population, it again drops to 10^{-6}. What has become of the large standardized variance calculated for the entire process?

Remember that for a given population growth rate, the maximum standardized variance is given by a process in which one individual reproduces the entire population, which is effectively the situation we consider when we look at the overall spread of a mutant gene from its origin to fixation. However, if the variance is not produced simply by death versus survival but rather by a more subtle difference in expected rate of increase, the variance involved may be much lower. In Haldane's case, the maximum standardized variance required per generation is only 0.025 and occurs in the middle of the process of replacement.

If we consider his higher estimate of selection "which would hardly be compatible with survival" (1957, 521) of $I = 4$, necessary for a replacement period of about 7.5 generations, this gives ratios of the rate of increase of mutant to ancestral type of about 50:1. In the first generation of replacement, this would give a standardized variance of about 0.0024; and in the middle of the replacement process, about 0.92. This is indeed rather larger, but certainly not ridiculously so. It corresponds to the situation where the selectively favored individuals have a relative rate of increase of about 1.96, the selectively disfavored of about 0.04. Our overview of figures on opportunity for selection (table 8.1) certainly shows many populations that have the requisite variance.

Thus, unlike Haldane's calculations of "selective deaths" that increase during a process of gradual gene replacement, when examined from the perspective of variance in relative rate of increase, the variance required per generation is much lowered if the process takes as few as 7.5 generations. This certainly accords more with one's intuition about the process. What about Muller's (1950) calculations? Are they similarly affected?

Like Haldane, Muller (1950) was concerned largely with mutation-selection balance, especially under increased mutation rate. Remember that Muller argued that one in five people die for genetic reasons. This is equivalent to 80 percent survival, or a standardized variance of 2/8 or 0.25. He invokes a Poisson process of genetic death, however, in which 0.20 is the probability of having a mutant lethal gene; the probability of being mutation free is thus a little higher (owing to multiple incidence of lethals) at 81.9 percent, so that the standardized variance involved is a little lower, about 0.22. Based on a maximum reproductive rate of three offspring per individual surviving to breed (six per couple), he argues that the highest rate we could possibly tolerate is a probability of a lethal mutation of about 1.0, for which a Poisson distribution gives only 36.8 percent of the population mutation free and surviving. This corresponds to a standardized variance of 63.2/36.8, or 1.72, which, on the face of it, does seem like rather a high value for humans to be able to afford. Although some studies of humans have found standardized variances up to 10.0 or more, others have found them to lie closer to 1.0. This is the expected number in a stable population (skipping subtleties for now). Given that genetic death is unlikely to be the major cause of death (discussed later), there is simply not enough variance to account for that associated with lethal genes, even if all such variance were involved in removing the lethals. This suggests that Muller might indeed have a point.

However, could there be a problem with the way Muller framed his argument? Is it really possible that almost one in five humans now dies (or at least formerly died) a genetic death? What does this even mean? Let's return to the basic population genetics model Muller uses and see what an analysis in terms of the standardized variance suggests.

GENETIC LOAD AND GENETIC DEATHS

As we have seen, Muller (1950) founded his calculations of genetic load on the idea of mutation-selection balance. He estimated that the average deleterious mutant persists about forty generations in humans, so that it must have a selective disadvantage of 2.5 percent, and that the average person carries eight such deleterious genes; hence, he calculates a 20 percent average disadvantage. However, Muller's model more explicitly assumes that each individual receives a variable number of deleterious mutations, each of which gives a 2.5 percent decrease in probability of survival, with an average of eight such mutations per individual. He thus envisions a Poisson distribution of deleterious mutations, with eight the mean number. As we saw, he concludes, "On the whole, those individuals who died the genetic deaths would simply be that fifth of them who, on the whole, carried a more detrimental assortment of these genes, and they would average only about one more detrimental gene per individual than the others" (142).

Now for a Poisson distribution, the variance is equal to the mean, so the variance of mutation number in this population would also be 8. What, then, is the variance in relative fitness (expected relative survival rate)? If one looks at the overall distribution of fitnesses, assigning to each class with a given total number of mutations f_t a fitness of $0.975 f_t$ (as Muller seems to have intended), the standardized variance in survival is only about 0.005. This estimate is about forty times lower than that which assumes a simple subdivision of the population into those that die a genetic death and those that don't. Where is the extra variance? It is, of course, found as variance *within* each of the classes. In fact, in the situation hypothesized by Muller, only 2.3 percent of the variance in survival is variance among genotypic classes carrying different numbers of mutations. This situation arises because the only way to get, for example, 90 percent survival within a class is to kill off 10 percent of its members, and this creates variance in survival within the class. What Muller has done here is to try to account for all variance in survival by the number of deleterious mutant genes one receives. But because each class necessarily must have internal variance in survival, much of the variation is within each class.

Where does this leave us? The real question is whether there is enough variance in survival in the population to remove the one mutant in five necessary to compensate for new mutations. The answer is yes—Muller was quite right that 20 percent of the mutants would be lost. But how can this be if so much of the variance is within classes of mutants? The solution lies in the observation made earlier that, for a given overall survival rate, the case of complete survival versus death gives the greatest standardized variance. Another way to look at this is that the greatest variance is observed if the classes one is interested in are precisely those that either survive or don't; if instead the classes have a certain expectation of survival, then much of the variation exists within classes. Nevertheless, the overall rate of change can be identical if the lowered variance among classes is compensated for by greater frequency of the classes.

A simple example should make this clear. If we have a population of 100, of which 98 are normal (survival rate $= 1.0$) and 2 are lethal mutants (survival rate $= 0$), then the mean survival rate is 0.98, and the variance in survival is 0.0196 overall (standardized variance in survival is $0.0196/0.98^2 = 0.0204$). The two lethal mutants are lost from the population, of course. Of the total variance in survival in the population, 100 percent is associated with the mutation. Now, consider a population of 100 consisting of 90 normals (survival rate 1.0), and 10 mutants, of which 2 die (survival rate 0.80). The total variance in survival is still 0.0196, but now the variance associated with the mutation is only about 18 percent of the total; the remaining 82 percent of the variance is within the mutant class. Nevertheless, the decreased variance between the mean survival of mutant and nonmutant classes is compensated by the greater frequency of the mutant class, so that the same number of mutants die, and the same overall variance is thus present. The lowered variance between mutant and nonmutant classes reflects our intuitive sense that there is less extreme differential survival between mutants and nonmutants in the second case. The relative survival rates are 1.02 and 0 in the first case, and 1.02 and 0.82 in the second. Thus, while classifying the death of mutants as "genetic deaths" may seem unwarranted, the general picture presented by Muller is correct.

THE MEASUREMENT OF TOTAL SELECTION
IN EXISTING POPULATIONS

Given the constraints on selection imposed by the standardized variance (opportunity for selection), what fraction of this is in fact evolution by narrow sense natural selection? Recall that within the framework established by Fisher (1930a), this fraction is the additive genetic variance in rate of increase (fitness)—that is, the fraction of the total variance that can be explained by the additive effects of alleles. What do estimates of this fraction look like? Ignoring for the moment issues with the underlying definitions, I would like to consider this question here.

Fisher (1930a) himself addressed the issue briefly in his discussion of "The Inheritance of Human Fertility." Here he compares the variance in number of off-spring among women to the mean, suggesting that the fact that the variance is higher than the mean (i.e., non-Poisson) argues for nonrandom factors at work. He infers this as well from the fact that half the population of Australia comes from families of eight or more: "It would be an overstatement to suggest that the whole of this differential reproduction is selective; a substantial portion of it is certainly due to chance, but on no theory does it seem possible to deny that an equally substantial portion is due to a genuine differential fertility, natural or artificial [?], among the various types which compose the human population" (190). In other words, Fisher posits (additive?) genetic variance in human reproduction equivalent to approximately half the total standardized variance, with the other half being due to chance.

Forty years later, in his thoughtful book on *The Genetic Basis of Evolutionary Change*, Lewontin (1974) came to a slightly lower estimate. He is specifically discussing the genetic (rather than additive genetic) variance:

> In a sexual population that is just replacing itself, with a Poisson distribution of off-spring, the *total* variance in offspring number is 2, so that the genetic variance in fit-ness is likely to be less than 1. In human populations, which are growing, and which have more than Poisson variance in offspring number, the total variance ranges from about 3 for stable Great Britain to 21 for rapidly growing Brazil. The genetic compo-nent of the variance certainly does not exceed 25 percent, from the available evidence, and is more likely to be on the order of 5 percent. (206)

More recently, Burt (1995) explicitly examined the question of the additive genetic variance in rate of increase (fitness) that exists in populations—that is, the variance associated with evolution by natural selection. Evidence from a variety of sources led him to conclude that the additive genetic variance in rate of increase over periods of a generation or less typically lies in the range from 0.0001 to 0.3. If we assume a standardized variance of 1, the additive genetic variance, or the rate of "increase in fitness due to natural selection," would fall in the range from 0.1 percent to 30 percent of this value.

Finally, an interesting paper by Kruuk et al. (2000) looked at the question of additive genetic variance in rate of increase in the context of heritability of num-ber of offspring produced in Clutton-Brock's well-studied Isle of Rum population of red deer *(Cervus elaphus)*. They estimated a mean number of offspring produced per individual (from birth) of 2.9 in females and 2.1 in males. The corresponding variances were 6.4 for females and 21.6 for males (standardized variances of 0.8 and 4.9, respectively). Of this variance, none was additive genetic variance in females, while 2 percent (not significantly different from 0) was additive genetic variance in males. Thus, there was no evidence for any evolution by natural selection occurring at all.*

In summary, although the total standardized variance sets an upper limit on additive genetic variance, and thus on evolution by natural selection, it is entirely unclear how much of this variance is actually evolution by natural selection. In spite of the large amount of effort devoted to studying natural selection in the wild (e.g., Endler 1986; Kingsolver et al. 2001), the simplest quantitative question—of the total variance in rate of increase in a given population, how much is evolution by natural selection?—remains largely unanswered. What is perhaps surprising is that this quantitative aspect of selection has so rarely been considered.

POPULATION GROWTH, SELECTION, AND STANDARDIZED VARIANCE

An interesting feature of these estimates of the fraction of the standardized variance in rate of increase due to selection is the general lack of interest one senses in any possible relation with overall population growth rate. Yet, as we have seen, population growth rate can greatly affect the standardized variance. And as Barrowclough and Rockwell (1993) pointed out, population growth rate varies across the life cycle, so that the measured standardized variance depends on the part of the life cycle considered (see Fisher 1939; Crow and Morton 1955). In particular, they noted that a population of organisms in which each adult produced exactly one offspring would have a standardized variance in lifetime reproductive success among adults of 0. On the other hand, if in this same population only 1 in 10 fledglings survive, then the standardized variance in lifetime reproductive success among fledglings is $9/1 = 9$. The standardized variance among gametes might be 10^8 for the same population. They conclude only that "if estimates of [variance in lifetime reproductive success] are to be compared among populations, taxa, field studies, and so forth,

---

* It is worth noting that these figures for heritability and additive genetic variance were arrived at by fitting a maximum likelihood model (derived from livestock breeding) to all of the data on correlations across the entire pedigree. These figures are thus averages from times with different populations, different population growth rates, and so forth, with up to eight generations in a single lineage over the study period. The implicit assumption is that absolute number of offspring per individual is a trait to be studied like any other, dependent primarily on genotype, rather than a property relative to a particular population over a specified time period. Nevertheless, in other studies Clutton-Brock and his colleagues have noted significant year-to-year variation in cohort reproductive success in this population, among both females (Albon et al. 1987, 1991) and males (Rose et al. 1998). Given this underlying year-to-year variation, it would make more sense to examine the population over a defined time period and only estimate parameters with respect to that period (see Coulson et al. 2006). The fundamental assumption of quantitative genetics, that the environment is independent of time, is belied by their own data, so it is not too surprising they found no additive genetic variance in rate of increase in their analysis.

the estimates must be obtained for a complete life cycle and with reference to the same end points of that cycle" (283).

Yet from the general relation between survival rate and standardized variance shown earlier in this chapter, it is clear that the maximum standardized variance is obtained during periods of population reduction. For example, if a marine invertebrate typically spawns 100,000 eggs, of which only one survives, the standardized variance in survival among the eggs is 99,999, while if each adult has a Poisson distribution of offspring, with expectation of 1, then the standardized variance in offspring number among adults is only 1. Now, of course that doesn't mean that more narrow or medium sense "selection" is going on during the first period of the life cycle than the second, in the sense that there is differential survival among genotypic and/or phenotypic classes, but it certainly does mean that there is more opportunity or scope for such selection to occur. There is no justification for neglecting this in favor of a complete life cycle from adult to adult, or egg to egg, because variance in rate of increase, if differential among genotypes, can cause evolutionary change whenever it occurs. It is the maximum, not the minimum, value of this quantity across the life cycle that is most important.

Not only can the standardized variance be enormous due to low rates of survival for early stages in the life history, but it will likewise be large if a population undergoes a crash in numbers. The standard argument is that there is no reason to believe there is more selective survival at times of declining population than at any other time: the variance in relative rate of increase among genotypic (or phenotypic) classes is generally independent of population growth, even if the overall variance is not. However, if there is a thousand times more "opportunity for selection" during a population crash than with a stable population, it at least appears possible that there is additional medium sense selection occurring as well (i.e., "selection of" genotypic or phenotypic classes). Such an idea has been promoted by a number of authors, going back at least to Alfred Russel Wallace's (1858) initial scenario in which selective replacement is tied to a cycle of population decline and subsequent recovery (cf. Bulmer 2005):

> Let some alteration of physical conditions occur in the district—a long period of drought, a destruction of vegetation by locusts, the irruption of some new carnivorous animal seeking "pastures new"—any change in fact tending to render existence more difficult to the species in question, and tasking its utmost powers to avoid complete extermination; it is evident that, of all the individuals composing the species, those forming the least numerous and most feebly organised variety would suffer first, and, were the pressure severe, must soon become extinct. The same causes continuing in action, the parent species would next suffer, would gradually diminish in numbers, and with a recurrence of similar unfavourable conditions might also become extinct. The superior variety would then alone remain, and on a return to favourable circumstances would rapidly increase in numbers, and occupy the place of the extinct species and variety. (58)

The nearly exclusive focus of Darwin and his successors on changes in frequency of types, rather than changes in their number, has certainly tended to limit any interest in the relation between changes in population size and natural selection. As I insisted in the introduction to this chapter, by separating the adaptedness of organisms from their existence, Darwin's formulation has thereby tended to obscure the real relations of organisms to their environment. As a parting shot, I would like to examine more closely the changes in population size and standardized variance that typically occur across the life cycle, and consider how these relate to the total standardized variance, or opportunity for selection, over a longer period of time.

PARTITIONING THE VARIANCE
IN RATE
OF INCREASE ACROSS
THE LIFE CYCLE

The relation between the standardized variance measured in juvenile and adult stages of the same population is complex, even when both are measured over a complete generation, and depends on the distribution of reproduction and survival. Fisher (1939) showed that for a Poisson model of reproduction with subsequent random survival, and a stable population size, the overall variance measured between more numerous (juvenile) stages of the life cycle would be almost exactly n times that found between less numerous (adult) stages, where $2n$ is the ratio between the more numerous (juvenile) and less numerous (adult) stages. And as noted previously, Barrowclough and Rockwell (1993) examined the case of a hypothetical stable population where each adult has 10 offspring, of which exactly 1 survives to adulthood; in this case the adult-to-adult standardized variance is 0, but the juvenile-to-juvenile standardized variance is 9.0.

In the general case given that we focus on a particular population over a particular period, it is clear that there must be some determinate relationship between the standardized variance over the entire period and that for various subperiods. In Crow's original paper on the "Index of Total Selection" (1958; see also Crow 1962, 1989a, 1989b), he suggested partitioning the total index into an index of mortality (I_m) and an index of fertility (I_f). His basic model was thus of a population that consists of two parts, one that undergoes a process of pure survival, one that undergoes a process of differential reproduction (these processes can be conceived of either as simultaneous, or with differential survival occurring first; to consider differential reproduction first, one must know the distribution of survival among offspring of reproducers). Crow showed that

$$I = I_m + \frac{1}{s}I_f,$$

where s is the proportion of the population that is included among those differentially reproducing.*

However, as noted, Crow's simple division of the standardized variance into mortality and fertility components does not work for the case in which we consider reproduction first if there is any variation in survival among families . In this case, the relation between the two measures is complicated by possible interactions (covariances) among rates of increase of each lineage between time periods.

In the same paper in which they rechristened the "index of total selection" the "opportunity for selection," Arnold and Wade (1984b) observed that Crow had assumed independence of the variation of the two phases in his model, and proposed an alternative partitioning of the standardized variance across the life cycle for cases where such independence doesn't occur. They noted that for the population as a whole, its rate of increase over an extended time period is simply the multiple of its rate of increase over a series of shorter time periods. They proposed calculating a separate mean and standardized variance of rate of increase for each subperiod considered. They recognized that in doing so, however, the measured standardized variance over each subperiod could not be used to find the standardized variance over the entire period, due to possible covariances among subperiods. Moreover, for each subperiod, the calculated mean and standardized variance of the rate of increase are for the population as it exists at the beginning of that subperiod, not for the population at the beginning of the entire period. Thus, all lineages that are already extinct are left out, and those still extant are weighted by the number of extant representatives. The effect is that one is calculating rates of increase and associated variances for different reference populations over each subperiod.

This defect in the Arnold and Wade approach was pointed out by Brown (1988), who proposed an alternative method of partitioning the standardized variance. Brown noted that if one deals with the *lineages* founded by members of the initial population, then the rate of increase of each lineage over a longer time period is simply the multiple of its rate of increase over shorter time periods (cf. Howard 1979). Thus, the standardized variance over the entire period is equal to the variance of the product of the rate of increase over each subperiod. Brown noted that the exact formula for the variance of a product had been given by Goodman (1960, 1962) and that it is readily applied to this case.

However, Brown used the variance formula not only for products that occur at successive periods of the life span, as in offspring production and offspring survival,

---

* Although he did not note this, this model is formally identical with a model previously developed by Crow and Morton (1955) for family-correlated survival in the context of genetic drift, because I_m is simply the number dying/number living $\left(= \frac{1-s}{s}\right)$ and I_f is the standardized variance among the parents who survive.

but also for those that are not successive in time, such as breeding life span and mean yearly fecundity. Such a procedure is mathematically valid, but its biological interpretation is difficult because the values one calculates for individuals will likely be affected by varying population growth rates (e.g., "good years" vs. "bad years"); at best, one is adding additional noise to the system. In any case, if the division is not made in time, it is clear that one is not dealing with a measure that applies to successive periods. If applied to members of different cohorts, so that one is not dealing with a single biological population through time but rather an average across a series of populations with different histories of population growth, interpretation can only be more difficult.

An even more significant problem with Brown's (1988) approach is that when lineages go extinct in an early subperiod, it is not clear how to assign a rate of increase to the lineage over a later subperiod. For any given lineage, it doesn't matter what rate one uses, because that rate will be multiplied by zero from the preceding subperiod. And as we saw earlier, rates of increase (realized fitnesses) themselves can simply be averaged across all living members of the population. But the treatment of extinct lineages does matter for one's calculations of variance in rate of increase at later subperiods, and thus the partitioning of the variance among subperiods. If one sets the rate of increase at zero for these extinct lineages, they will contribute to the variance at that later subperiod, which seems counterintuitive. Alternatively, one might set them to the mean rate for all extant lineages, but this results in a similarly counterintuitive situation in which an extinct lineage is assigned a nonzero rate of increase that depends on the situation of the entire population. Brown dealt with this problem by excluding all nonbreeders from consideration, thus effectively treating the population as consisting of a part that undergoes a process of survival and another that reproduces, much as Crow (1958) had before him.

Finally, there is the issue of how to interpret the results arrived at; while the opportunity for selection can indeed be decomposed this way, the biological interpretation of the results obtained is not clear. This is perhaps why Brown's (1988) method of decomposing the standardized variance has not been as popular as Arnold and Wade's (1984a, 1984b).

A simple alternative approach that deals with the problem posed by noncomparable reference populations in the approach of Arnold and Wade (1984a, 1984b), as well as the problem posed by extinctions of lineages in early subperiods in the approach of Brown (1988), is to always examine the population from the same initial time, but with different ending times. The contribution of each subperiod to the overall variance then can be assessed by the change in the total variance as one adds successive subperiods into consideration. An example of such a breakdown is shown in figure 8.3. In this method, it is immediately clear when the events of a given subperiod increase the standardized variance among lineages and when they decrease them (see the negative value in the last column of fig. 8.3). But once again,

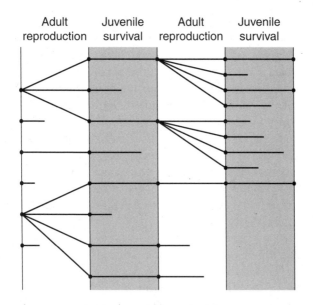

| | Adult reproduction | Juvenile survival | Adult reproduction | Juvenile survival |

Initial population of adults
(values to end of each subperiod)

$\bar{R}$	1.33	0.83	1.50	0.50
Abs var	2.56	1.47	8.58	0.58
St var	1.41	2.11	3.81	2.32
Δ St var	—	0.70	1.70	-1.49

Initial population of juveniles
(values to end of each subperiod)

$\bar{R}$	0.63	1.13	0.38
Abs var	0.23	1.69	0.48
St var	0.60	1.34	3.44
Δ St var	—	0.74	2.10

FIGURE 8.3. Diagrammatic representation of lineage proliferation and survival across two discrete, asexual generations. The figures give mean rates of increase ($\bar{R}$) and absolute and standardized variances in these rates of increase for two different initial populations: adults and juveniles. Note that the values calculated depend on the initial population chosen. Note also that standardized variance can decrease if survival and reproduction in successive periods are negatively correlated.

note that all figures calculated are relative to the initial population examined; if one begins with juvenile stages that undergo an initial subperiod of survival, rather than adults that undergo an initial subperiod of reproduction, one calculates different figures because one is examining a different set of lineages.

Understanding these straightforward and rather obvious relations of the standardized variance in rate of increase (opportunity for selection) to population growth rates across the life cycle would seem critical to better understanding the relation of natural selection to the life cycle in particular and to organismal ecology in general. Yet the tendency of evolutionary ecologists to focus on the adaptive significance of reproductive and life history traits, and of population geneticists to focus on differential fitness among adults (together with its associated genetic load and genetic drift), has left these rather obvious relations almost entirely unexplored.

· · ·

As we will see, the concept of genetic load was one of the key contributing factors to the rise of the "neutral theory." However, most incarnations of the genetic load concept have been illegitimate, due to the same blurring of the distinction between absolute and relative rates of increase implicit in Wright's metaphor of the adaptive landscape. In particular, Haldane's measures of the "cost of selection" in terms of depression of mean fitness and the number of "selective deaths" involved in gene replacement are no more directly tied to any real feature of a population than is the adaptive landscape. Nevertheless, it has been clear since Crow's work some fifty years ago that there are indeed quantitative limits to selection, in the simple form of the total standardized variance in rate of increase (opportunity for selection). We have seen further that the total standardized variance should, and does, vary with population growth rate, and that times of population decline are times of high standardized variance. The proportion of this variance that constitutes evolution by natural selection, and whether this proportion varies with population growth rate, remains entirely unclear.

With this basic grounding in the mathematics of Mendelian population genetics, we are now better prepared to examine the distinction between natural selection and genetic drift. Evolution by both natural selection and genetic drift involves differences in rate of increase among classes of genetic lineages, but these two processes are currently given rather different roles in evolution. What is the basis for this distinction?

9

Natural Selection and Genetic Drift

Their Role in Evolutionary Change

The customary uses of the term Natural Selection are not only vague, but discrepant, and please imagine what theoretical physics would be like if we not only all talked about the second law of thermodynamics but meant materially different things by the phrase.

—R. A. FISHER, 1930, LETTER TO E. B. WILSON

If we turn to any modern textbook of evolutionary biology (e.g., Ridley 2004; Futuyma 2005; Freeman and Herron 2007), we find agreement on some of the basics of evolutionary theory. In particular, the general view of evolution is that it is a process that involves three main "forces" or processes: mutation, natural selection, and genetic drift. Unlike mutation, both natural selection and genetic drift involve differential survival and reproduction among classes of lineages. Nevertheless, "adaptive evolution" and "adaptations" are ascribed exclusively (or almost so) to natural selection, not to genetic drift (or mutation), much as Fisher, Wright, and Haldane did in the beginning years of the synthesis. To know what this means, however, depends critically on how natural selection and genetic drift are defined—if we don't know what they are, we certainly can't decide which evolutionary change to ascribe to each of them. Unfortunately, a critical examination of such introductory textbooks immediately makes it clear that no agreed-on definition of natural selection or drift exists, but rather several competing definitions, each used depending on the circumstance.

For example, two key questions in the definition of natural selection have to be (1) does it include inheritance or merely work with inheritance?, and (2) does it involves differential survival and reproduction of classes of entities, *due to the features* by which we define them (narrow sense selection), or merely differential survival and reproduction of classes of entities (medium sense selection)? In other words, does selection include or exclude the hereditary "response to selection" (see Endler 1986, 14)? And are we concerned with what Sober (1984) called "selection for" features or merely "selection of" features (Endler 1986, 50)? These distinctions clearly matter.

Nevertheless, we find a good deal of variation in thought on just these points. For example, Ridley (2004) defines selection as *inclusive* of heredity and as a *causal* process based on relative adaptedness: "The process by which the forms of organisms in a *population* that are best adapted to the environment increase in frequency relative to less well adapted forms over a number of generations" (686). By contrast, Futuyma (2005) defines it in a way that *excludes* heredity and is apparently *noncausal*: "The differential survival and/or reproduction of classes of entities that differ in one or more characteristics." However, he adds the clarifying remark: "To constitute natural selection, the difference in survival and or reproduction cannot be due to chance," and he also notes that "usually the differences are inherited" (550). In other words, Futuyma is at pains to indicate that natural selection must indeed be a causal process, but not necessarily one in which the causes are the features we use to distinguish the classes. And although heredity is not part of the process, it usually goes along with it. Finally, Freeman and Herron (2007) define natural selection simply as "a difference, on average, between the survival or fecundity of individuals with certain phenotypes compared with individuals with other phenotypes" (803). This is truly noncausal and independent of heredity, though the phrase "on average" is perhaps intended to point in the direction of a cause. And these are just the glossary definitions; a thorough examination will find much variation in usage even within a single book.

Yet in spite of such lack of clarity over the precise definition of natural selection, Ridley, Futuyma, and Freeman and Herron are united in their ascription of the adaptedness of present species to natural selection, not random genetic drift. Given this ascription, one might wonder if we can get a more explicit idea of precisely how these two processes differ. In chapter 7, we saw how one can measure rates of increase in Mendelian populations. But because natural selection (at the level of both average effect and average excess) is defined only as a probabilistic outcome, this leaves open the possibility that random differences in rate of increase among genetic lineages, with no causal basis, may occur. It is such differences, of course, that are contemplated under the concept of genetic drift.

In this chapter, I want to further explore the distinction between natural selection and genetic drift. This is critical to our project because the distinction between selection and drift is fundamentally the distinction between what was *supposed* to happen and what *did* happen. Yet whenever one speaks of what is supposed to happen, there is obviously great danger of lapsing into a teleological perspective. In particular, as I have insisted, if one merely *assumes* that what did happen is what was supposed to happen—as is the case when evolutionary changes are ascribed to natural selection alone, without any evidence to support the ascription beyond the end state reached—one is guilty of teleological determinism. Moreover, unless it is at least possible that other evolutionary processes could be responsible for the changes seen, the ascription of such changes to natural selection has no empirical content.

To be able to explain a given evolutionary change by (narrow sense) natural selection, we need to have a *causal* account of any differential survival and reproduction occurring, as well as showing that such differential survival and reproduction was responsible for the evolutionary change. In such a causal account, genetic drift appears as random noise around the values determined by the interaction of a defined set of organisms and their environment (i.e., within a deterministic model of the evolution of a population). It is in the context of such a model that Fisher (1922a, 1930a) first developed the theory of stochastic processes in population genetics, subsequently elaborated on by Wright (1931) to form the well-known "Wright-Fisher model" of genetic drift.

I begin by examining the distinction between natural selection and genetic drift at the most basic, physical level—as part of the standardized variance in rate of increase we observe in populations. In this context I look at some simple computer simulations to further explore the relationship of population growth rate, the opportunity for selection, and genetic drift. This makes it clear that the distinction between selection and drift depends on certain assumptions about the distribution of offspring number within the population.

To see where these assumptions came from, I give a brief overview of the historical origin of the concept of genetic drift in the work of Fisher and Wright. This paves the way for a consideration of the neutral theory and tests used to distinguish selection from drift at the molecular level. Finally, I examine some of the complexities of molecular evolution, involving such features as hitchhiking, meiotic drive, variable selection, mutational bias, recombinational bias, jumping genes, and horizontal gene transfer. What I hope to show is that while the distinction between natural selection and drift seems clear in principle, it is much more difficult to make in practice. Even mutation may sometimes look like selection or drift when considered from an alternative perspective.

WHAT'S REALLY GOING ON?

How do natural selection and genetic drift differ at the most basic, physical level? The general sense one has is that natural selection is a causal process, or perhaps the result of a causal process, whereas genetic drift is "random." Yet both natural selection and genetic drift involve differential survival and reproduction among classes of lineages. As we have seen in the last chapter, such differential survival and reproduction can best be examined quantitatively within the context of the standardized variance in rate of increase (opportunity for selection), which sets the upper limit for the rate of selection or drift that can be occurring in a population. This quantitative context provides a good place to begin our examination of the distinction between natural selection and genetic drift.

Consider an initial population of zygotes for which we record their state at two loci, with two alleles at each locus (*AABB*, *AaBb*, etc.), and count their offspring zygotes in the next generation. Within the population, some individuals will have no offspring, some one, some two, and so forth. As we saw in chapter 7, it is a simple matter to calculate the average rate of increase across the entire population. One can also calculate average rates of increase for each of the nine genotypic classes we have identified. Both individual and average genotypic rates of increase can be made into relative rates of increase by dividing throughout by the average of the whole population.

Now consider the variation in rates of increase we see within this population. As we saw in chapter 8, this can be measured simply as the total variance in rate of increase. Considered purely descriptively, some of this will be variance among genotypes, some will be variance within a given genotype, and the total variance will be the sum of these components. This partitioning is identical to the familiar partitioning in the analysis of variance, where the total sum of squares can be split into the sum of squares among and within groups. If there is variation among genotypes in rate of increase, then the fraction of the total variance associated with genotypes (among groups variance/total variance) can easily be calculated.

How do we decide whether such variation among genotypes in survival and reproduction is natural selection or genetic drift? To do this, it is clearly not enough to simply measure values associated with particular genotypes; one must try to come up with a *causal* account of variation in rate of increase. In particular, it seems reasonable to think that to the extent the genotypes are themselves causing (or are at least statistically associated with the causes of) among-group variance, then the among-group variance reflects natural selection; to the extent they are not, it reflects genetic drift.

Let us first consider the case where the variation among genotypes is substantial. In that case, it is a simple matter to test the hypothesis that there is a nonrandom association between genotype and rate of increase. Nevertheless, even in this case we have not identified the causes of differential rates of increase but merely shown that the differences are too great to be explained by chance and thus must have some cause. It is quite possible that other loci statistically associated with the ones we are examining are in fact the true cause of differential rates of increase, or that environments statistically associated with the genotypes are the cause of differential rates of increase. Moreover, if we try to separate out the "natural selection" component of the among group variance from the "genetic drift" component, we will find it impossible—the best we can hope for is to have a certain confidence interval for the fitnesses (expected rates of increase) of the genotypes and thus for the proportion of the standardized variance due to selection.

Now consider the case where the among-genotype variance is not too great to be explained by chance. Does this mean that it is all "genetic drift"? Not necessarily, because it is entirely possible that real, small differences in rates of increase exist among genotypes but cannot be detected as significantly different (and this is in the context of a hypothetical population, where we have perfect information on rates of increase).

The general point is that while natural selection and genetic drift are conceptually two extremely different things, it is very difficult to distinguish them precisely at the level of any real population (even a model one) in terms of what is actually happening to individuals or in terms of any measurements one might hope to make. When one considers that both are in fact identical processes in terms of what is actually happening (differential survival and reproduction) and that their causes can be the same (as in the case of genotype-environment interaction or a phenotypic distribution that doesn't mirror the one expected for the allele frequencies present), this is not so surprising.

If we compare the situation to another where one posits stochastic processes in nature, radioactive decay, it is clear that in this latter case it *is* possible to measure the actual rate and use it to estimate an underlying expected rate, precisely because one is dealing with an isotope of presumed constant nature, and the process is insensitive to environmental variation. Yet in evolution we are dealing with a situation where it is never clear we are dealing with quite the same "isotope" (the individual organism), and its rate of increase is extremely sensitive to the environment. This point was obvious to Fisher at the outset, and we should recall in this context his insistence that statements of probability in studies of populations pertain "to the hypothetical population sampled, and depend only upon its nature and circumstances. The inexactitude of our methods of measurement has no more reason in statistics than it has in physics to dim our conception of that which we measure. These conceptions would be equally clear if we were stating the chances of death of a single individual of unique genetic constitution, or of one exposed to an altogether transient and exceptional environment" (1930a, 23–24). The conceptual distinction is indeed clear, but we must remember that in this case our inability to specify the environment precisely means that it is hopeless to disentangle the effect of varying environments and varying genetic natures. Only by assuming a constant (average) environment can we ascribe variation either to innate genetic causes or to stochastic deviations from them. In the real world, while we *conceive* of "fitnesses" or expected rates of increase as physical properties like radioactive decay rates, which can be used to construct a dynamic, stochastic model of the evolution of a population, such "fitnesses" can only be *measured* retrospectively (as Fisher was quite aware) and thus can never be cleanly separated from genetic drift.

Yet this is only a general perspective. How do these general relationships work out numerically? This is most easily seen in the context of some simple models of populations, in which we know all the parameters. By examining such models, we will be able to better understand the difference between natural selection and genetic drift.

MODEL POPULATIONS

In the Wright-Fisher model of drift, the general situation contemplated is a diploid population in which each parent has a binomial or (in the limit) a Poisson distribution of expected offspring number, but with a different mean expected number of offspring (fitness) for each genotypic class. To explore the relation between variance in rate of increase and "fitness," I constructed a simple computer simulation in which a population was generated by randomly combining alleles at two loci (random mating, no linkage). With two genetic loci containing two alleles each, there were nine possible genotypes. Each had a specific expected number of offspring (fitness) and an actual number drawn from a binomial distribution around this number. If we assign the same expected rate of increase to all genotypes in this simulation (pure "genetic drift"), we see that the total variance in rate of increase is generally close to the mean fitness or rate of increase, as one would expect for a Poisson distribution. As we saw (box 8.1), since this is true for absolute rate of increase, the variance in *relative* rate of increase is equal to the mean absolute rate of increase, divided by its square; in other words, the standardized variance equals the inverse of the mean absolute rate. Thus, for a population that is growing by a factor of 10, the standardized variance is only about 0.1; while for one shrinking by a factor of 10, the standardized variance is about 10.

How much of the variance in rate of increase is associated with our genotypic classes? In a stable population (mean rate of increase of 1.0), the Poisson variance in rate of increase is about 1.0. In a typical run of the simulation, about 0.1 percent of this variance in a population of ten thousand is associated with the genotypes. In a population of only one hundred, about 10 percent of the variance is associated with the genotypes. The increase in variance is an increase in genetic drift, since (by hypothesis) no variation should exist between genotypes except by chance.

If the population has fitness (expected rate of increase) differences among genotypes, what becomes of variance in rates of increase? Consider a simple case of a recessive lethal at one of our two loci. In this case, the fitness of homozygotes is 0. If we adjust the fitnesses so that the mean rate of increase remains 1.0, then we find that the variance in rate of increase in a population of ten thousand increases from about 1.0 to about 1.33. The variance associated with genotypes is about 0.33, showing that the additional variance is indeed simply added to the existing variance. In this case, about 25 percent of the variance in rate of increase is associated with the genotypes.

What happens if we change the growth rate of the population but keep the same fitness ratios? As we have seen, the overall variance in relative rate of increase is greater in decreasing populations. Nevertheless (by our hypothesis of constant fitness ratios), the variance associated with the genotypes remains the same, so that a decreased fraction of the variance is associated with these genotypes. For a population growth rate of 0.15, only about 5 percent of the variance in rate of increase is associated with these genotypes. Conversely, in a growing population, the situation is reversed, so that in a population with a growth rate of 2.25 and these fitness ratios, about 42 percent of the variance in rate of increase is associated with these genotypes.

Finally, it is worth returning to the relation between standardized variance and traditional measures of genetic variance as modeled and measured by quantitative geneticists. Remember that Fisher (1918, 1930a, 1941) divided phenotypic variance into genotypic and environmental variance. Within the genotypic (now usually called genetic) variance, he defined genetic (now usually called additive genetic) and dominance components.

Let's examine the basic distinction between genotypic and environmental variance, which sum to give phenotypic variance. As just noted, considering the entire population of a diploid, outbred species, it is unlikely we will find two individuals with the same genotype. All are unique. Presumably, almost all of the loci have variation with some effect on rate of increase. In this situation, all variance in rate of increase is genotypic variance, in the sense that it is variance among genotypes. What became of the environmental variance? What we have come up against is a problem in the way environmental variance is conceived of in quantitative genetics. At the level of individual value, we assume that each individual has some genotypic value but that environmental causes result in individual deviations from this value. This accords with the common observation that identical twins are very similar but not truly identical, that there is phenotypic variability within inbred lines, and so forth. Variation among individuals is thus (reasonably) considered partially due to genetics and partly due to environment. But if every genotype is unique, what does it mean to specify a particular expected rate of increase for it and to regard deviations from this value as environmental noise? How could we ever hope to measure the genotypic value, when we have no replication of genotype? This is the general problem of interpreting "heritability" in the context of natural populations, where there is no ability to randomize across environments as in experimental situations (see Feldman and Lewontin 1975; Jacquard 1983; Kempthorne 1997).

In this context, it pays to remember Fisher's dictum just quoted. Although an individual in a population is genetically unique, we are presumably to regard its value (contribution to variance in rate of increase) as a sample from a distribution of possible values for that individual in a population with the same "nature and circumstances." The same nature here means the same genotypes and phenotypes, the same circumstances means the same general environment. We are to imagine rerunning the tape of this population repeatedly, and the distribution of values for that individual genotype in an infinite series of runs will determine the mean (expected) value.

While we can't rerun the tape in practice, we can consider each individual to represent a sample (of one) from such a distribution. Then for the population as a whole, the distribution of values might be expected to closely resemble the distribution expected if each individual is randomly assigned one value from the distribution for its genotype, and this resemblance should be greater the larger the population. In fact, our observation of the distribution of values depends in turn on the distribution of this distribution across our infinite series of populations. How close are the values we observe for such quantities as population growth rate, rate of increase of genotypic classes, and so forth, to those expected values? This is the fundamental question at the heart of the division of rate of increase into natural selection and genetic drift. As we have seen, the division depends ultimately on the assumption that the form of the distribution for each individual is known: it is a binomial or (in the limit) a Poisson distribution.

Given this assumption, we can see how much variation results from genetic causes, how much from environment. For example, in a run of my simulation with a population of 100, the total phenotypic variance in rate of increase was 2.10, the variance among the nine genotypes was 1.06, and the variance in genotypic value (expected rate of increase or fitness) was 0.60. At the level of individual alleles at one of the two loci, mean absolute rate of increase was 1.48 for A and 0.98 for a, compared with expected rates of 1.22 and 0.77. Most of this difference was tied to a greater–than-expected population growth rate: based on the distribution of genotypes, a growth rate of 0.91 was expected, but the actual growth rate was 1.14. At the level of relative rates of increase, the difference is much less apparent: the actual relative rates are 1.30 for A and 0.86 for a, while the expected relative rates are 1.34 and 0.84. The difference between expected rates and actual rates is, by definition, genetic drift. Interestingly, if we look at variance in relative rate of increase between alleles (variance in average excess), we find that expected variance for this population is 0.053, but actual variance is 0.043. In other words, genetic drift happens to be reducing the rate of evolution in this population.

To make one further point about drift with respect to this population, it is important to note that so far we have only dealt with one of the two possible sources of drift based on variation in organismal survival and reproduction (as opposed to drift due to segregation and differential survival during the gametic phase). In particular, we have effectively considered each individual in the population to represent a sample of one from a Poisson distribution of rate of increase for that genotype. However, we have not dealt with the fact that the distribution of geno-types is other than one might expect based on allele frequencies, mating preferences, and so on. In this case, we have assumed random mating. With allele frequencies of 0.32 and 0.68 at the A locus, under random mating we would expect relative rates of increase of the A and a alleles of 1.26 and 0.88, respectively, rather than the 1.34 and 0.84 we saw earlier as the expected value for the actual distribution of geno-types, or the 1.30 and 0.86 that we happened to observe in this particular popula-tion. Another possible comparison would be with the population expected based on "parental" allele frequencies—namely, the ones used to determine the distribu-tion of genotypes in the simulation, which had allele frequencies of 0.30 and 0.70, rather than the 0.32 and 0.68 that occurred in the run of the simulation.

Thus, even in this ideal situation, where we have assumed a population with completely known genotypes, fitnesses (expected rates of increase), frequency distributions of expected rate of increase on an individual level (Poisson), and mating scheme (random), it is rather difficult to decide what the theoretical population or "universe" is that our actual population is a sample of. In reality, the situation is much more complicated, with the distribution of individual rate of increase (realized fitness) perhaps the biggest unknown. For example, it has re-cently been argued that the Poisson assumption is unreasonable for many broad-cast-spawning marine invertebrates, which have a much higher variance in rate of increase due to effects of stochastic environmental variation on fertilization suc-cess (e.g., Hedgecock 1994; Hedrick 2005; Eldon and Wakeley 2006). Finally, one wonders what can it even mean to talk about averaging across all possible envi-ronments, with a certain probability of meeting such environments, when this is presumably a function of all other parameters of the system (distribution of geno-types, phenotypes, environments, etc.).

The upshot once again is that even in such a population, the distinction between natural selection and genetic drift appears to be impossible to make cleanly. This is true even without bringing in all of the nontraditional sources of evolutionary change we will examine later in this chapter, involving random change or change for other reasons than effects on organismal survival and reproduction. It does not seem too extreme to say that natural selection as a causal agency (narrow sense nat-ural selection) can be defined rigorously in a theoretical sense, but not in any op-erational sense, due to the sheer number of different genotypes possible and the difficulty of establishing what one means by "expected rate of increase" and "ex-pected distribution of environments" for each one. This is not a new realization. Recall that Fisher (1930a) noted that "the actuarial information necessary for the calculation of the genetic changes actually in progress in a population of organisms, will always be lacking; if only because the number of different genotypes for each of which the Malthusian parameter is required will often, perhaps always, exceed the number of organisms in the population" (44).

At the same time, to understand what is happening in populations at a descriptive level—namely, at the level of actual (not expected) rates of increase—seems by no means impossible. Nor does it seem impossible to isolate particular causal influences on these rates of increase. It is this more limited goal, rather than the complete determination of the expected evolution of a population, that one might realistically hope to achieve. Nevertheless, it also seems reasonable to believe that in general what one does observe will be fairly close to what one might expect to observe under some expected distribution of genotypes and environments, and the more so the larger the population observed.

With this general introduction to the physical meaning of genetic drift, I would like to turn to a brief historical overview of the origin of the concept. We will see that although the mathematical development indeed justifies calling this the Wright-Fisher or Fisher-Wright model, the theory of drift played a very different role for Fisher than it did for Wright. This point again is connected to the question of how we want to regard our models of populations: are they merely heuristic aids to understanding what is happening in the real world, or are they explicit representations of real relations, of which the appropriate parameter values can be determined for real populations?

THE HAGEDOORNS, FISHER, AND
THE ORIGINS OF GENETIC DRIFT

Because both Darwin and Wallace had made natural selection a mechanism that acts in accordance with the laws of probability, rather than completely deterministically, a role for chance in evolution was implicit from the beginning. However, with the rise of Mendelism, the statistical nature of the Mendelian ratios made it inevitable that chance should now prominently enter evolutionary theory, where it had previously been a minor factor (except in its role in the "chance" origin of variants). Hagedoorn and Hagedoorn (1921) took advantage of this opportunity to call the ubiquity of selection into question with their infamous book on *The Relative Value of the Processes Causing Evolution*, in which they suggested that random factors might account for nonadaptive species differences (Provine 1986; Beatty 1992).

Fisher's general opposition to any significance for random factors in evolution is well known. In 1921, he reviewed the Hagedoorns' book for *Eugenics Review*. He noted that the authors believed "that the random selection of individuals of one generation to become the parents of the next generation is a more important factor in reducing the variability of a species than is the natural selection of advantageous characters, while they also believe that crossing is a far more important factor in increasing variability than mutation. The whole question is worthy of a thorough discussion, but the authors evidently lack the statistical knowledge necessary for its adequate treatment" (467–468). He went on to briefly examine this question himself:

> By taking simple examples of species of *only two or three individuals* it is easy to show that by random selection the variability will be rapidly reduced. It by no means follows that the reduction will be equally rapid in an interbreeding group of 2 or 3 million individuals. The present reviewer has examined this particular point, and finds that in the absence of mutation or crossing, a perceptible reduction in variability will take place in a number of generations equal to four times the number of the species breeding. In the case of great seasonal fluctuations in number, it is only fair to take the number of interbreeding individuals at its lowest point, but even so the number of interbreeding individuals in any group will seldom be of less than five figures. (468)

And the following year, long before Wright had published anything on evolution, we find Fisher already quite opposed to the idea that random survival could be of any evolutionary import:

> The interesting speculation has recently been put forward that random survival is a more important factor in limiting the variability of species than preferential survival. The ensuing investigation negatives this suggestion. The decay in the variance of a species breeding at random without selection, and without mutation, is almost inconceivably slow: a moderate supply of fresh mutations will be sufficient to maintain the variability. When selection is at work even to the most trifling extent, the new mutations must be much more numerous in order to maintain equilibrium. (Fisher 1922a, 323–324)

Nevertheless, in the same paper, Fisher constructed a probabilistic model for genic survival that showed that random factors are important to consider with respect to the survival of individual mutant genes. Fisher's basic model derives from an analogy with the gas laws (Provine 1986, 241), and his statistical distribution of gene frequencies is clearly analogous to the Maxwell-Boltzman distribution for the velocities of molecules of an ideal gas (see Morrison 2000). Biologically, the situation considered was an annual plant, such as a field of cross-fertilized grain. In such a situation, each adult individual "is the mother of a considerable number of grains and the father, possibly, of an almost unlimited number" (1922a, 325). Each ovule or pollen grain has a very small chance of survival, and Fisher thought it reasonable that the probability distribution for the number of ovules or pollen grains destined to become mature plants in the next generation could be approximated by the Poisson distribution, with the mean of the distribution representing the population growth rate.

This model is the historical source for the concept of genetic drift; it suggested that the early stages of survival of a mutant are determined largely by random factors, but the eventual fate of a mutant depends almost entirely on its selective advantage (average excess). As he summarized his findings some eight years later, in the *Genetical Theory of Natural Selection*, "the range of selective advantage which may be regarded as effectively neutral is . . . extremely minute, being inversely proportional to the population of the species. Since it is scarcely credible that such a perfect equipoise of selective advantage could be maintained during the course of evolutionary change, random survival, while the dominant consideration in respect to the survival of individual genes, is of merely academic interest in respect to the variance maintained in the species" (1930a, 118).

Nevertheless, in a 1931 letter to E. B. Ford commenting on Wright's review of the *Genetical Theory*, Fisher notes, "He thinks I have overlooked a major factor in the effect of random survival in small isolated colonies, but though I see it may be of special importance in some cases, . . . I do not appreciate how it can generally favour a more rapid progress in *adaptive* modification. Probably he will develop the view more fully later, when it will be possible to judge better how much weight should be given to it" (Bennett 1983, 199). Fisher's apparent open-mindedness in this letter is interesting in light of the extended controversy Fisher and Ford were to engage in in subsequent years with Wright on just this point (see Provine 1986, 420–437). It is clear there was nothing in Wright's further work that made Fisher feel the need to reevaluate his initial position.

THE "SEWALL WRIGHT EFFECT"

In contrast, the testimony of systematists that most differences among species were nonadaptive led Wright (1932) to conclude that "The principal evolutionary mechanism in the origin of species must thus be an essentially nonadaptive one" (364). He used accidents of sampling in small subpopulations to explain nonadaptive differences between races, species, and even genera (Wright 1931, 1932; see Provine 1986). His background in animal breeding, in particular, had led him to see great value in inbreeding (with concomitant loss of genetic diversity) as a way to reveal differences among stocks, which could then be selected among and crossbred to restore vigor (e.g., Wright 1923; see also Wright 1978). As we have seen, this was the core of the shifting-balance theory.

Because random genetic drift through accidents of sampling became known as the "Sewall Wright effect" and this drift played a crucial role in his shifting balance theory of evolution, it is only natural to assume that this is the major role that Wright saw for random factors in evolution. In his 1931 paper, which had its origin in Fisher's (1922a) work, random drift due to restricted population size indeed appears to be the only random factor in evolution; all other factors are systematic pressures, determined by particular (constant) coefficients. The evolutionary role of random drift in this initial version of the shifting balance theory is to permit nonadaptive differentiation of local populations, exploring gene combinations that would be unlikely to arise in a large population, and thereby increasing the overall rate of evolution of the species. However, while this does appear to represent Wright's initial conception, his views evolved significantly over time (Provine 1986; Beatty 1992).

In 1948, Wright wrote a paper "On the Roles of Directed and Random Changes in Gene Frequency in the Genetics of Populations" as a response to Fisher and Ford's paper of the previous year, which denied any important role for random drift in the fluctuations in frequency of the *medionigra* gene in the moth *Callimorpha* (formerly *Panaxia*) *dominula* (Provine 1986; Beatty 1992). Here he vigorously defended his shifting balance theory as one that was based not on random changes in gene frequency but instead on "the simultaneous treatment of all factors by the inclusion of coefficients measuring the effects of all of them on gene frequency in a single formula" (Wright 1948, 281)—that is, in a dynamic model of evolution. Moreover, he pointed to random fluctuations in selection as a factor with the same evolutionary import as sampling error and noted that he had suggested something like this in a passage in the 1931 paper. Of course, as Fisher and Ford (1947, 1950) pointed out, such fluctuations in selection apply equally to small and to large populations and thus argue against the evolutionary significance of population subdivision that Wright had always insisted on.

Nevertheless, apparently beginning at this point, Wright came to formulate a new classification of evolutionary factors, first displayed in his influential 1949 paper on "Adaptation and Selection." There his classification of the "Modes of Immediate Change of Gene Frequency" was as follows:

1. *Systematic change* (Δq determinate in principle).
   A. Pressure of recurrent mutation.
   B. Pressure of immigration and crossbreeding.
   C. Pressure of intragroup selection.
2. *Random fluctuations* (δq indeterminate in direction but determinate in variance).
   A. From accidents of sampling.
   B. From fluctuations in the systematic pressures.
3. *Nonrecurrent change* (indeterminate for each locus).
   A. Nonrecurrent mutation.
   B. Nonrecurrent hybridization.
   C. Nonrecurrent selective incidents.
   D. From nonrecurrent extreme reduction in numbers.

Finally, he concluded, "Back of these processes are all factors that cause secular changes in mutation rates, conditions of selection, size and structure of the population and possibilities of ingression from other populations" (1949, 369).

Wright's classification is striking. Random processes now included fluctuations in selection, mutation, immigration, and crossbreeding, not just accidents of sampling. Yet what can such random fluctuations mean, if not seen against the background of a presumed first-order constancy of these parameters? While in 1931 Wright had assumed constant parameter values as a mathematical convenience, he here seems to have reified his model of the evolutionary process to the point that he takes his assumed constant parameter values as evolutionary modes distinct from fluctuations or secular trends in these values. In any case, it is clear that random factors have come to have an entirely different meaning for Wright than they did in 1931 and 1932. Thus, in a 1956 paper on the "Classification of the Factors of Evolution" (see also Wright 1978, 444–446), Wright concludes, "The directed processes may be described as producing a cumulative *steady* drift in gene frequency, complementing the term *random* drift that has been used for the cumulative effects of the random processes. The common restriction of "random drift" or even "drift" to only one component, the effects of accidents of sampling, tends to lead to confusion" (1956a, 18). One might wonder on which side of this usage the confusion lies.

What I have tried to show in this brief historical overview is that the concept of random genetic drift as an evolutionary force or process explanatorily equivalent to natural selection derives from Wright's perspective, even though the mathematics had first been developed by Fisher. This is because Wright was occupied in trying to construct a dynamic, if probabilistic, mathematical theory of evolution. Where Fisher had given up on a complete dynamical theory of evolution at the outset, considering it beyond human possibility to ever measure the relevant variables, and instead focused on understanding what natural selection does as a predictable *part* of the evolutionary process, Wright was always focused on developing just such a comprehensive theory, however inadequate it might be. It is no wonder they failed to understand each other.

DRIFT AND THE NEUTRAL THEORY OF
MOLECULAR EVOLUTION

Of course, the major place where random genetic drift now enters evolutionary theory and practice is in studies of molecular evolution, where the consensus is that much nucleotide substitution that occurs is due to random drift. At the same time, phenotypic evolution is largely considered exempt from such drift and instead due to natural selection. In the remainder of this chapter, I would like to examine the evidence for both of these views.

The early history of the neutral theory has been reviewed by Lewontin (1974) and Kimura (1983). Kimura (1983, 25) tells us that the three major features which led him to propose the "neutral mutation-random drift hypothesis" were the approximate uniformity of rate of amino acid substitutions across different lineages, an apparent randomness in the pattern of substitutions, and a high overall rate of substitution. The high rate of substitution was considered a problem because of the arguments made by Haldane (1957) on the "cost of natural selection" (i.e., genetic load associated with substitution of amino acids by natural selection; see chap. 8). In fact, this impossibly high "substitutional load" was the prime argument initially used by Kimura (1968) in proposing the neutral theory. On the face of it, this argument is rather strange, since neutral substitution must involve precisely the same variation in rates of increase as selective substitution. That Kimura didn't see this is at the very least an interesting historical anomaly.

In any case, the main argument now made for the predominance of neutral changes in molecular evolution is the observation that evolution is fastest at those nucleotide and amino acid sites that are functionally unconstrained—that is, at sites that we expect changes to make little difference to the overall functioning of the organism. Thus, at the amino acid level, substitution rates are much higher at areas outside the active site of enzymes than within them; at the nucleotide level synonymous substitution rates are generally higher than nonsynonymous, and extremely high nucleotide and amino acid substitution rates are seen in pseudogenes—those duplicated genes that have apparently lost any specific function. The relative lack of change in functionally important parts of proteins is such a standard feature of protein evolution that evolutionary conservation of sequence is used to predict such regions (Simon, Stone, and Sidow 2002).

When examining nucleotide and amino acid substitution rates, the neutral theory functions as a null model, with which the actual substitution rates in different regions of the genome can be compared. Given a particular effective population size and the assumed binomial distribution of offspring gene number probability for any given gene, the mean absolute rate of random drift can be predicted. In practice, the effective population size is not known, and all that can be done is to see whether a rough estimate of effective population size is consistent with the prediction of evolutionary rate based on genetic drift. Nevertheless, the vast quantity of molecular data that can now be relatively easily accumulated has stimulated the development of tests to distinguish evolutionary change due to genetic drift from that due to natural selection, based on patterns of genetic variation within and between populations. Let us take a look at some of these tests.

MOLECULAR TESTS OF DRIFT AND SELECTION

By assuming that survival of neutral alleles is a binomial process and that mutations in a certain region (e.g., a pseudogene) or of a certain type (e.g., a silent substitution) are neutral, substitution rates in other regions or of other types can be compared; and if they differ significantly from this neutral expectation, then this can be taken as evidence for nonrandom survival or selection. This approach is the basis for the two fundamental tests applied to molecular evolution: the McDonald-Kreitman (MK) test and the test of the ratio of nonsynonymous to synonymous substitutions (dN/dS). An excellent review of these tests is provided by Eyre-Walker (2006).

The dN/dS (or K_a/K_s) test (Yang and Bielawski 2000; Eyre-Walker 2006) is perhaps the simplest test for natural selection based on patterns of variation in natural populations. This test compares the rate of nonsynonymous (dN) to synonymous (dS) substitutions within a gene (where the substitutions are inferred by comparing two or more species). The null hypothesis here is always that synonymous and nonsynonymous sites evolve at the same rate, while reduced evolutionary rate at nonsynonymous sites is taken as evidence of negative selection (selection against deleterious mutants), and increased evolutionary rate is taken as evidence of positive selection (selection for advantageous mutants). Special interest in recent years has attached to estimates of selection at single amino acid sites (codons), where inclusion of a large number of sequences allows tests of the dN/dS ratio even for such single codons (Kosakovsky Pond and Frost 2005; Eyre-Walker 2006). In practice, one must account for such issues as unequal mutation rates (e.g., transition/transversion ratios) and codon bias (Yang and Bielawski 2000), but after attempting to do so, many examples of genes or codons that appear to have undergone episodes of positive selection have been found. Nevertheless, these tests can certainly give false positives, as indicated by conflicts between the results obtained with different tests (e.g., Nunney and Schuenzel 2006; Sorhannus and Kosakovsky Pond 2006).

The MK test (McDonald and Kreitman 1991) compares the numbers of polymorphisms within populations of a species to the number of substitutions between two species at two kinds of sites, at one of which variation is assumed to be neutral. This is done most commonly for synonymous versus nonsynonymous sites. The specifics of the test are not important here, but the assumptions of the underlying model are worth examining, as these are even more restrictive than those for the dN/dS test. These include complete neutrality of synonymous substitutions, a distribution of nonsynonymous mutations into three (constant fitness ratio) classes (strongly deleterious, neutral, and strongly advantageous), and the rarity of advantageous mutations (Eyre-Walker 2006). As Eyre-Walker pointed out, this test is also dependent on a constant population size (or at least a known history of population size) and is fairly sensitive to deviations from constancy. In fact, in a recent maximum likelihood study comparing *Drosophila simulans* with *D. melanogaster* and *D. yakuba*, the MK test suggested that 40 percent of nonsynonymous substitutions were driven by positive selection, supporting an earlier study by Smith and Eyre-Walker (2002) that put the figure at 45 percent, but lack of variation in this value among loci suggested that this apparent selection was likely an artifact of demographic history (Welch 2006).

While I have found it necessary to note certain shortcomings of tests for the relative role of drift versus selection in molecular evolution, lest we be too prone to accept them at face value, let us assume for a moment that such shortcomings could be resolved. In other words, let us assume that we could determine precisely, for each nucleotide difference in a comparison of sequences, which had been fixed by positive selection, which by genetic drift. What exactly would this mean? Again, we are back to many of the issues discussed in a more general context earlier. First, here *genetic drift* means explicitly genetic drift of neutral alleles. In particular, these alleles must be neutral not just at one point along the phylogeny but throughout the phylogeny. Likewise, positive selection here means a certain fixed selection coefficient on each branch of the phylogeny (or at least an average coefficient that can serve in its place). So, already we see there is little room in this formulation for substitutions that fluctuate from "advantageous" to "neutral" to "disadvantageous" at different points along each branch.

Second, what one is measuring is clearly "selection of" the substitution, not "selection for" the substitution. In other words, one is looking at the average excess in fitness, not the average effect on fitness of the alleles. As I will discuss further later in the chapter, other factors besides selection and drift such as linkage (and associated "genetic hitchhiking") and mutational bias can also cause rates of substitution at a site to differ significantly from the neutral substitution rate.

Finally, even if one were to assume that all the substitution rates that are significantly greater than the neutral rate are in fact due to selection *for* the specific amino acid change that occurred, one still wouldn't know when in the life cycle or at what level this selection was acting—it could have been during gametogenesis, gamete survival, fertilization efficiency, or during embryonic development or adult life.

In summary, while the prospect of robust statistical tests distinguishing selection from drift based on nucleotide diversity among populations is appealing, in practice one is not really asking the question that it appears most people would like to ask: to what extent does selection for specific changes due to their effect on survival and/or reproduction account for sequence evolution? Instead, one is only able to ask the question: to what extent does nonrandom survival and reproduction versus random survival and reproduction account for sequence evolution?

My last comment in this section has to do with a matter of terminology. In the study of substitution rates, all differences from the neutral rate are assumed to be due to selection, and positive selection is equated with "adaptive evolution." Thus, in his review of methods for detecting selection, Eyre-Walker (2006) considers that he is examining the question of "The Genomic Rate of Adaptive Evolution." He concludes optimistically: "For over 30 years, scientists have debated, often bitterly, the relative contributions of genetic drift and adaptation to evolution at the molecular level. Although we are still some way from resolving the controversy, it is clear that we should know the answer within the next few years" (574). Eyre-Walker here clearly considers that all nonrandom survival and reproduction are due to selection and hence, by definition, are "adaptive." Drift and adaptation (by natural selection) are the only two alternatives in play. Nevertheless, he is forced to confess his mystification when looking at evolution from a quantitative standpoint:

> *Drosophila melanogaster* and *D. simulans* are estimated to differ from one another by at least 1.3 million adaptive differences; even if we focus on the protein-coding complement of the genome they appear to have ~110 000 adaptive amino acid differences.... And yet these species are almost identical morphologically. What do all these adaptive differences do? It might be that many of them are involved in the physiology and ecology of these species, something that we know remarkably little about. And it might be that some of the adaptive substitutions are a consequence of arms races between host and parasite. But it is still difficult to comprehend how so much adaptive evolution can be going on. It might be that we just have no idea how complex the environment really is and how it is constantly changing in ways that always challenge an organism to adapt. (574)

Or perhaps we are seeing changes that do not reflect the kind of "adaptive evolution" that Eyre-Walker envisions.

PROBLEMS IN PARADISE

In spite of the apparent simplicity of approaches to studying the causes of molecular evolution based on among- and within-population diversity, in particular, in assigning molecular changes either to drift or to selection, it is well known that many other phenomena can affect patterns of nucleotide diversity. I will spend the remainder of this chapter examining a few of these challenges to the simple contrast of drift versus selection in a little more detail. My goal is not to discredit attempts to study the causes of molecular evolution, but rather to note the complexity of the genetic situation that we are now aware of, a complexity that was not envisioned at the time when Fisher, Wright, and Haldane established the basic theory of population genetics.

Hitchhiking

One of the most widely acknowledged complications to the simple equation between substitution rates significantly different from neutral expectation and selection is the phenomenon of "genetic hitchhiking" or "selective sweeps," and its opposite, "background selection." Genetic hitchhiking (Maynard Smith and Haigh 1974) is the association of alleles due to linkage, so that an allele can be carried to a high frequency due to linkage with an allele at a different locus that is experiencing "positive selection," meaning of course selection *for* the change. As already pointed out, the hitchhiking allele is itself experiencing positive selection of another sort, selection *of* the change. The pattern of reduced variation in a region due to such hitchhiking is the associated phenomenon of a selective sweep of all variation in the region. By contrast, the phenomenon of background selection (Charlesworth, Morgan, and Charlesworth 1993; Charlesworth, Charlesworth, and Morgan 1995) involves a decrease in frequency of an allele due to association with deleterious alleles (alleles selected *against*) at linked loci; this also produces a region of reduced variation, and much attention has focused on how to distinguish these two potential causes of reduced variation (see, e.g., Inman and Stephan 2003). Such hitchhiking and background selection are obviously at the mercy of recombination, which breaks down linkage.

At first one might think that hitchhiking and background selection are really just two sides of the same phenomenon, with hitchhiking focusing on those selected alleles with relative rates of increase greater than one, and the alleles linked to them, while background selection focuses on the alternative selected alleles with relative rates of increase less than one, and the alleles linked to them. However, they are really two ends of a continuum, because with hitchhiking, one is dealing with a selectively favored rare allele, which is therefore increasing in frequency, while with background selection, one is dealing with selectively disfavored rare alleles, with their frequency being maintained by mutation-selection balance. That these are two ends of a continuum can be seen if we think about the process of a selective sweep. At the beginning of the process, the favored allele is rare and linked alleles increase by hitchhiking; but at the end the process the disfavored alleles are rare and are maintained at a low frequency by mutation-selection balance, which likewise holds down the frequency of linked mutants.

Gillespie (2000, 2001, 2004) has noted that at a population level, hitchhiking and background selection will have a similar effect to genetic drift in reducing overall variability; he calls this effect "genetic draft" as effectively "neutral" alleles get pulled along with their linked selectively favored or disfavored alleles. In an interesting argument, he suggests that this may explain the relative independence of levels of standing variation in a population and population size, which is contrary to expectations based on at least some versions of the neutral theory. In Gillespie's view, small populations undergo increased drift, which reduces variation, and large populations undergo increased draft (because selection is more effective in large populations), which reduces variation; the net result is little difference in levels of standing genetic variation across a wide range of population sizes. While the jury is still out on this, a recent paper (Bazin, Glémin, and Galtier 2006) supports the effect of genetic draft at least for mitochondrial evolution, which involves very little recombination.

Meiotic Drive

Another exception to the simple contrast of selection and drift involves meiotic drive. Meiotic drive (Sandler and Novitski 1957) is the selection that occurs due to distortion of normal Mendelian ratios in diploid populations; loci experiencing meiotic drive do not show normal Mendelian inheritance; rather, heterozygotes consistently produce sperm or eggs with a preponderance of one allele (reviewed by Burt and Trivers 2006). The two classic examples are the *t* locus of the house mouse *Mus musculus* and the *SD (segregation distorter)* alleles of *Drosophila melanogaster*. While this is clearly selection (and can be selection in the narrow sense), it is not the sort of selection commonly envisaged in studies equating positive selection with "adaptive evolution," although in most molecular tests it will be indistinguishable from changes due to differences in organismal survival and reproduction. In their recent review, Taylor and Ingvarsson (2003) note that the meiotic drive systems that have been identified share a number of features, such as sex linkage and deleteriousness, and they ask whether these characteristics are necessary features of such systems or just a biased sample due to the ones that we are likely to notice. They conclude in favor of the latter view, suggesting that meiotic drive may be a much more common phenomenon than usually allowed. Chevin and Hospital (2006) have proposed that such transmission ratio distortion (TRD) may not only be more common than previously realized but account for much hitchhiking of linked alleles.

Molecular Drive

Molecular drive is a term introduced by Dover (1982, 2000, 2002) to describe a variety of mechanisms (gene conversion, unequal crossing over, slippage, transposition, and retroposition) that can result in spread of a particular variant throughout a multigene family, effectively homogenizing such a family. The resulting pattern of "concerted evolution" of multigene families is clearly real for some, such as the rDNA genes in which it was first identified. However, its prevalence has been questioned. For most such families, more recent work has suggested that a birth and death process among repeats accounts for within-family evolutionary change (Nei and Rooney 2005). Nevertheless, examples of concerted evolution and molecular drive continue to turn up (Johannesson et al. 2005) and appear to be particularly prevalent in the noncoding repetitive DNA sequences found throughout most genomes (e.g., Sharma and Raina 2005; Rudd, Wray, and Willard 2006).

Given that the molecular mechanisms underlying concerted evolution exist, how are we to regard them? In other words, is molecular drive a new evolutionary "force" or process in addition to mutation, selection and drift, as Dover argued strongly, or is it merely a form of one or more of these processes? If we consider what is actually happening during unequal crossing over, to pick just one example, we find that a repeat appears more frequently in one of the homologous chromosomes and less frequently in another. From one perspective, this is merely a mutation that happens to affect two homologous chromosomes in a correlated way. At a population level, this can only matter if one or both of the two mutant homologous chromosomes produced increase in frequency in the population—and this can happen randomly (drift) or due to cause (selection). Alternatively, the same unequal crossing over event can occur repeatedly, which would constitute an example of "mutation pressure" in standard population genetic terms. Considered from this point of view, it appears that no new evolutionary principle is in play.

Conversely, from the viewpoint of continuity of information, focusing on the lineage of the duplicated allele, one could regard the initial unequal crossing over event as a kind of *intragenomic selection* within this individual cell for (or of) the duplicated DNA section in one daughter chromosome, and against it in the other. Even from this point of view, intragenomic selection can only do so much without a population-level process to increase the frequency. Mutation pressure, or repeated mutation to the same form, could be such a process and would constitute the population-level expression of intragenomic selection. However, under this perspective, molecular drive turns out to be an example of meiotic drive, in which the distribution of alleles among gametes produced does not simply represent a random sample of the parental allelic distribution.

In summary, whether considered under either the "mutational" or "intragenomic selection" point of view, molecular drive does not seem to deserve to be enshrined as a distinct evolutionary factor, as Dover has argued. Unequal crossing over, gene conversion, and the like, can clearly be incorporated into the standard framework (Burt and Trivers 2006). Nevertheless, Dover's broader point is certainly correct: this sort of intragenomic selection is not traditional Darwinian selection at the organismal or gametic level, and does appear to play a role in the evolutionary process.

Fluctuating Selection

Another complication that affects attempts to assign molecular evolution to drift or selection is the undoubted fact that selection coefficients (average excesses) of alleles are not constant but vary in time and space. In fact, it is one of the gravest problems with the standard models that each mutant form is supposed to have a fixed relative fitness—to be classifiable as "deleterious," "neutral," or "advantageous." The artificiality of this assumption is obvious, but the simplicity of models that assume this constancy, as compared to those with variable selection coefficients, continues to encourage their use. The appearance of silent sites and pseudogenes as (almost) neutral in terms of function has tended to support this view, so that many people now think that it makes sense to talk of "neutral mutations" and even of "deleterious mutations" and "advantageous mutations" as if these were properties of the mutation itself.

The most vocal heretic on this count has again been John Gillespie, who was almost alone for many years in continuing to emphasize the important effects of temporally fluctuating selection coefficients on population-level parameters, particularly with respect to the maintenance of variation. In his 1991 book on *The Causes of Molecular Evolution*, Gillespie provides an interesting historical review of work with fluctuating selection coefficients (228–230). His account may be summarized as follows.

In 1948, Wright published a paper "On the Roles of Directed and Random Changes in Gene Frequency in the Genetics of Populations."* However, Wright made an error by not including a variance term in his drift coefficient, thus concluding that variable selection would only drive variation from the population. Kimura (1954) followed Wright in this error and concluded that the allele with the greatest arithmetic mean fitness would prevail. In 1955, Dempster published an analysis of a haploid model that corrected this error and showed that the genotype with the greatest geometric mean fitness would prevail. Nevertheless, in both the Wright/Kimura model and Dempster's model, variation was reduced. Finally, in 1963, Haldane (in his retirement in India) published a paper with Jayakar examining a diploid model and showed that polymorphism could be maintained if the geometric mean fitness of the heterozygote was greater than that of the homozygotes. In the 1970s, four population geneticists—Gillespie, Jensen, Felsenstein, and Ewens—realized the error in Wright's diffusion and supported Dempster's conclusion, leading to the general realization that fluctuating selection could maintain variation (reviewed in Felsenstein 1976). Nevertheless, many population geneticists to this day (i.e., 1991, when Gillespie was writing) fail to recognize that fluctuating selection can maintain variation. Gillespie finds it "interesting" that a simple mathematical error could have had such long-lasting effects.

The heart of Gillespie's book is the development of a class of models that he calls SAS-CFF (stochastic additive scale–concave fitness function) models, involving diploid populations with random variation in selection coefficients. His goal is to show that much of the amino acid variation in populations explained by the neutral theory of Kimura (1983) as a transient phase in the genetic drift of neutral substitutions can be explained better as the result of fluctuating selection coefficients that lead to an average heterozygote superiority, thus helping maintain variation. He bolsters his argument by discussing evidence that almost all changes to proteins have some effect on their kinetic properties or stability and by noting the adaptive nature of at least some amino acid substitutions in proteins, such as high-altitude adaptations in hemoglobins.

---

* This is the same paper discussed earlier, written in response to Fisher and Ford's (1947) criticisms of the evolutionary significance of random drift.

I am not competent to judge the relative merits of Kimura's and Gillespie's models as a description of molecular evolution, although it is clear that both depend on a number of assumptions that can be only partially justified. What is relevant to us here is that a completely neutralist model (Kimura's) and a completely selectionist model (Gillespie's) can produce similar distributions of allele frequencies, so that it is difficult to decide between them on the basis of evidence from standing variation in natural populations. Moreover, as Gillespie stressed, a theory based on fluctuating selection coefficients is certainly more realistic than one with constant selection coefficients. Nevertheless, these models have received little attention, owing to the difficulty of determining exactly what parameters to use for the models (the problem of "specification") and their mathematical difficulty. This situation has now begun to change, and a number of recent studies have begun to explore the potential of fluctuating selection to explain diversity in natural populations (e.g., Dean 2005; Huerta-Sanchez, Durrett, and Bustamante 2008).

Mutational Bias

The analysis of DNA sequences in the search for selection has made it clear that a null model must not only incorporate random propagation (genetic drift) but also account for biases in the mutations that occur. From the earliest days of molecular sequencing, it has been apparent that at the molecular level, all mutations are not equally likely. One of the first forms of mutational bias to be discovered was transition-transversion bias, with an excess of transitions (purine > purine or pyrimidine > pyrimidine) over transversions (purine > pyrimidine or vice versa) commonly being observed (Jukes 1977; Kimura 1980). Rosenberg et al. (2003) have shown that the apparent rate variation among lineages in this ratio is mostly an artifact of error in estimation, and the ratio is consistently about 4 across a variety of mammalian species. To find this consistency, however, they had to remove the effect of CpG dinucleotide hypermutability.

CpG hypermutability is the tendency of CG dinucleotides to mutate to TG (Bird 1980). This tendency is correlated with levels of DNA methylation, because the 5-methylcytosine (5mC) is primarily found in mCpG dinucleotides that have a much higher (tenfold in mammals; Rosenberg et al. 2003) tendency to mutate to TpG than do unmethylated CpG dinucleotides (Bird 1980). This CpG hypermutability means that GC-rich regions of the genome mutate at higher rates than do AT-rich regions.

One area where mutation pressure has been generally acknowledged as a key evolutionary factor is in the phenomenon of codon usage bias (e.g., Powell and Moriyama 1997). Because of the degenerate nature of the genetic code, most amino acids have several different codons that code for them. However, these codons are not used equally frequently. In fact, the frequencies of use vary both among taxa and among genes. A general rule is that the overall codon usage bias reflects the overall GC content of the genome, but highly expressed genes tend to depart from this significantly in the direction of the isoaccepting tRNA pool available. This effect is particularly prominent in microorganisms (Ikemura 1985). The effects of expression level have generally (and reasonably) been attributed to constraints associated with efficiency of translation, but the correlation with overall GC level has been attributed to biased mutation (see Necşulea and Lobry 2006).

A comprehensive study of thirty-seven nematode species concluded that overall GC content was the major factor explaining codon usage bias (Cutter, Wasmuth, and Blaxter 2006). Constraints associated with translation efficiency only appeared to be important in free-living species, which the authors attributed to large population size and consequent increased effectiveness of selection. Interestingly, in mammals such an effect is less evident, and in a thorough study in humans, no evidence for such an effect could be found (Urrutia and Hurst 2001). The overall picture we are left with, then, is of mutation pressure that is variably effective across species and regions of the genome, depending on the level of functional (selective) constraint on the sequences.

Recombinational Bias

Another factor affecting patterns of molecular evolution is recombinational bias. Rates of recombination vary across the genome: certain areas are "recombination hotspots," others are "recombination deserts." In a survey across the entire human genome, Myers et al. (2005) found more that twenty-five thousand such hotspots. Interestingly, these hotspots seem to be extremely labile in evolutionary time, with little conservation between humans and chimps (Ptak et al. 2005) and much interindividual variation in human populations (Jeffreys and Neumann 2005). While such hotspots have obvious implications for patterns of linkage disequilibrium, they can also affect patterns of nucleotide evolution. Birdsell (2002) provided evidence that in yeast, biased mismatch repair is responsible for biased gene conversion in regions of high recombination, increasing their GC content. Such mismatch repair, he proposed, could counter an overall mutational bias toward AT. In a recent paper examining processed pseudogenes, Khelifi et al. (2006) argue that in mammals, a tendency for biased gene conversion in regions of high recombination rate can explain much of the evolution of GC content and may also help explain the GC rich isochore structure of mammalian genomes. In contrast, a thorough examination of the relation of recombination to genetic diversity in humans by Spencer et al. (2006) found only weak evidence for biased gene conversion as the primary factor controlling GC content. Nevertheless, the biochemical evidence for biased gene conversion associated with recombinational hotspots cannot be doubted (e.g., Jeffreys and May 2004). From a larger perspective, such gene conversion can be viewed as either selection or mutation, depending on one's perspective (see my earlier comments on molecular drive). Interestingly, a recent study suggests that gene conversion is more important than recombination in the generation of haplotype diversity (Morrell et al. 2006). Moreover, meiotic drive of specific alleles may commonly occur in association with gene conversion at recombination hotspots (Jeffreys and Neumann 2005).

Jumping Genes

Since their discovery by Barbara McClintock in the 1940s, transposable elements have represented a challenge to traditional Mendelian genetics. Their evolution also doesn't fit the traditional population genetics model based on point mutation in individual, protein-coding genes. In 1980, soon after the appearance of Dawkins's book *The Selfish Gene* (1976), two papers appeared proposing that DNA might itself be "selfish": Doolittle and Sapienza's "Selfish Genes: The Phenotype Paradigm and Genome Evolution," and Orgel and Crick's "Selfish DNA: The Ultimate Parasite." Estimates suggest that more than 35 percent of the human genome is made up of repetitive DNA derived from active or inactive transposable elements. From the initial hypothesis that transposable elements multiply selfishly, as "parasites" oblivious to the benefit to their "host" organism, has come a recognition that transposable elements can come to serve many roles at the organismal level, especially over evolutionary time (Kidwell and Lisch 2000, 2001). For example, Kidwell and Lisch propose that the shuffling of immunoglobulin genes that forms the basis of adaptive immunity in mammals arose from a mechanism of excision derived from a transposable element. As with other mechanisms of DNA evolution discussed here, transposition of such elements is "selection" at the level of DNA but "mutation" at the level of the organism. For an excellent recent review of transposable elements and other "selfish genetic" phenomena, see Burt and Trivers (2006).

Horizontal Gene Transfer

An even greater challenge to the basic system of Mendelian inheritance is the phenomenon of horizontal transfer of genes, which may occur within populations (lateral gene transfer), between closely related species, or between organisms as different as bacteria and plants (Syvanen and Kado 2002). Although quantitatively it is apparently a relatively rare event in prokaryotes, usually accounting for no more than 2 to 3 percent of the genome (Koonin et al. 2002; Kurland, Canbeck, and Berg 2003; Ge, Wang, and Kim 2005; Kurland 2005) and even rarer in eukaryotes (Andersson 2005), there is no doubt that it occurs and is of significant importance in evolution, particularly among microbes (Novozhilov 2005). As with other challenges to the simple contrast of selection and drift at the level of individual organisms, such horizontal transfer can be seen either as mutation or as selection/drift, depending on the focus of one's attention (gene or organism).

DRIFT AND MUTATION PRESSURE
IN PHENOTYPIC EVOLUTION

I began this chapter by noting that three main evolutionary processes are admitted by the synthetic theory: mutation, selection, and drift. At the molecular level, the mechanisms of mutation are known, as is the range of potential genotypes that can be realized. We have seen that accurate models of molecular evolution depend on knowledge of the biases of the mutation process, and their relation to others factors such as recombination. Moreover, we have seen that linkage, through hitchhiking and background selection, can result in an effect that Gillespie calls genetic draft, in which favorably and unfavorably selected alleles draw linked alleles with them. Finally, genetic drift of neutral alleles is acknowledged by all to be important for at least some regions of the genome. Given the acknowledged role of these factors in molecular evolution, the question then becomes: Do such factors have any effect at the phenotypic level, the level of bodies, with their physiology and behaviors?

The standard textbook view (e.g., Ridley 2004; Futuyma 2005; Freeman and Herron 2007) is that all significant phenotypic change is due to selection, with drift responsible only for neutral changes with no presumed effect on the phenotype or occasional minor nonadaptive changes. Likewise, in the initial proposals of the neutral theory, Kimura (1968) and King and Jukes (1969) didn't envision any effect at the phenotypic level, and this attitude was still maintained by Kimura in his 1983 book. However, by 1991, three years prior to his death, Kimura was promoting a rather different view: "If the neutral theory is valid so that the great majority of evolutionary changes at the molecular level are controlled by random genetic drift under continued input of mutations, it is likely that selectively neutral changes have played an important role in the origin of life and also in phenotypic evolution" (5972). Under an acknowledged influence from Gould's book *Wonderful Life* (1989), Kimura here envisions a four-stage process of evolution. In the first stage, a species is "liberated from the preexisting selective constraint," as happened to mammals when the dinosaurs died. In the second stage, a wide range of phenotypically neutral variation accumulates by mutation and drift. In the third stage, the "latent selection potential" of some of the neutral mutants is realized in the exploitation of new niches. Finally, in the fourth stage, "intergroup competition and individual selection lead to extensive adaptive evolution, creating a radically different taxonomic group adapted to a newly opened ecological niche." There are obvious similarities to Wright's vision of shifting balance on an adaptive landscape, a comparison that Kimura himself is also drawn to make:

> In conclusion, I would like to emphasize the importance of random genetic drift as a major cause of evolution. We must be liberated, so to speak, from the selective constraint posed by the neo-Darwinian (or the synthetic) theory of evolution. Wright, in his later years, used to claim that he had never attributed any significance to random drift except as an agent to bring about shift of adaptive peaks. As shown in Provine's recent book, however, Wright in his papers of the early 1930s used to attach much more weight to random drift. Personally, I was mainly influenced by Wright's earlier papers, so that he is truly the forerunner in whose footsteps I have followed. I admire him very deeply. (5972)

Kimura is certainly not the only one who has emphasized the possibility of neutral evolution at the phenotypic level. In the wake of the initial proposal of the neutral theory, there was a brief spate of papers looking at the possibility of neutral evolution at the phenotypic level. By far the best known is the work of Lande (1976), who developed a quantitative genetics model to examine the relative roles of selection versus drift (see also Lynch and Hill 1986; Lynch 1990). Using data on the evolution of mammalian teeth, he showed that rates of phenotypic evolution commonly observed in the fossil record could easily be accounted for by drift, rather than selection, even assuming rather large effective population sizes.

We have seen that Gillespie (2000, 2001, 2004) used the phenomena of genetic hitchhiking and background selection to create an argument for "genetic draft" due to selection on linked genes as a significant evolutionary process/force. Gillespie (2000) apparently does not consider this genetic draft to be visible at the phenotypic level, but he does insist that "the stochastic effect of linked substitutions as captured in the pseudohitchhiking model is remarkably like genetic drift" (918). Given this similarity, there is no good reason to think that it could not also cause phenotypic changes (as long as they were "neutral"). Thus, even under Gillespie's model, there is no need to interpret the phenotypic changes we see in the fossil record or infer from contemporary organisms as due to selection for those changes. Once again, we find that the distinction between "selection of" and "selection for" in Sober's terminology, or average excess and average effect, in Fisher's, is critical.

Finally, the discovery of a hodgepodge of mechanisms of selection and/or biased mutation at the molecular level—including meiotic drive, biased gene conversion, unequal crossing over, transition/transversion ratio distortion, CpG hypermutability, transposable elements, and horizontal gene transfer—has led to an increased awareness of the importance of actual mechanisms of mutation in understanding the evolution of genomes. A similar realization for phenotypes has been less widespread but is beginning to happen. One notable proponent of this view has been Stoltzfus (1999), in a paper with the intriguing title "On the Possibility of Constructive Neutral Evolution." His basic argument is the same one I have just made here: "The only necessary 'degeneracy' in the concept of neutrality is with respect to fitness: neutral evolution is a transition between states with approximate parity of fitness, there being no restriction on how a given degree of fitness is achieved, so that changes in phenotypic and 'functional' characters are fair game, including everything from morphological changes documented in the fossil record to molecular changes that alter enzyme activities" (170). Utilizing this perspective, he examines four cases in which new functions at the organismal level have arisen from molecular mechanisms that do not appear to have been selected for at the organismal level—RNA pan-editing, gene scrambling in ciliates, the evolution of the eukaryotic spliceosome, and the retention of duplicate gene loci. He concludes that

> for most biologists, features that are complex or coordinated, that figure prominently in the biology of an organism, and that can only have arisen by a long series of changes will "appear to be adaptations." The common assumption . . . is that such traits arise by natural selection, usually by the classic model of a series of successive, small modifications, each of which is beneficial for some reason relating to the "function" or current utility of the trait. Clearly, the traits addressed in the case studies above would qualify as "complex and intricate." They also "appear to be adaptations" in the sense of eliciting proposals of hidden adaptive benefits.
>
> In the models outlined here, complex and intricate traits arise, not by the classical model of beneficial refinements but, instead, by a repetition of neutral steps. The fundamental sequence of events is that a novel attribute appears initially as an excess capacity and later becomes a contributor to fitness, due to a neutral change at some other locus that creates a dependency on it. (176)
>
> Interactions among evolving sites, excess capacities, and biases in the production of variants may bring about the evolution of complex and aptive features, without the necessary involvement of selective allele replacements. (179)

With this it is hard to argue.

* * *

In this chapter, I have tried to survey evolutionary biology as it has come to exist today, to understand the contrasting roles given to selection and drift, the two evolutionary "forces" involving differential survival and reproduction established by the synthesis. Given that most theories of phenotypic evolution depend on natural selection only, it is of special interest to see how that concept is defined. While it is easy to understand Fisher's frustration with the lack of a consistent definition of natural selection in 1930, the same situation can only be more frustrating today. In fact, not only is there no consistent definition of selection available, but we have seen that many processes at the molecular level can be viewed as drift, selection, or even mutation, depending on one's point of view.

Moreover, at the phenotypic level, drift, draft, and mutation pressure are all quite similar in their effect: they promote changes that are consistent with survival, though not necessarily better in any way. In other words, all evolutionary change must be consistent with the satisfaction of the conditions for existence, but the potential mechanisms of change are diverse. Among this diverse set of mechanisms, natural selection for the specific change that actually occurred (to the extent that one can even determine what this means) is not privileged in any way—linked selection, natural selection at other levels, mutation pressure, and random drift are all potential explanations (cf. Lynch 2007).

More generally, our review of population genetic theory in this part of the book has shown that there has been much lack of clarity on the relation between a *causal* and a *descriptive* account of evolution. Natural selection is supposed to be a causal mechanism of evolution, but in attempting to explain the evolutionary changes we can reconstruct, we will rarely (if ever) be able to offer such a causal account. To put this another way, fitness, the fundamental mathematical concept of the population genetic theory of natural selection, is defined as a value determined by the environment, so that it can form part of a forward-looking, dynamic, stochastically deterministic account of evolution. Within this forward-looking account, genetic drift appears as those differences in rate of increase not explained by the environment.

But, as Fisher realized at the outset, we are not really in a position to establish such a forward looking account. Instead, he defined fitness retrospectively, as the number of descendants present at some later time. Within this framework, we can construct forward-looking models to understand the interaction of various parameters, but we should not confuse our models with reality. The parameters of the models cannot be determined *theoretically* for any populations; they can only be estimated *empirically* and thus *retrospectively*. In other words, once again, we see that adaptedness ("fitness") cannot be used to determine existence; instead, we must insist that overall adaptedness is identical with continued existence, because this is the only explicit and universal measure of adaptedness we have.

By focusing on what is supposed to be happening in a population, rather than what is actually happening, population genetic theory has been constructed in such a way that the relevant parameters are almost impossible to measure. Nevertheless, in the previous chapters we have seen that the mathematics of differential survival and reproduction can fairly readily be examined in a purely descriptive way in terms of the rate of increase of genetic lineages. We have seen further that the variance in relative rate of increase, or the standardized variance (opportunity for selection), is a number that is easily calculable and that has both an empirical and theoretical connection to population growth rate, and thus to mean rate of satisfaction of the conditions for existence. This descriptive framework is what I call "medium sense" selection; it allows us to ascribe evolutionary changes to differential survival and reproduction, but not to establish the causes of such differential survival and reproduction.

Yet, of course, it is the causes of such differential survival and reproduction that are fundamental to the theory of evolution by narrow sense natural selection. And it is because of the connection between adaptedness and natural selection that *past* natural selection is used as an explanation of *present* adaptedness. With our solid grounding in the measurement of overall adaptedness (realized fitness), we are now prepared to critically examine this connection. More generally, we are prepared to see how the principle of the conditions for existence can provide the foundation for a more rigorous, nonteleological conceptual framework for understanding evolution.

PART FOUR

The Conditions for Existence and Evolutionary Biology

10

Adaptedness, Natural Selection, and the Conditions for Existence

I regard it as unfortunate that the theory of natural selection was first developed as an explanation for evolutionary change. It is much more important as an explanation for the maintenance of adaptation.

—G. C. WILLIAMS, 1966, *ADAPTATION AND NATURAL SELECTION*

We have also here an acting cause to account for that balance so often observed in nature—a deficiency in one set of organs always being compensated by an increased development of some others—powerful wings accompanying weak feet, or great velocity making up for the absence of defensive weapons; for it has been shown that all varieties in which an unbalanced deficiency occurred could not long continue their existence. The action of this principle is exactly like that of the centrifugal governor of the steam engine, which checks and corrects any irregularities almost before they become evident; and in like manner no unbalanced deficiency in the animal kingdom can ever reach any conspicuous magnitude, because it would make itself felt at the very first step, by rendering existence difficult and extinction almost sure soon to follow.

—A. R. WALLACE, 1858, *ON THE TENDENCY OF VARIETIES TO DEPART INDEFINITELY FROM THE ORIGINAL TYPE*

The rather long and convoluted path we have taken from the Greeks through Cuvier and Darwin and on to Fisher, Wright, Haldane, Muller, and Kimura has now finally put us in a position to better consider the meaning of the term *adaptation* as used in evolutionary biology. After all, *adaptation* is both a key term in natural theology, expressing the relation between means and ends taken as evidence for a designer, and a key term in Darwin's theory of evolution. And of course, the association between natural selection and adaptation is axiomatic for many evolutionary biologists.

ADAPTATION VERSUS ADAPTEDNESS

To begin with, I must make a distinction made many times before, the distinction between adaptation as a process and adaptation as a state of being, or adaptedness (Bock and von Wahlert 1965; Williams 1966; Lewontin 1978, 1984; Lauder and Rose 1996). *Adaptedness* means a certain appropriateness or fit between the various aspects of an organism's morphology, physiology, lifestyle, and environment. *Adaptation* refers to a change in morphology, physiology, or lifestyle, maintaining such appropriateness in the face of a change in the environment. For the moment, I want to focus entirely on adaptedness; we will return to the process of adaptation to changing environments later.

When we talk of adaptedness, we are generally talking about the relation between organisms that have a particular lifestyle and morphological or physiological features that appear to be correlated with that lifestyle. We say that woodpeckers are adapted to their lifestyle of drilling holes in trees by long stiff tails used for bracing themselves on trees, a stout beak used for drilling holes in the bark, and a long tongue used for probing these holes for insects. This example makes it clear that adaptedness refers to individual features of the organism that have functions, relative to a particular lifestyle. As we saw in examining Fisher's (1930a) geometric model of adaptedness (in chap. 7), one is implicitly considering a series of alternatives, of which the one observed is more appropriate than others. We assume that a woodpecker with a weak tail and a wimpy beak would have a harder time living the life of a woodpecker. In fact, we assume that woodpeckers with weak tails and wimpy beaks would, on average, have a lower rate of increase than ones with the fine appendages they now have, everything else being equal. Adaptedness consists in the appropriateness of a particular feature for a particular lifestyle, as judged by the ability of the organism to satisfy its conditions for existence while living that lifestyle (overall adaptedness of the organism).

As I noted in my discussion of Fisher's treatment, this view certainly does not constrain one to think of the lifestyle as somehow necessary and thus as a goal toward which evolution must be directed. To use his example, we can decide that accuracy of focusing light is the standard we wish to compare eyes by; and if we do so, then an eye that focuses light less accurately is clearly less well adapted to the lifestyle of the organism than one that focuses better, but this entirely depends on the organism we consider. It may be true that for an owl an eye that focuses accurately is critical, but perhaps for a snail, it might be less well adapted than one that maximizes the light gathered. When we speak of adaptedness, we are really speaking of a particular organism in its particular environment, living its particular lifestyle, and how any deviation in any aspect of this relationship would impact the ability of the organism to survive.

This view of adaptedness is quite similar to the "ahistorical" definition of adaptedness promoted by Reeve and Sherman (1993; see also Frumhoff and Reeve 1994;

Leroi, Rose, and Lauder 1994; Lauder 1996). Reeve and Sherman define adaptedness (their "adaptation") of a particular feature on the basis of three components: a set of phenotypes, a measure of "fitness" (in my terms, rate of increase), and a clearly defined environmental context. Given this definition, a feature is an "adaptation" or what might be better called an *adapted feature* if it causes organisms bearing it rather than one of the specified alternatives to have the highest rate of increase in the given environmental context.

Recall that Cuvier, in his introduction of the principle of the conditions for existence, focused on just such coadaptedness of organismal morphology and lifestyle in a particular environmental context. In his words, "the different parts of each being must be coordinated in such a way as to render possible the existence of the being as a whole, not only in itself, but also in relation with other beings," and his favorite example of such correlation was the sharp claws and teeth, short gut, and so forth characteristic of carnivorous animals, features that certainly would seem to qualify as adapted features under Reeve and Sherman's definition.

What is important to emphasize here is that this definition of an adapted feature is relative to a phenotype, not a genotype. When we talk of adaptedness, we are necessarily talking about phenotypes. The whole question of the genetic basis for such adapted features is thus pushed aside, with the implicit assumption that the genetics will take care of itself. Where genetics really enters is in constructing the phenotype set, because the underlying assumption is that alternative phenotypes will not differ except for the particular feature under consideration. Such an assumption is problematic, but it seems difficult to avoid if we don't know everything about the genetics of the organisms and want to focus on phenotypic traits. The assumption is not just that the alternative phenotypes are genetically possible but also that they are possible with no important effect on any other features of the organisms. Traditionally, the alternative phenotypes considered are those that involve simple change of size or traits that appear in closely related organisms. Thus, for Cuvier, it was reasonable to look at mammalian features adapted to carnivory because they could easily be compared with features in mammalian herbivores, and their appropriateness for carnivory was readily apparent.

ADAPTEDNESS OF WHAT?

The previous discussion has shown that overall adaptedness can only be measured by survival, or satisfaction of the conditions for existence. Given this link to survival, as measured by rate of increase, there are several possible ways to examine the adaptedness of particular phenotypic features of organisms. First, the *overall adaptedness* (realized fitness) of individuals can of course simply be averaged across a class of organisms distinguished by some particular feature. In doing so, however, one is examining only medium sense selection—that is, the average adaptedness of this class of organisms. To consider this narrow sense selection, one must assume that no excluded features that have important effects on survival are correlated with the one used to divide the groups. This is true in spite of well-intentioned attempts to remove such correlations statistically by measuring a series of variables (Lande and Arnold 1983).*

Second, when it comes to features of organisms, one can consider them *adapted features* to the extent that their effect is to boost the rate of increase of an organism, relative to other features (of the phenotype set). This effect can be measured most directly by experimental manipulations of the phenotype (Sinervo and Sinolo 1996).

Finally, a third possible meaning of adaptedness has to do with the relation of a feature to a particular biological role (Bock and von Wahlert 1965) or function. Given that a feature has a positive effect on rate of increase, how can we try to explain this? By its improved *performance* at some task (or tasks) that it encounters during the organism's life span—in other words, by improved function. Thus, adaptedness has at least three meanings, which are not equivalent to each other.

To see how this works out in practice, let's consider a simple example of adaptedness—namely, that which is relative to organismal survival (rather than reproduction or survival at some other level) and in which the relevant environmental context can be reasonably postulated. Lizard toe fringes have evolved in a number of unrelated lineages of lizards, generally (although not universally) in association with sand dwelling (Luke 1986). Carothers (1986) tested the hypothesis that toe fringes improve running performance in the Mojave fringe-toed lizard *Uma scoparia* by comparing running performance before and after trimming the fringes off. He measured maximum velocity and acceleration on level sand platforms and found a 15 percent decrease in both; on platforms with a 30-degree incline, maximum velocity decreased by 9 percent, acceleration by 22 percent. In his control comparisons on rubber mats, the lizards with trimmed fringes did as well as (in fact slightly, though not significantly, better than) their wild counterparts. Thus, judged by this standard, the phenotype of having fringes is better adapted to running on sand than the one without (table 10.1).

However, other standards could be used. Perhaps what really matters to the lizards about running performance is the energetic cost of locomotion, and this is reduced by toe fringes during normal cruising. Moreover, it is possible that locomotion is not even the most significant biological role of toe fringes; Carothers notes that Stebbins (1944) had suggested shimmy burial as an important biological role of the fringes.

---

* In the case of populations with overlapping generations (most populations in nature), it is not even clear how phenotype should be defined, because one is necessarily looking at phenotype across a series of lineages at different points in their life cycle.

TABLE 10.1 Adaptedness

Organism (Population)	Environmental Context	Trait (Alternatives)	General Effect of Trait (Compared with Alternatives)	Hypothesized Major Context of Trait Use (Biological Role)	Function/Criteria of Functional Performance in This Context	Other Effects
			COARSE-GRAINED ADAPTEDNESS			
Lizards	Sand dunes	Fringed toes (not fringed)	Increases survival rate	Running on sand	Increased speed, acceleration Decreased metabolic cost/distance	Increased cost of growth Increased digging efficiency
Humans	World with light	Eyes (no eyes)	Increases survival rate	Vision (light/dark discrimination, orientation to environment, detection of food, detection of mates)	Increased light detection Increased color discrimination Increased object discrimination Increased acuity Increased field of view	Increased cost of growth Increased potential for injury Regulation of melatonin production Production of social signals
			FINE-GRAINED ADAPTEDNESS			
Lizards	Sand dunes	Fringed toes (1% longer/shorter, 1% stiffer/floppier, etc.)	Increases survival rate	Running on sand	Increased speed, acceleration Decreased metabolic cost/distance	Increased cost of growth Increased digging efficiency
Humans	World with light	Eyes (1% thicker/thinner lens, 1% more abundant/less abundant cones in retina, 1% greater/lesser field of view, etc.)	Increases survival rate	Vision (light/dark discrimination, orientation to environment, detection of food, detection of mates)	Increased light detection Increased color discrimination Increased object discrimination Increased acuity Increased field of view	Increased cost of growth Increased potential for injury Regulation of melatonin production Production of social signals

When we try to connect adaptedness to running on sand to survival of the lizards themselves, we run into additional issues. The effect of toe fringes isn't limited to increased running performance. We might expect some negative effects—for example, that there is increased cost of growing the fringes. We might also expect additional positive effects. Even if Stebbins is wrong that improved performance in shimmy burial is the most important effect of fringes, it might be *an* effect. What we must assume, if we are to count toe fringes as having positive effects on survival and attribute this effect on survival to improved running speed and acceleration (aiding in predator avoidance and prey capture), is that all of these other effects are of secondary importance.

Finally, if we want to connect this increased survival of the lizards with toe fringes to the underlying genetics, we must assume that there is no genetic correlation between toe fringes and any other features, so that nature presents us with lizards differing only in this feature. Such an assumption is clearly unrealistic. We are left with a black box—the unknown relationship of genotype to phenotype—which we must assume will behave "as if" the feature of interest (toe fringes) was an isolable part.

In short, in any given case, even a simple one in which we can show experimentally that the feature of interest indeed performs better at a particular task than an alternative, as in Carothers's (1986) elegant work, we must make a number of simplifying assumptions to go from the performance of a particular function to the survival of the organism as a whole. This is merely the point I have made more generally: we can't provide a causal explanation of the survival of an organism on the basis of its individual features ("adaptations"), because its survival can't be determined on the basis of its features. Such an explanation is a teleological deterministic one. Instead, we can only provide conditional teleological explanations of survival: running fast can be a condition for existence of the lizards in the context of both predator avoidance and prey capture, and fringes aid in satisfying this condition for existence.

ADAPTEDNESS, ADAPTATION, FUNCTION, AND NATURAL SELECTION: HOW ARE THEY RELATED?

The debate over the meaning of adaptation (Lauder and Rose 1996) has numerous parallels with the debate over the meaning of function discussed in chapter 2. In fact, as we have just seen, the concept of adaptedness and that of function are closely related. Consider again the example of the lizard toe fringes. Carothers showed that they have a positive effect on running speed and acceleration, and hypothesized that this effect increased survival by its role in predator avoidance and prey capture. In the terminology I developed earlier, this is merely asserting that a major function of the toe fringes is aiding in running on loose sand. This is because the comparison (the phenotype set considered) is between lizards with toe fringes and those without, as with any functional ascription. Thus, to assert a particular function of a feature is to assert what might be called a "coarse-grained adaptedness" of the feature—having it is better than not having it.

Consider a second example of adaptedness of a feature: the human eye (table 10.1). As with lizard toe fringes, the eye is often considered an "adaptation." We can again define a phenotype set (eyes, no eyes) and see how vision is improved (by any number of functional criteria) by the presence of an eye. We can also understand how vision can be a condition for existence, even in modern human societies. Thus, we can say that a major function of the eye is vision.

However, the concept of the adaptedness of a feature is not identical to the concept of its function. Functional ascriptions implicitly refer to a phenotype set that consists only of organisms with the feature and those without. In contrast, adaptedness can be defined with reference to a variety of phenotype sets (Reeve and Sherman 1993). In particular, it usually refers to a set including phenotypes that differ merely quantitatively from the one observed, as well as alternative phenotypes that are observed in closely related species. We can thus consider a fine-grained adaptedness as well as a coarse-grained one identical with function. It was this fine-grained adaptedness of the organism that Fisher (1930a) highlighted in his geometric model of adaptedness.

The basic point of Fisher's model is that given a lizard with toe fringes, we can ask whether a lizard with toe fringes 1 percent longer or shorter, or 1 percent stiffer or flimsier, would perform better at running on sand or would survive at a higher rate. Likewise for the eye, we can ask whether a human with a slightly thicker or thinner lens, slightly more or less abundant cones in the retina, slightly greater or smaller field of view, would perform various tasks as well or would survive at a higher rate. Thus, a trait either has a function or it doesn't, but its performance at a specific task is a relative matter.

How does this adaptedness relate to natural selection (in the narrow sense)? At one level, the relation is quite clear: variation in the relative adaptedness of phenotypes *is* the causal basis of natural selection. Given that there are some optimum toe fringe lengths for all lizards in this particular population, if less than optimum toe fringe lengths occur in some lizards in the population, they will be "selected against," while the lizards with better adapted toe fringes will be "selected for." But of course, as already noted, this is a very naïve view of the relationship of genotype to phenotype. In other words, we have assumed that we can simply carve the organism up into traits, each of which can respond independently to selection. In doing so, we deny the fundamental unity of the organism, living its lifestyle, in its environment.

If we think about what is actually happening in a population, we see that each individual has its own lifestyle. When we assess adaptedness, then, it can only be relative to the unique lifestyle of that individual organism. Although we can take averages across a population, measuring the average adaptedness of organisms to their own lifestyle is not the same thing as measuring their adaptedness to an average lifestyle, particularly if organisms tend to adopt lifestyles suited to their abilities. This of course they do, not only because of behavioral adjustments to a particular set of features, but also due to the simple fact that if a lizard can't run fast on sand, it won't, and it will necessarily do something else (in quantitative genetics terms, this creates a genotype/phenotype/environment correlation). So when one is examining variation in adaptedness in a natural population, one is indeed examining natural selection, but natural selection in an environment that differs for each individual and that can never be fully ascribed to variation in the adaptedness of a particular phenotypic trait. By contrast, experimental manipulations of phenotype can give us good information on the adaptedness of specific traits, but with only a tenuous connection to genetics.

To summarize my argument so far, in the study of adaptedness of individual features or traits, we are essentially carving the organism/lifestyle/environment continuum at the boundary between that trait and the remainder of the continuum and assessing its fit to the remainder. By assuming a particular critical functional role (or roles), we can test the performance of the feature with respect to that function (or those functions), but it is the overall fit that is the real issue. This is the background we must understand to Fisher's geometric model of adaptedness, unless we are to regard the optimum as a fixed goal of evolution.

How does the adaptedness of a trait relate to its evolutionary history? In particular, how does it relate to the *process* of adaptation? This topic has been at the center of the debate over the definition of adaptation, and the proper way to study it, from the beginning of the current incarnation of this debate (e.g., Williams 1966; Lewontin 1978, 1984; Gould and Lewontin 1979; Gould and Vrba 1982; Sober 1984). In Gould's convenient phrase, the debate has been framed as one of "current utility" versus "historical genesis." Those who have supported the "historical genesis" definition of an adapted feature have generally required that natural selection for the performance of some task or set of tasks has resulted in the trait of interest having its current form, as opposed to its previous form. Thus, in the terminology developed by Gould and Vrba (1982), a trait with current utility is only an "aptation." If it was created by natural selection for its current function, it is an "adaptation." If it was created by some other process (or by natural selection for some other function) and then co-opted for its current use, it is an "exaptation." Thus, the status of a feature as an adaptation becomes a matter of history.

This view of adapted features as history (parallel to Larry Wright's views on function; see chap. 2) led in the late 1980s and early 1990s to the proposal of numerous phylogenetic methods of analyzing adaptations (Coddington 1988; Baum and Larson 1991; Harvey and Pagel 1991). The fundamental assumption of such analyses is that a correlated transformation of a particular character and a particular feature of the lifestyle or environment, if repeated independently across the phylogenetic tree, is evidence for the adaptedness of the derived state of the character to the derived environment, as well as for its status as a historical adaptation that originated under the influence of selection. This latter claim in particular excited a great deal of protest (Reeve and Sherman 1993; Frumhoff and Reeve 1994; Leroi et al. 1994; Lauder 1996; see also Fisher 1985), and the paper by Reeve and Sherman that we have used as a guide to adaptedness was one response. The objectors pointed out the significant disconnect between any long-term historical process of evolution and the current situation of a population, arguing for a definition based on current utility, rather than an inferred past process of evolution, as a more operational concept of adaptation.

How might one consider the process of adaptation without involving natural selection (in the narrow sense)? To even ask such a question may appear rather heretical, when adaptation and natural selection for most authors go hand-in-hand. However, given the difficulties with precisely defining natural selection (in the narrow sense) detailed in the previous part of this book, it seems foolish to connect the process of adaptation with natural selection by definition. What I really want to ask, of course, is how to connect the process with natural selection in the broad sense—namely, with the conditions for existence.

From the perspective of the conditions for existence, the process of adaptation can only mean a change in lifestyle or mode of adaptedness, since overall adaptedness (rate of increase) does not generally increase in evolution. Thus, adaptation as a process generally refers only to changes in the adaptedness of individual features of organisms, not changes in the adaptedness of the organisms themselves. In parallel with the two ways of looking at the adaptedness of specific features, adaptation with respect to those features might mean one of two things. First, it might indicate relative improvement in the effect of the feature on organismal survival (and reproduction), compared with what would obtain if the ancestral form was retained. It is often assumed as a corollary that the derived state of the feature would do less well than the ancestral one in the old environment/lifestyle. Second, adaptation might mean improved performance of the feature at some particular task that is critical for survival of the organism in the new lifestyle/environment, compared with the ancestral (plesiomorphic) form.

Thus, for example, in a lizard population that changes evolutionarily from one in which most of the individuals spend most of their time on relatively hard substrates, to one in which most lizards spend most of their time in sandy washes, adaptation as a process means an integrated change in the environment (little sand > lots of sand) and lifestyle (not running on sand often > doing so), of which the change in form of foot is one necessary condition (in the conditional teleological sense). The environment, the lifestyle, and the form of the foot all change, with the result being that adaptedness is maintained. As Cuvier saw, the functional integration of the organism, its lifestyle, and its environment limits the possibility of evolutionary change. However, this doesn't mean that evolution is impossible, as Cuvier thought, only that functional integration is necessarily maintained during evolutionary transformation.

Given this noncausal definition of the process of adaptation of individual features, how does natural selection (in the narrow sense) fit in? Consider the evolution of lizard toe fringes. Our initial population consists of ancestral lizards with some genetic variation in the roughness of the foot scales, and our final population of lizards has toe fringes similar to those seen today, also with genetic variation. In both cases we will assume that an examination of the population would show that the average existing state of the foot was better than the alternative states at that time. It is only reasonable to think, then, that the same would therefore be true of all the intermediate populations, at least to a first approximation. If we imagine the ancestral population living adjacent to a rising mountain chain or nascent dune field, we can also imagine the members of the population spending more and more time exploring the sandy washes or dunes as they became capable of doing so (and as resources to exploit appeared). As they moved into the washes or dunes, one might also expect that their toe fringes were elongating. Natural selection in the narrow sense would certainly have been involved throughout this process in maintaining the adaptedness of the lizards to their lifestyle, but there is no reason to assume that the process of change itself couldn't involve some fraction of genetic drift, genetic draft, correlated selection, and the like.

Imagine one year sometime in the middle of the process. At this point, lizards are generally spending half of their time on the dunes and half in the area alongside the dunes, with some spending more, some less. Their fringes are also generally of intermediate length. In this particular year, the dune insects the lizards forage on are rare, and those lizards who invested energy in growing long fringes do poorly, compared to how they would do with shorter fringes, all else being equal. However, among these lizards long toe fringes are genetically correlated with other features that happen to do well that year, and the toe fringes increase in length. The following year, the situation is reversed. The lizards running on the sand have a banner year, and among them, those with long fringes do best (on average). The overall length of the fringes again increases in the next generation, this time clearly due to natural selection. In a third year, the relative advantage of long and short toe fringes exactly balances, but by chance (genetic drift) toe fringes happen to lengthen. Now vary these scenarios for ten thousand years, with the end result being that the fringes have lengthened. Even if the lengthening of the toes was favored by natural selection overall, is it really correct to say that natural selection (in the narrow sense) was the cause of their lengthening, when some of it was due to other factors? If we are to allow other factors in evolution, isn't it true that any changes due to such other causes, like those due to natural selection, must be consistent with the conditions for existence? And isn't the integration of lifestyle and morphology just such a condition for existence, as Cuvier insisted?

My point is not to argue for some speculative account of the origins of this trait, but to note that if the only evidence we have bearing on the question is that the toe fringes are useful today, there is no reason to think that their use is the exclusive reason for their "historical genesis." This was the main point of Gould and Lewontin's (1979) original critique, and a valid one. But the solution is not to try to decide on a particular historical cause for each feature, and to consider natural selection by definition the explanation for all "adaptations." If we do so, we necessarily accept a teleological view of natural selection aiming from a past unadapted state to a present adapted state, when overall adaptedness is necessarily maintained throughout the process, whatever the causes.

Lizards with toe fringes live on sand dunes, lizards without toe fringes live on dirt and rocks, and those with partial toe fringes live part of the time in sandy washes and around the edge of dune fields. None of this requires natural selection (as commonly understood) to be the exclusive factor that gave lizards on sand dunes their fringes, though it does suggest that selection is maintaining the length of toe fringes in each of these situations. I am not arguing that natural selection (in the narrow sense) was *not* involved in the origin of toe fringes, and I am not stating categorically that during the process lizards with longer fringes were not frequently surviving at a higher rate than those without, and precisely because of the length of their fringes—but instead arguing that this is merely a belief based on plausibility, not on any concrete *evidence* that we could ever hope to gather.

But what of adaptation to a changed environment, rather than a shift in morphology and lifestyle occurring in an integrated fashion over evolutionary time. Can't this be studied in existing populations? Doesn't this give us concrete evidence of the role of natural selection in evolution, as an important factor favoring changes that suit organisms to changed conditions? Without yet entering into the question of whether adaptation to a changing environment is an adequate model for all evolutionary change, let's examine some putative examples of such a process and see what they can teach us.

EMPIRICAL STUDIES OF EVOLUTION: BACTERIA, PEPPERED MOTHS, AND DARWIN'S FINCHES

The evolutionary process is notoriously difficult to observe, customarily occurring in a time frame that exceeds the research life of most scientists. However, there are several favorable situations in which the active evolution of populations can be monitored, including those with extremely short generation time (unicellular organisms, especially bacteria, and viruses), those with clearly identifiable variants (Mendelian polymorphisms, recognizable either phenotypically or molecularly), and those in which individuals can be followed throughout their life span (plants, insects, birds, and mammals being the best studied examples). Let us begin with bacteria.

Bacteria in Chemostats and in the Wild

Experimental evolution in microbial systems is one of the most exciting areas of evolutionary biology today (see Elena and Lenski 2003). These systems allow one to observe evolution in replicate populations over hundreds or thousands of generations, with extensive control of the environment, initial population, and population growth rate. Beginning with early work arising from the development of continuous culture chemostats (Kubitschek 1970), the field took off in the early 1980s with the work of Dykhuizen and Hartl (1981; see also Hartl and Dykhuizen 1981; Hartl, Dykhuizen, and Dean 1985) and Hall (1982). Since then, a large number of workers have explored this fascinating set of systems. Most prominent among them have been Richard Lenski and his colleagues, who have been running an ongoing analysis of evolution in *E. coli* for more than twenty thousand generations (Lenski 2004). Let's take a look at some of the conclusions that have emerged from this remarkable study.

The basic setup of the experiment is a daily serial transfer; populations grow exponentially upon transfer until the glucose supply is exhausted, at which point they enter stationary phase until the next transfer. The total population expands from about 5×10^6 cells at the beginning to 5×10^8 cells at the end of a cycle, from which 5×10^6 cells are taken to start the next cycle. The mean absolute rate of increase is thus about 100/day for the increasing phase and 0.01/day for the decreasing phase. There are approximately 6.6 generations per day. One thing to note initially is that the population decrease phase of the experiment occurs "randomly"; in other words, there is (by design) no tendency for particular variants to survive at higher rates than others. Because no death occurs in the system other than this culling associated with dilution of the culture during transfer, all competition between cell lines is due to differences in growth rate, not survival. These populations are entirely asexual and thus clonal.

Of special interest for us, Lenski and his colleagues have paid particular attention to the evolution of "fitness" in these cultures. How do they measure this fitness? By competing two strains (distinguished by a neutral marker) in common culture, under conditions identical to those of the long-term experiment. The strains are thus competing in terms of net growth rate until the glucose is depleted, which depends on both how quickly they emerge from stationary phase and how fast they grow when they do. They measure fitness as the ratio of the logarithms of the absolute rates of increase. Thus, if a culture is started with two strains, each at 2×10^6 cells, and one increases to 3×10^8 cells, while the other increases only to 1×10^8 cells, the absolute rates of increase are 150 and 50 per day, respectively. The logarithms of these rates are 5.01 and 3.91, so the relative "fitness" of the better type is 5.01/3.91 = 1.28 (Lenski 2004). In the formalism I developed earlier, this corresponds to a situation in which the relative rate of increase of the better type is 1.5; that of the worse, 0.5. In their actual experiment, Lenski and his colleagues found that compared with the ancestral population ("fitness" defined as 1.0), mean "fitness" increased with time. After twenty thousand generations, mean fitness of the derived populations was about 1.7, a 70 percent increase in net growth rate, and was rather consistent across populations. In our terms, this is equivalent to an absolute fitness of 180 per day for the derived type and 20 for the ancestral type, an increase of nine times. Was there any change in actual numbers of cells, though? In fact, there was: total cell number at the end of stationary phase *declined* about 40 percent over the first ten thousand generations of the experiment. Thus, although the derived cells can outcompete their ancestors, total population size declined under the conditions of the experiment (i.e., the overall rate of increase was less than 1.0). At the same time, however, average cell size approximately doubled, so that overall biovolume produced increased by about 20 percent, which suggests an increase in efficiency of energy utilization.

If we take a step back and consider the general situation of these populations, it is clear that they are evolving in response to a novel environment, one characterized by a single energy source, extreme temporal fluctuations in the availability of that source, and a grim reaper that sweeps through the population randomly on a regular basis. Evolution by natural selection indeed results in increased growth rate and yield under the novel conditions, which Lenski considers an improvement in "fitness." Nevertheless, overall adaptedness (absolute rate of increase) averages slightly less than one over the time course of the experiments, resulting in the decreased overall population size seen. Thus, even in this case of adaptation to a novel environment, the improvement in performance seen clearly does not do anything to the overall fitness of the populations, even though the relative "fitness" of the population with respect to the ancestor continually increases in a transitive way.

What is striking about this experiment is the degree of general similarity in the results across the replicates at the level of such phenotypic features as cell size and growth rate. Yet some other phenotypic features, such as cell shape, evolved differently across the populations. What genetic changes are associated with these phenotypic differences? Lenski and his colleagues have found that one feature that varies among replicates is mutation rate. Four of the twelve populations developed a mutator phenotype, with increased mutation rate due to changes in the DNA repair mechanism. He argues that this mutator allele (which varies among populations) swept to fixation by hitchhiking with a selectively favored allele at another locus; the frequency with which this occurred is owing to the greater likelihood that a mutator will show a favored mutation. When tested in environments other than that in which they had evolved, derived populations generally (though not universally) did worse than their ancestors, a situation that Lenski attributes to "ecological specialization" due to antagonistic pleiotropy. The only substrate tested that all populations could not grow on, however, was D-ribose. It turns out a mobile genetic element in the ribose operon was associated with this loss in all populations; it was subject to a high mutation rate and also appeared to have a 1 to 2 percent greater rate of increase. In this case, then, we see mutation pressure and selection cooperating to produce a parallel change across all populations.

Schneider and Lenski (2004) reviewed the role of such insertions in overall genetic evolution of experimental bacterial populations. They found that such mutations were not only widespread but highly divergent among populations. Many, if not all, of the genetic changes known in such systems are associated with mobile genetic elements that function as insertion sequences. Point mutations appear to be rare in this long-term experiment, aside from the mutator lines (Lenski, Winkworth, and Riley 2003). Nevertheless, when four genes carrying a set of four IS150 "candidate mutations" were surveyed across the replicate populations, there was a strong tendency for all populations to show nonsynonymous point mutations in the same genes, suggesting that narrow sense natural selection had favored particular physiological changes in these genes (Woods et al. 2006). The take-home message from all of this elegant work is that natural selection is indeed operative in increasing growth rate and efficiency in a novel environment. Nevertheless, overall rate of increase is not increased but in fact may even decrease. In addition to selection, mutation pressure in the form of mobile genetic elements and hitchhiking of mutator alleles along with favored mutations appear to be the major mechanisms of change.

Two other studies of experimental evolution in bacteria are worth noting. MacLean and Bell (2002) studied evolution of *Pseudomonas* grown on BIOLOG plates consisting of ninety-six wells, with each well (except one control) containing a different carbon compound, as well as a carbon and nitrogen supplement that was progressively reduced over the course of the experiment, requiring increased reliance on the carbon source present. They ran their experiment for 100 transfers to new plates, each representing about 11 doublings, or 1,100 generations altogether. At the beginning of the experiment, the ancestral strain could grow in only about half of the wells. Population growth was assessed by absorbance measurements in the wells after twenty-four hours of growth, thus representing a measurement of overall yield. What is striking about their results is that under the conditions of the experiment, bacteria were unable to evolve the ability to utilize thirteen of the novel carbon sources. Nevertheless, variants that could utilize all these sources appeared in other wells, as a "correlated response to selection." Thus, we see novel functions appearing not due to any direct "selection" for them but due to changes occurring for other reasons.

Finally, in a twenty-eight-day experiment on glucose-limited chemostat cultures of *E. coli*, Maharjan and colleagues (2006) found that considerable diversity of "ecotypes" arose within the population, even though the environment was apparently homogenous. Moreover, of the ecotypes tested in direct competition with the ancestral form under chemostat conditions, eight of ten had higher growth rates, but the other two did not. They concluded that "sharing of a niche by a large number of diversifying members of the same species is a feasible evolutionary strategy. A single fitness solution, or survival of the fittest, is not the only answer in a competitive environment" (517). In other words, there is not a single solution to a "problem" posed by the environment, but instead a broad range of solutions by which individual organisms can satisfy their varying conditions for existence.

Peppered Moths

The origin of industrial melanism in the peppered moth *Biston betularia* is a classic example of evolution in action. Moreover, Kettlewell's famous selection experiments of the 1950s have long provided one of the best examples of the process of natural selection, in a context where it could reasonably be argued to be responsible for a particular evolutionary change. They have thus been at the center of recent controversy: not only are they a favored target for creationists (e.g., Wells 2000; see Coyne 2002; Rudge 2002), but a book-length popular exposé appeared a few years ago, arguing that Kettlewell fudged his data (Hooper 2002; see Rudge 2005).

In his very first paper on natural selection, Haldane (1924) introduced the peppered moth (his *Amphidasys betularia*) to the stage of evolutionary biology. He noted that the melanic form first appeared in Manchester in 1848 and had "completely ousted" the recessive variety by 1901. Assuming that the change was from less than 1 percent to more than 99 percent of the population in fifty years, Haldane calculated a minimum average selective advantage of the melanic form over the recessive form of about 50 percent, assuming complete dominance. This corresponds to relative rates of increase of about 1.5 and 0.99 at the beginning of the process, 1.2 and 0.8 when the two forms are equally common, and 1.01 and 0.67 at the end. In our terms, when the forms are equally common, the standardized variance associated with differences in this feature is 0.2^2, or 0.04. Haldane reasonably considered this a "not very intense degree of natural selection" (1924, 26).

The problem of melanism was briefly examined by E. B. Ford in his 1937 review of "Problems of Heredity in the Lepidoptera." In Ford's initial view,

> melanic forms have spread in industrial areas owing, primarily, to selection for characters other than colour. The action of the genes producing melanism as one of their effects may sometimes give the organism a physiological advantage. That such favourable factors have not become widely established may be due to the handicap of black coloration which, in normal circumstances, would render some species very conspicuous. On the other hand, melanism, as such, may at least be no longer a drawback in the blackened countryside of many manufacturing districts, in which, furthermore, the number of predators may be reduced. Here then, the insects may be able to avail themselves of the other benefits conferred by these genes. (487)

Ford's caution was due to much evidence that larval or adult "hardiness" might differ between the two forms (see Ford 1940) and that there might also be differences in time of emergence or other features, aside from any action of predation on adults.

Kettlewell's first studies of the problem were published in two 1955 papers. The first (Kettlewell 1955a) examined the possibility that the *typical* and *carbonaria* moths could choose an appropriate background to rest on, finding evidence to support this hypothesis. The second (Kettlewell 1955b) was the notorious release and recapture study, run following the protocol devised by Fisher and Ford (1947) for population estimation. Kettlewell had raised a large number of moths and released them, initially in an aviary with a resident nesting pair of great tits, then in a polluted wood near Birmingham. He spent considerable effort trying to document bird predation on the moths and assess to what degree it was selective. Finally, he used a scale of conspicuousness to humans (himself) to assess how well the moths matched their background. His results can be summarized as follows: as judged by humans, *carbonaria* was less conspicuous than *typical* or *insularia* (an intermediate morph) on oaks in the polluted woods near Birmingham, though more conspicuous on birches. In contrast, in an unpolluted area of Devon, *carbonaria* was much more visible than *typical* on the heavily lichen encrusted background. His aviary experiments showed that the great tits would not only eat resting moths but did so in an order corresponding to their visibility to humans, supporting the idea that our judgment of conspicuousness corresponds to that of birds. Finally, in his release of moths in the polluted wood near Birmingham, he recaptured 149 of 630 released (using both light and scent traps), of which his recovery rate was 27.5 percent for *carbonaria*, 13.5 percent for *typical*, and 17.4 percent for *insularia*. In our terms, this corresponds to relative rates of increase of 1.16, 0.57, and 0.74.

Kettlewell considered alternatives to differential predation as an explanation, including differential attraction to the traps, differential dispersal from the area, or differential survival, and argued effectively against each. Moreover, he was able to gather some observations on bird predation that supported the idea that birds were not only feeding on moths but selectively eating the more conspicuous *typical* form. He concluded that "the effects of natural selection on industrial melanics for crypsis can no longer be disputed" (341).

The following year, Kettlewell (1956a) published a second paper, detailing a further series of experiments. He had repeated his previous experiments near Birmingham and found quite comparable ratios of returns (relative rates of increase), though the overall return rate was higher. He had also conducted a similar mark-recapture study in an unpolluted area with abundant lichen in Deanend Wood, Dorset. Here he found that the *typical* form was quite inconspicuous, while the *carbonaria* form was quite conspicuous. He gave up on experiments to test the rate at which individuals disappeared from resting places because he kept losing sight of his *typical* forms (because they were so well camouflaged). Here he had a return of 4.7 percent and 13.8 percent for his *carbonaria* and *typical* forms, respectively. In our terms, this corresponds to relative rates of increase of 0.51 for the *carbonaria* and 1.51 for the *typical*.

This new set of experiments was accompanied by a series of films made by the famed behaviorist Niko Tinbergen, recording birds eating moths; Tinbergen and Kettlewell's observations directly confirmed the increased predation by birds on the more conspicuous morph in both Birmingham and Dorset. Nevertheless, his conclusion here is rather interesting: although he strongly supports the role of predation in the relative advantage of the two morphs in the two localities, he is quite willing to consider that other factors may also be involved: "The difference in cryptic coloration alone could be responsible for the rapid spread of the Industrial Melanics. There are also, however, other character and behaviour differences between them and their *typical* forms. These are at present the subject of investigation" (Kettlewell 1956a, 301). These factors no doubt included the larval viability and behavioral differences discussed previously by Ford (1937, 1940).

A third major paper by Kettlewell appeared in 1958. Here he looked at clines in frequency of the morphs across Britain based on his own records and those from a network of correspondents. He found a general correlation of melanism with industrial areas, though melanics were also common in rural areas of eastern England, a pattern that Kettlewell attributed to prevailing southwesterly winds carrying pollution from central England.

The work by Kettlewell inspired a large number of studies on the phenomenon of melanism in *Biston betularia*, examining both the natural history and the genetics of the phenomenon (e.g., Clarke and Sheppard 1963, 1964, 1966; Lees 1968; Cook, Askew, and Bishop 1970; Bishop 1972). In 1973, Kettlewell reviewed this burgeoning field in a book with the interesting title *The Evolution of Melanism: The Study of a Recurrent Necessity*. What Kettlewell meant by this phrase is spelled out in his introduction: "The hypothesis I wish to present is that dark and light forms of many organisms have possessed advantages or disadvantages under varying environments since the inception of life on this planet" (7). This seems an eminently reasonable hypothesis.

Further studies in the years since the appearance of Kettlewell's book have called into question some aspects of Kettlewell's work. For example, criticisms have been raised about the choice of tree trunks as resting sites for moths (reviewed in Majerus 1998; Cook 2000). The increased density in his releases has been an issue as well. Both of these, however, are factors that Kettlewell (1955b) himself freely acknowledged as likely heightening the differential relative returns observed. Since the 1950s, there has been a decline in the frequency of the *carbonaria* morph in the wake of clean air legislation, but the decline has not always been well correlated with an increase in the abundance of lichens (Grant, Owen, and Clarke 1996; Grant et al. 1998).

The genetics has also turned out to be somewhat complicated, particularly with respect to the *insularia* morph (reviewed by Majerus 1998). Moreover, Ford's old ideas about different larval viabilities have been supported by more recent analysis (Creed and Lees 1980). In spite of such criticisms, however, there is clear evidence that differential survival and reproduction (differential rates of increase) are responsible for the change in morph frequencies over time, and that bird predation is an (if not necessarily the only) important factor in this change (Majerus 1998; Cook 2000, 2003).

How are we to view this classic example of natural selection in action? Even a textbook such as Ridley (2004) is quite clear on the fact that the increase in frequency of the melanic morph represents a case of selection for possession of the melanic allele, not necessarily for any particular phenotypic effect of that allele (such as increased crypsis). Nevertheless, the case is often presented as a simple one fitting the following model of natural selection: the environment changes (smoke pollution darkens trees), and the now poorly adapted organism, under the influence of natural selection, adapts to the change.

As we have seen, the initial view of Ford (1937; but cf. 1964) was more subtle, as was indeed that of Kettlewell (1955a, 1955b, 1956a, 1956b, 1958, 1973). Ford suggested that the melanic form has an innate physiological advantage, which in unpolluted situations is outweighed by its conspicuousness, resulting in low frequencies of the melanic form in populations. In the presence of air pollution, however, the relative fitnesses of the two forms is reversed, and the melanic is able to increase. In the present situation of declining frequency of the melanic morph (occurring in parallel in both Europe and North America; Grant et al. 1996, 1998; Cook 2003), the balance has clearly shifted the other way. In all of this we see natural selection functioning not as a process *producing* adaptedness but as a process *maintaining* adaptedness in the face of environmental change (as emphasized by the fine-scaled geographic and temporal adjustment that occurs). The environment has changed, the moths have changed, and the moths are still here. The genes underlying melanism, and the melanic moths themselves, can clearly satisfy their conditions for existence better in a polluted that a pristine environment; we can investigate possible reasons for this fascinating phenomenon—end of story.

Darwin's Finches

The group of birds known as "Darwin's Finches" includes thirteen species on the Galápagos Islands off the west coast of Ecuador and one in the Cocos Islands some seven hundred kilometers to the northeast. Since the early 1970s, Peter and Rosemary Grant and their colleagues have been conducting an extensive and elegant series of studies of these finches, focusing in particular on the populations of two species, the medium ground finch *Geospiza fortis* and the cactus finch *G. scandens*, on the small Galápagos islet of Daphne Major (Grant 1999). There is far more to their studies than can be discussed here; I would like to focus on only four of their papers in particular.

In 1984, Price, Gibbs, and Boag published a study with Peter Grant that demonstrated "recurrent patterns of natural selection in a population of Darwin's finches" using Lande and Arnold's (1983) newly published method for analyzing selection in natural populations in a quantitative genetics framework. They began by affirming that "the adaptive significance of morphological traits can be assessed by measuring and identifying the forces of selection acting on them" (Price et al. 1984, 787). Their goal was specifically to assess the adaptive significance of body size and of beak size and shape. They found that during droughts, seed availability dropped and mortality increased. With other measured variables held constant, Lande and Arnold's method detected significant selection for increased weight, increased beak depth, and decreased beak width during the period of the 1977 drought (by far the worst, with only 15 percent survivorship across the period for the sampled population). They concluded that "results of the study support an adaptive interpretation given to beak size and shape in this population of Darwin's finches" (788–789). Nevertheless, beak width was negatively correlated genetically with body size and beak depth, so that estimated selection for decreased beak width opposed selection for increased depth and overall size, and there was little net effect on the evolution of the population. A follow-up study (Gibbs and Grant 1987) found that selection for large body size in the medium ground finch during droughts was reversed during periods of abundant rainfall, a situation the authors referred to as "oscillating selection."

Some years later, the Grants examined the variation in rate of increase (fitness) among their finches in a paper titled "Non-random Fitness Variation in Two Populations of Darwin's Finches" (Grant and Grant 2000). They examined data for both species for the entire twenty-four years from 1976 to 1999, but focused in particular on the 1978, 1981, and 1983 cohorts, for which data sets were largest and most complete. Fitness was measured as recruits per breeding adult, or "the number of offspring per parent that survive to breed" (Grant and Grant 2000, 132).

As in their earlier studies, they found large year-to-year variation in breeding activity correlated with rainfall patterns. They also again found large variation in breeding success of cohorts. For example, of the 1976 cohort of *G. fortis*, only 1 of 376 fledglings banded eventually bred (due largely to a terrible drought the following year). Of the 1978 cohort, 89 of 222 fledglings banded (40 percent) eventually bred. The distribution of fitnesses across individuals was nonrandom in the sense that there was much greater variance in number of recruited offspring and number of fledglings produced than one would expect for a Poisson distribution. Unfortunately, they report the standardized variance in recruits and fledglings, but not the variance and mean themselves, so that a possible connection with population growth rate (as seen in chap. 8 for other species) cannot be assessed.

There was some relationship between properties of the offspring and their success in becoming recruits, but this was weak and variable among cohorts and sexes. There was also no evidence that number of recruits produced per individual was heritable. They conclude that variation in realized fitness (number of recruits per breeding adult) "is neither detectably heritable nor entirely random" (135). They suggest this nonrandomness arises through the effect of variation in longevity on variation in breeding output: those individuals who manage to survive a long time can produce an inordinate number of offspring.

They end this interesting paper by again emphasizing the importance of environmental fluctuations in causing the lack of association between life history and morphological measurements, on the one hand, and rate of increase (fitness), longevity, and fledgling production, on the other. This they ascribe to the variation in association of such traits with different periods of the life history in such long-lived (up to fifteen years) birds. The timing of "bad" (low rate of increase) and "good" (high rate of increase) years with respect to the life history of the cohort is particularly critical. In a population in which environmental fluctuations are so extreme that half of the years see no reproduction at all, this does not strike one as surprising.

The implications the Grants derive from their results are that "although morphological traits contribute to fitness and are heritable, they do not change unidirectionally because they are selected in opposite directions, and in different combinations, under fluctuating environmental conditions" (137). This statement echoes the message of the earlier study by Gibbs and Grant (1987) and is identical with the viewpoint I have been promoting here: natural selection acts to maintain adaptedness under fluctuating environmental conditions, not to produce adaptedness. Phenotypic changes result from this selection, as well as other factors, but there is not necessarily any net change in phenotype over time.

The final paper I'd like to discuss is the Grants' 2002 overview of the lessons learned from their fieldwork, titled "Unpredictable Evolution in a 30-Year Study of Darwin's Finches." In looking at the finch populations over these thirty years, they see clear evidence for evolutionary change in principal components corresponding approximately to body size, beak size, and beak shape. Much of this they can ascribe to natural selection that they have measured, but the unexplained portion they attribute to introgression of genes from the ground finch into the cactus finch populations. This introgressive hybridization is thus driving the evolution of the cactus finch in a direction different from that dictated by selection. Overall, they conclude that "reversals in the direction of selection do not necessarily return a population to its earlier phenotypic state. Evolution of a population is contingent upon environmental change, which may be highly irregular, as well as on its demography and genetic architecture" (709–710). In other words, though selection (differential survival and reproduction) is important as a cause of evolutionary change, it is ultimately the interplay of such selection with other processes and with the phenotypic state of the population that determines the path of evolution. Because both the environment and the state of the population are constantly changing, evolution is unpredictable. Nevertheless, adaptedness has been maintained: the finches are still there.

. . .

In the three cases I have just examined in detail—experimental evolution of bacteria, peppered moths, and Darwin's finches—I have tried to show that there is no reason to interpret the role of (narrow sense) natural selection in evolution as anything other than a factor maintaining adaptedness in the face of environmental change and associated changes in the mode of adaptedness (lifestyle). Because the environment is constantly changing, so is the direction of selection. In fact, in Darwin's finches, environmental changes cause the direction of selection (in the previous time period) to be maladaptive. To evolve, the conditions for existence must be satisfied, but evolutionary change results not only from natural selection for that specific change but from a wide variety of other processes (mutation pressure, hybridization, etc.).

With this general view of evolution at the level of existing populations now somewhat better supported, it is time to step back and take a broader view. Given this picture of the role of natural selection in existing populations, what role should we ascribe to natural selection and to the conditions for existence in macroevolutionary explanation?

11

How to Talk about Macroevolution

Adaptation to environment is not something which has arisen during the course of organic evolution. To be capable of evolutionary change, organisms must be viable, and to be viable they must already be adapted to environment (and to some degree, specialised). While therefore the innumerable diverse forms of adaptively specialised organisms which have appeared in the history of the globe owe their existence to an evolutionary process, the factors and causes of which may be discovered, the fact remains that the very earliest forms of life, to exist and persist at all, must themselves have been adapted to their environment, for this is the very condition of their existence.

—E. S. RUSSELL, 1945, *THE DIRECTIVENESS OF ORGANIC ACTIVITIES*

Darwin's fundamental error was to erect into a whole theory of evolution— on an implicitly teleological basis of "advantage" and "dis-advantage"— Natural Selection and its corollary, Adaptation.

—LEON CROIZAT, 1962, *SPACE, TIME, FORM: THE BIOLOGICAL SYNTHESIS*

All of the discussion up to this point has really just been preamble to the question that is the historical source of my own interest in the problem of adaptation: how should natural selection enter into explanations of macroevolutionary changes? What I have in mind here are such traditional issues in my own field of evolutionary morphology as the evolution of the eye and the diversification of vertebrate forelimb structure. Ever since Darwin, most people supporting natural selection as the major agent of evolutionary change have felt that it should, and does, provide an explanation of such evolutionary transformations. But what aspects of these transformations are explained by natural selection? And of all the versions of natural selection discussed so far, what version (or versions) is doing the explaining? These questions are the subject of the present chapter.

THE EXPLANATORY ROLE OF NATURAL SELECTION: THE MECHANISM AND THE PRINCIPLE

There are three major versions of natural selection I have identified in previous chapters (along with many debates about precise definitions within them): broad sense, medium sense, and narrow sense (table 11.1). *Broad sense selection* is satisfaction of the conditions for existence. Any organism, organismal feature, genetic or cellular lineage, or species is being "selected for" in this sense if it survives over a given time period; it is being "selected against" if it dies. Broad sense selection is necessarily a selection of *individuals* (at whatever level one is considering). It can be measured by absolute or relative individual lineage rates of increase (see chap. 7).

By contrast, *medium sense selection* is differential survival and reproduction among genotypic or phenotypic *classes* of individuals, over a given time period. It is thus the *average* degree to which this particular class is satisfying its conditions for existence, compared with others. This is Sober's (1984) "selection of" classes, or Fisher's (1930a) "average excess in fitness" of alleles (if random effects are included). It is what is frequently measured (however imperfectly) in natural populations, where it is called "mean realized fitness." It can be measured by the mean absolute or relative rate of increase of the class.

Finally, *narrow sense selection* is differential survival and reproduction among genotypic or phenotypic classes of individuals, *caused by their distinguishing*

TABLE 11.1 Modes of Selection

Mode of Selection	Mechanism of Evolution?	Evolutionary Phenomena Explained
Broad sense Satisfaction of conditions for existence by individuals	Yes	Explains general internal and external functionality of cells, organisms, species
Medium sense Differential satisfaction of conditions for existence by classes of individuals	Yes	To the extent that mutation and immigration can be excluded as explanations, explains all hereditary evolutionary changes
Narrow sense Differential satisfaction of conditions for existence by classes of individuals caused by their distinguishing features	Yes	Explains subset of hereditary evolutionary changes due to effects of distinguishing features on survival of classes
		Explains how lineages are able to maintain adaptedness of individual features in the face of continual changes to individual and its environment

characteristics. It is differential *ability* to satisfy their conditions for existence among genotypic or phenotypic classes, due to their distinguishing features. This is Sober's (1984) "selection for" properties of classes or Fisher's (1930a) "average effect on fitness" of alleles. This is what most people mean when they think of natural selection, which is certainly intended to be a *cause* of evolution, not just a description. It is also closest to what Darwin and Wallace were thinking when they speak of averages and probabilities with which their principles will act.

In the introduction to this book, however, I made another distinction—that between natural selection as *mechanism* and as *principle*. How does this distinction relate to the three versions of selection? Let's begin with narrow sense selection. Since differential ability to satisfy the conditions for existence can clearly be a causal factor in the differential survival or reproduction of genetic or phenotypic classes of organisms, narrow sense selection can indeed be viewed as a probabilistic mechanism of evolution (with the caveat that the actual definition of what one is measuring is not entirely clear; see chaps. 7 and 9).

Medium sense selection is clearly also a mechanism of evolution, in the sense that it can explain changes in a population over time. In fact, it explains such changes better than narrow sense selection, because it incorporates all sources of differential survival and reproduction, both causal and random. It is not tautologous, however, because an alternative explanation for changes in frequency of genetic classes in a population is repeated mutation, or mutation pressure (phenotypes are subject to other effects as well). In an evolutionarily closed population (no immigration), all variants arise by mutation. Differential survival and reproduction, on the one hand, and repeated mutation, on the other, are competing hypotheses for the increase in number or frequency of a type after its origin. In an open population, immigration is an additional possible explanation both for the origin and for the increase in number or frequency of a variant type.

Finally, satisfaction of the conditions for existence by individuals, or broad sense selection, is also a mechanism of evolution, in the sense that whether an individual (organism, genetic or cellular lineage, species, etc.) survives or not is an event that is part of the evolution of the population within which it is examined. The medium (and narrow) sense selection of classes of individuals is made up of such broad sense selection. However, it is important to note that any given individual that survives (or doesn't) could be considered a member of a number of different possible classes of individuals, depending on our point of view. This relativity of survival to our definition of classes is a key distinction between broad sense selection and medium (and narrow) sense selection.

In short, all three versions of natural selection can count as a mechanism of evolution, though in different ways. What about explanation? What can we explain by each? A central thesis of this book has been that the principle of the conditions for existence (broad sense selection) can account for the "adaptedness" and "functionality" of organisms, their complexity and apparent "designedness," apart from any historical process of selection. Not to put too fine a point on it, if organisms weren't adapted and complex, they wouldn't exist. Moreover, they are not in any way more adapted and complex than they need to be to simply exist, because mean overall adaptedness is simply population growth rate, and population growth rate is generally about 1.0.

Therefore, at a first level of analysis, the adaptedness and functionality of organisms are accounted for by the principle of the conditions for existence, just as Cuvier (and before him Empedocles, Epicurus, and Lucretius) recognized. When we think of the evolutionary process as a whole, the conditions for existence of individuals are the range of *boundary conditions* within which the process must proceed, if the process is to proceed at all. There is no exception to the rule that those lineages that continue to exist in evolutionary time are those that are satisfying their conditions for continued existence. The playing out of this principle in populations over time can explain the maintenance of adaptedness in the face of environmental change, as Wallace insisted some 150 years ago. If under changing environmental conditions some varieties in a population survive better than others, and this variation is heritable, then these varieties come to make up a larger part of the population. If overall adaptedness is maintained, then the population (by definition) continues to exist; if not, then it doesn't. What is left, then, to be explained by medium sense and narrow sense selection?

Let us begin with medium sense selection. As already noted, the differential survival and reproduction of classes of individuals is one of the two main causes of evolutionary change in evolutionarily closed populations. At the level of individual genetic lineages, when we know that a population is closed, and we know that a single mutation has swept (however slowly or fitfully) through a population, we know that it has been selected in this sense. In other words, we know that its mean relative rate of increase has been greater than 1.0. This explains (accounts for) its increase in frequency.

At the organismal, phenotypic level, however, the evolutionary effect of one phenotypic class satisfying its conditions for existence better than another depends critically on the genetics of the situation. The classic (but nonetheless valuable) example of this is in a sexual, diploid population with heterozygote superiority in rate of increase at a single locus. With constant ratios of rates of increase of the three genotypes, in this case there is a stable equilibrium value, in which both homozygotes and the heterozygote continue to exist, in spite of the differences in their rate of increase (Fisher 1922a). Thus, even with medium sense selection, explaining changes in frequency of types within a populations requires that we know more than just the rate of increase of the types.

What about narrow sense selection, or what most would consider natural selection proper? As we have seen, this can also explain changes in number or frequency of types in a population, given that we can show that these changes in number or frequency are indeed due to the *effects* of the features we are using to distinguish the types, as they relate to a specified environment, rather than simply a correlated feature or random chance. However, in the absence of such a demonstration—which is exceedingly difficult even in current populations—attribution of any change in frequency or number of a type to narrow sense selection necessarily involves a teleological determinism. In other words, one simply assumes that the observed end state was selected for. Given the many alternative possible explanations for increase in frequency or number of a type, this is clearly unjustified.

Thus, to take the example we will pursue at greater length at the end of this chapter, if we assume that the wings of birds have been "molded by selection" to their present form (as my introductory zoology text suggests), we are implicitly assuming that all of the changes in wing shape that occurred in the ancestry of the lineage were directly selected for, rather than being, for example, neutral changes that occurred by genetic drift, changes that occurred by genetic draft or correlated selection, or changes induced by mutation pressure. In short, it is precisely in the context of such macroevolutionary transformations that we find the mechanism of narrow sense natural selection least useful to invoke, because there is no principled way to rule out alternative explanations for the observed phenomenon.

But given the epistemological limits on our knowledge of past populations, and the teleological determinism involved in invoking selection to explain evolutionary transformations, does this mean that narrow sense selection is of no relevance to macroevolutionary explanation at all? This position seems a rather unhappy one to take, because it implies that there is no way to relate causal processes seen at the population level to broader patterns of evolution. I would argue that at the macroevolutionary level, narrow sense natural selection (differential satisfaction of the conditions for existence by classes of individuals, due to their distinguishing features) is indeed extremely important. It is so, however, not because it explains all "adaptive" evolutionary transformations, but because it is a necessary part of the explanation for the *continued existence of characters in populations.*

Recall that in my discussion of function in chapter 2, I pointed out that the function of a part explains its own existence in two ways: (1) by explaining the life of the organism, and thus its own continued existence, and (2) by explaining the continued existence of the part as a feature of the organisms in a population, given certain assumptions about the underlying genetics. It is the first of these explanations that was highlighted by Cuvier, the second by Darwin and Wallace.

Clearly, if an organismal feature continues to exist in a population, in the face of continual change due to mutation, changing environmental and genetic background, and so forth, it is satisfying its conditions for continued existence. It can do so due to the functions of the feature in individual organisms (narrow sense selection), or for other reasons, as in the classic example of heterozygote superiority just noted.

But when a complex phenotypic feature persists over millions of years, it seems reasonable to think that any genetic correlations involved in maintaining it would have broken down, so that its continued persistence can only be ascribed to its necessity for survival. The continued presence of a heart in vertebrates for over five hundred million years can hardly be ascribed to genetic correlations.

To return to the example of the dorsal plates of stegosaurs touched on earlier, while we may debate what the functions of these plates were, the belief that they must have had *some* function can be grounded in the belief that if they didn't, they would (relatively) soon have disappeared from the population or, in fact, would never have been able to appear in the first place. This is what Wallace (1858) meant in the epigraph to the previous chapter, when he pointed out that the principle of differential survival and reproduction based on cause (narrow sense selection) explains the maintenance of adaptedness, just as the centrifugal governor on a steam engine explains its maintenance of a certain speed. This governor, in his words, "checks and corrects any irregularities almost before they become evident; and in like manner no unbalanced deficiency in the animal kingdom can ever reach any conspicuous magnitude, because it would make itself felt at the very first step, by rendering existence difficult and extinction almost sure soon to follow" (62). Wallace clearly had in mind both the ability of a variety to satisfy its conditions for existence in an absolute sense, and its ability to do so relative to other varieties in the population.

So what to make of the distinction between selection as mechanism and as principle? We have seen that all three versions of selection can be considered mechanisms of evolution, or change in populations over time. Narrow sense selection is clearly operative, but to an extent difficult to determine even in living populations under study. Medium sense selection, or simple differential survival and reproduction, can often be demonstrated, and it is most easily connected to evolutionary change in populations. If differential survival and reproduction—rather than mutation or immigration—can be assumed responsible for evolutionary transformations, it does indeed explain them, but it itself is unexplained without a known causal basis for such differential survival and reproduction. Finally, broad sense selection, or satisfaction of the conditions for existence by individuals, is also a mechanism of evolution; indeed, it makes up both medium sense and narrow sense selection. It is the underlying reality, from which our division of individuals into classes creates the differential survival and reproduction (selection) we can measure.

As a principle of interpretation, however, natural selection is something different than it is as mechanism. The principle of the conditions for existence, as proposed by Cuvier, holds that the internal and external harmony of organisms is explained by the fact that if such harmony didn't exist, the organism wouldn't, either. As I have suggested, this principle takes the ancient materialist doctrine of Empedocles and

Epicurus, developed as a criticism of theism and the argument to design, and turns it into a guide for the study of life. From Cuvier's perspective, the world operates as a harmonious whole, and every part of it that exists does so in virtue of the fact that it satisfies its conditions for existence. These conditions can be elucidated, which constitutes a scientific understanding of the organism, but of a different kind than that afforded by physical laws.

The Darwin-Wallace principle, as I have called it, expands on that of Cuvier by adding to his principle dealing with the conditions for the existence of *organisms,* a principle dealing with conditions for the continued existence of *characters* in *populations.* Once one accepts evolution, one clearly needs both the Darwin-Wallace principle *and* Cuvier's principle to deal with the situation of evolving populations. The principle of the conditions for existence is a guide to the physiology of organisms; the principle of the conditions for existence of characters in populations (under the influence of natural selection and other evolutionary factors) is a guide to the genetic physiology of populations. Both provide only for what I have termed conditional teleological explanations, rather than deterministic ones.

TELEOLOGY AND THE TERMINOLOGY OF SELECTION

Before looking at macroevolutionary explanations in general, it will be useful to critically examine some of the terminology associated with the use of natural selection in these explanations. Much of this terminology has a decidedly teleological flavor, as when one talks of the "selective advantages" directing the course of evolution, the "force" of natural selection, or selection "pressures." The problem here is that the pressure or force is frequently derived from the difference between the present state and an imagined improved future state, or between the present and an imagined less improved past state (as discussed in chaps. 2 and 7). Nevertheless, it is difficult to believe that there is not some real subject under discussion when these terms are used, which might perhaps be expressible without such teleological overtones. Let's consider some examples of the use of the terms *selective force* and *selective pressure* from the current literature and see how far the twin principles of the conditions for existence of individuals, and of characters in populations, will take us in seeing what these terms are really expressing.

Selective Force

For the term *selective force,* I found only two major uses (with some minor variants within them). The first of these is essentially the distribution of (expected) rates of increase across alleles at a locus. For example, Ohta (2002), in the context of the nearly-neutral theory, states that "at the time of speciation, the magnitudes of drift may get large and the *selective force* may change" (16135, emphasis added—as I do for the term under discussion in all quotes in this section). Likewise, Suzuki and

Gojobori (1999) speak of the "*selective force* at single amino acid sites" (1315) and distinguish this from the "*cause* of the selective force" (1324), making it clear that the selective force itself is essentially the distribution of the average excess in rate of increase across the alleles at a locus.

The second use is of more interest in the present context, as it is a more metaphorical one. It generally involves a particular biological context and the form that selection will take with respect to a particular feature in that context. This selection is considered to account for the origin of the feature. For example, in looking at inducible defenses in the water flea *Daphnia*, particularly in microstructure of the carapace, Laforsch et al. (2004) noted that "inducible defenses are common strategies for coping with the *selective force* of predation in heterogenous environments" (15911). Here we have a particular biological context—predation; a particular feature—the inducible strengthening of the carapace; and a form that selection will take—namely, favoring the induction of a stronger carapace when predators are present. Underlying this account, however, is an evolutionary scenario: the origin of this trait (inducible toughening of the carapace) is associated with selection for this trait. We implicitly imagine poorly protected *Daphnia* swimming around, vulnerable to nasty predators, when along comes a selective force (nasty predators) that selects those that respond appropriately (by eating those that don't). As discussed earlier, such an underlying assumption is necessarily teleological, in that the only evidence to support this hypothesis is that this trait evolved, and it currently has the function of predator defense. What is the alternative?

In most of these uses of the term *selective force*, what is really going on is that one is constructing an evolutionary scenario, in which the Darwin-Wallace principle (the conditions for the existence of characters in populations) is used to frame an account of how a particular feature can continue to exist in a population. Thus, a suitable equivalent to *selective force* in such contexts is simply *function* or *functional significance*. For example, rather than saying "inducible defenses are common strategies for coping with the *selective force* of predation in heterogenous environments," we might simply say "inducible defenses are commonly found in heterogenous environments, where they *function* to reduce predation." This phrase should be understood to mean both (1) having inducible defenses (as opposed to not having them) can be a necessary condition for the continued existence of organisms exposed to variable predation, and (2) the evolutionary maintenance of particular inducible defenses depends on this function.

A slight further modification encompasses not just the maintenance but the origin of this trait. In the first ancestral *Daphnia* that happened to grow a thicker carapace in the presence of a predator and was henceforth protected, protection from the predator was a function of the induced thicker carapace. And at every point along the evolutionary path leading to the condition we see today, the same was undoubtedly true. Thus, the entire evolutionary scenario can simply be replaced by the statement that "inducible defenses commonly evolve in heterogenous environments, where they function to reduce predation (when predators are present)."

Let's try out this translation of the language of "selective force" on some other examples. Anderson-Nissen et al. (2005) looked at molecular variation in bacterial flagellin in the context of receptors for flagellin on mammalian cells. They found that Toll-like receptors could recognize some but not all subclasses of flagellin and proposed that "flagellin receptors provided the *selective force* to drive the evolution of these unique subclasses of bacterial flagellins" (9247). Here the biological context is flagellin receptors on host cells, the character is flagellin structure, and the form that selection would take is to favor variant forms of flagellin that are still functional although not recognized by the receptors. The translated statement would run as follows: "these unique subclasses of bacterial flagellins evolved in the context of flagellin receptors; they function to permit motility while avoiding receptor activation." Has there been any loss of content, other than the *assumption* that narrow sense selection drove the evolutionary process?

Aagard et al. (2006) looked at the proteins that make up the egg coat of abalone eggs. They note that "identifying the *selective forces* acting on egg coat proteins is a preliminary step in establishing their potential contribution to the speciation process" (17302). Again, this statement is rather easily translated as "identifying the functional significance of egg coat proteins is a preliminary step in establishing their potential contribution to the speciation process."

Blackledge, Coddington, and Gillespie (2003), looking at the form of spider webs, argued that "predation by wasps exerts a directional *selective force* that favours the building of complex, three-dimensional webs" (15). Translation: "complex, three-dimensional webs function to reduce wasp predation."

Finally, Tabak et al. (2006) asked, with respect to the origin of peroxisomes, "could the segregation of lipid biosynthesis in the ER from degradation in the peroxisomes have been a *selective force* to create a new organelle?" (1653). The latent teleology in such a question should be clear by now—the origin of a new organelle is attributed to a postulated function that the new organelle is going to have. Preferred translation: "could the segregation of lipid biosynthesis in the ER from degradation in the peroxisomes have been an important function of the newly formed organelle (and thus help explain its persistence)?"

Selective Pressure

The term *selective pressure* is often used in ways quite similar to *selective force*, although—judging from my online searches—it is much more popular and is used in a wider variety of contexts. For example, we find it being used in situations where it is effectively synonymous with fitness (expected rate of increase) differences, as when Brunet et al. (2006) speak of the "*selective pressure*, as measured by the Ka/Ks ratio" (1810). Here positive selective pressure is identified with positive amino acid substitution rates (compared with the synonymous substitution rate), negative pressure with negative substitution rates. Under standard theory, such rates will obtain in situations of positive and negative fitness effects of variants.

As with *selective force*, the term *selective pressure* is often used in the context of an historical change induced by natural selection. This is again interpretable as the function or functional significance of a feature, meaning both contribution to the ongoing survival of the containing organism or lineage, and to its own ongoing survival. Thus, Zhang (2006), in discussing convergent evolution of amino acid sequences in digestive RNAses of leaf monkeys, argues that "the parallel substitutions are not attributable to chance alone and therefore must have been driven by a common *selective pressure*" (819). This argument is easily restated as "must have a common functional significance." The functional significance is something that can be investigated, while the selective pressures (whatever they were) occurring during the sweep of the mutation to fixation are lost to us.

A similar use is provided by Heffner, Koay, and Heffner (2006), in the context of the auditory responses of bats. They state, "The main source of *selective pressure* for this high-frequency hearing in mammals has been to enable them to use the two high-frequency sound-localization cues, namely the pinna cues that result from the directionality of pinnae, and the binaural spectral-difference cue" (18). Again, this is easily restated as a simple function statement: "High frequency hearing in mammals functions mainly in sound localization, using the two high frequency cues available." In such contexts, natural selection is being brought in and paraded around essentially for show—no predictions follow from the first statement that don't follow as well from the second.

An example focusing on lack of change comes from a recent examination of homeobox gene cluster evolution in vertebrates by Mulley, Chiu, and Holland (2006). They argue that "the presence of a ParaHox gene cluster in amphioxus, humans, mouse, and *Xenopus* implies there has been a strong *selective pressure* to maintain physical linkage of the three homeobox genes, otherwise inversions and translocations would have dispersed these genes during the half-billion years since the divergence of cephalochordates and vertebrates" (10369). Translation: this "implies that physical linkage of the three homeobox genes has an important function, otherwise inversions and translocations would have dispersed these genes during the half-billion years since the divergence of cephalochordates and vertebrates."

A related, but not identical, use of the term *selective pressure* is the environmental context in which a trait functions and therefore in which differential performance of a trait occurs. Thus, in discussing adaptation of forms of the visual pigment opsin to different lifestyles, including nocturnality, Bowmaker and Hunt (2006) note that "animals have evolved their visual sensitivity to match aspects of their photic environment, and it is likely that the primary adaptive *selective pressure* is the spectral range and intensity of daylight" (R484). An alternate version of this statement might simply be "it is likely that the primary functional context for visual sensitivity is the spectral range and intensity of daylight."

These authors go on to discuss the convergent loss of one type of cone photoreceptor in seal and cetaceans: "Since the closest terrestrial relatives of the seals and whales (carnivores and hippopotamus, respectively) possess both LWS and SWS1 opsin genes, the mutations in the SWS1 gene in these two distinct orders of marine mammals illustrates convergent evolution, suggesting a common *selective pressure*. However, it is not clear what that *selective pressure* was. Most fish have retained short-wave cones and, as described above, water transmits primarily blue/green light, so the loss of SWS1 cones is somewhat counter intuitive" (R488). Here again, one might simply replace *selective pressure* with *functional significance*.

Selective pressures are also used to explain the occurrence of traits in one population and not another. When the only evidence is that such traits occur in one population rather than the other, this is simple teleological determinism. For example, Pausas, Keeley, and Verdú (2006) suggest that there has been "stronger *selective pressure* for fire persistence traits in California than in the Mediterranean Basin, such as higher frequency of lignotubers, serotiny and fire-induced germination by smoke or charred wood" (32). The evidence cited for such pressure is that such traits occur at higher frequency in California than the Mediterranean.

However, the most common use of the term *selective pressure* is slightly different from the major one we have seen for *selective force*. This use also relates to a particular biological context and to the form that selection will take with respect to a particular feature in that context. Rather than looking at the feature as a realized result of the evolutionary process or the hypothetical first appearance of the feature, this use focuses on the ongoing evolution of the feature. Selective pressure is imagined here as a factor leading to a *tendency* for evolution in a particular direction to occur, given that it is not opposed by other factors. Once again, we see selective pressure being brought in as part of an evolutionary scenario. As noted, this is using the Darwin-Wallace principle to explain the persistence (or lack thereof) of traits in populations.

A particularly nice example of such usage is the following one from Sawyer and Malik (2006): "host genes are under constant *selective pressure* to limit the success of mobile elements, whereas mobile elements are under constant *selective pressure* to evade these limitations" (17614). In this context, "selective pressure" points to a situation in which, everything else being equal, a particular phenotype will be more successful at satisfying its conditions for existence than another. In any actual situation, of course, such a selective pressure may be opposed by numerous factors, including other "selective pressures."

Rather than a simple function, such cases appear to deal with overall adaptedness (rate of increase) of variant types. Applied to the example just given (Sawyer and

Malik 2006), one might restate the passage as follows: "everything else being equal, host genes that consistently limit the success of mobile elements will survive at a higher rate than those that don't, while mobile elements that consistently evade these limitations will survive at a higher rate than those that don't." In other words, this is really just pointing out that a particular evolutionary situation exists, one that Sawyer and Malik call "genetic conflict." What do we make of this situation?

Sawyer and Malik (2006) make a great deal of it. From this situation of "genetic conflict," they infer that two functional classes of molecules of the host are especially likely to interact with the invading viruses and other mobile elements: (1) nucleases that provide direct protection from the mobile elements and (2) parts of the transcriptional and translational machinery of the host that are utilized by the mobile elements. By comparing two species of yeast, *Saccharomyces paradoxus* and *S. cerevisiae*, which diverged about five million years ago, they are able to estimate K_a/K_s (dN/dS) ratios across a wide variety of genes using a "sliding window" strategy. They find two functional classes of genes that have undergone significant nonsynonymous change: those involved in meiosis and those involved in nonhomologous end joining (NHEJ). They propose that NHEJ genes are "under positive selection to impede the integrative success of the Ty LTR-retrotransposons that populate the yeast genome." Why? Because mutant phenotypes of these genes are universally associated with changes (increases or decreases) in Ty retroposition rate. They further propose two models for direct interaction between the NHEJ proteins and Ty elements: a "defensive" one in which the NHEJ proteins are used by Ty elements to locate double-stranded breaks for integration, and an "offensive" one in which the NHEJ proteins actively alter Ty elements, preventing their integration. They find evidence for each, depending on the protein considered.

Let us accept Sawyer and Malik's hypothesis of what is going on here and consider what it means. In particular, let's look at this situation from the standpoint of the *stability* of amino acid composition of the NHEJ proteins. From this point of view, the presence of Ty elements is an environmental factor reducing the stability (degree to which the conditions for existence are satisfied) of the ancestral amino acid composition, by making it vulnerable to changes that allow for maintained function in repairing double strand breaks, while reducing the Ty retroposition rate. The system will continue to evolve as long as Ty elements are successful in occasionally transposing and the organisms (cells, etc.) containing them continue to exist. What we see here is the Darwin-Wallace principle (the principle of the conditions for continued existence of features in populations) being used to predict that under a situation in which the main players are the NHEJ genes and the Ty elements, we expect to see rapid ongoing change of each. What we don't know, of course, from this scenario is whether Ty elements will eventually win out and sweep through the genome, whether they will be successfully killed off, or whether the situation will continue indefinitely much as it has.

I have spent some time dissecting this example because it is a rather nice one that shows how, by discovering the existence of a particular evolutionary situation, one can elucidate functions that would otherwise not have been clear, and how these functions can reciprocally help us understand the evolutionary situation. This is, in fact, the major role of such evolutionary reasoning in biology in general, where the Darwin-Wallace principle is used as an interpretive guide to the diversity of life around us.

Rather than the coevolutionary situation discussed by Sawyer and Malik, where a continued evolution was expected, a more common situation is to see conflicting selection pressures used to explain the continued maintenance of features. Warringer and Blomberg (2006) were interested in explaining yeast protein size. They begin with the assumption that, all else being equal, protein size should be reduced: "in the light of the *selective pressure* to minimize biosynthetic costs by reducing the size of highly expressed proteins" (3). However, this assumption creates a problem for them, because many yeast proteins, including highly expressed ones, are quite large. How do they solve this problem? By looking for another selective pressure increasing protein size. In particular, they look for functions that large proteins can do that small ones can't. They conclude that multiple protein-protein interactions may be such a feature, and thus "the *selective pressure* to maintain a large protein size at least partially may be a *selective pressure* to entertain more protein-protein interactions" (4). Here the Darwin-Wallace principle (the principle of the conditions for existence of characters) is being used to explain the maintenance of a feature (large protein size) in a population.

Finally, I must note a kind of combined use of the term *selective pressure*. This is one that includes both a functional scenario and a quantitative aspect. For example, Tam et al. (2005) worked on antibiotic resistance in bacteria. They note that "as the drug exposure increased, a higher *selective pressure* was imposed on the bacteria" (422). Here *selective pressure* clearly means differential survival and reproduction due to the application of the antibiotic, with a quantitative aspect because more antibiotic equals more selective pressure.

In short, we see that the term *selective pressure*, like *selective force*, is generally used to talk about narrow sense natural selection as part of an evolutionary scenario explaining the continued existence (or lack thereof) of traits in populations. The strength of such explanations—when unjustified, teleological assumptions about selection as the cause of inferred evolutionary change are removed—is that they are essentially local explanations, about the persistence of traits at each point in evolutionary history. Thus, we are assuming no more about past populations than we can examine in present populations. We simply do not have the information to reconstruct the precise evolutionary factors involved in the origins of traits, but if we can predicate a certain situation that obtains at all points in the evolutionary trajectory, we have a general explanation for the observed maintenance (or not) of a trait. This explanation is, of course, a conditional teleological one.

This overview has been meant to show how the metaphor of natural selection as a force or pressure moving the evolutionary process from a poorly adapted past state to a better-adapted present or future state is brought into many discussions of the role of natural selection in macroevolution. I have tried to demonstrate that the invocation of a "force" or "pressure" of natural selection, when connected to a particular historical transformation, almost inevitably results in a teleological determinism. This does not mean that narrow sense natural selection has no role in our explanation of evolutionary transformations; we instead can look to narrow sense natural selection as a factor maintaining adaptedness through such transformations. This distinction is clearest when one considers that many uses of the term *selective force* or *selective pressure* can simply be replaced by *function* or *functional significance*. However, I have also tried to show how such teleological language is in fact frequently tied to the use of natural selection as a principle of explanation— in particular, a principle explaining the persistence of characters in populations over time. It is in this role that natural selection indeed is an essential principle for the interpretation of macroevolution.

With this brief introduction to the problems of thinking and talking about macroevolution, I'd like to turn to a more substantive aspect of macroevolution: the role of "constraints" in the evolutionary process. The study of constraints has been one of the most controversial, but nevertheless burgeoning, areas of evolutionary biology. What do such constraints mean from the perspective of the conditions for existence?

CONSTRAINTS: BY WHAT AND ON WHAT?

The notion that there are constraints on the evolutionary process has a long history, going back, on one hand, to Darwin's invocation of "laws of growth" as influences on form and, on the other, to Wallace's invocation of natural selection as the "governor" preventing unbalanced deficiencies of form. The modern use of the term *constraint* likewise appears to have two sources: in the work of paleontologists and functional morphologists, particularly those in Europe, and in the discovery that functionally important proteins and regions of proteins evolve at a slower rate than functionally unimportant ones. Let us consider this latter use first.

The notion of "functional constraints" on the evolution of protein structure emerged from some of the earliest comparative studies of amino acid sequences (e.g., King and Jukes 1969; Dickerson 1971), which showed quite clearly that the rate of change of amino acid sequence was inversely proportional to the likelihood of deleterious effects of sequence change on protein function. Thus fibrinopeptides evolve fast, cytochromes slow, and the active sites of enzymes evolve slower than internal structural regions. Similar patterns are now known at the nucleic acid level, where "silent" substitutions typically occur much more frequently than amino acid–replacing ones.

The term *functional constraint* in the context of protein evolution appeared at least by 1974, in a paper by Kimura and Ohta entitled "On Some Principles Governing Molecular Evolution." These patterns were, of course, happily claimed (Kimura and Ohta 1974; Kimura 1983) as evidence for the "neutral theory." The idea of such functional or selective constraints is now a fundamental one in the study of molecular evolution, and an assumed negative correlation between evolutionary rate and functional constraint is used to predict functionally important regions of proteins (e.g., Simon et al. 2002).

What are such constraints by? What are such constraints on? Clearly, in the envisaged scenario of protein evolution, such functional constraints are constraints on protein *evolution* (as opposed to natural selection), by the *function* which the protein (or part of protein) plays in the organism. How does narrow sense selection enter this scenario? Primarily as a negative factor, rejecting all functionally inferior forms and thereby maintaining overall adaptedness. This is again using the Darwin-Wallace principle (the principle of the conditions for existence of characters in populations) to explain evolutionary trajectories: those sequences that are critical for survival are retained longer than those in which changes consistent with survival are possible, and the rates are proportional to the functional constraint. Interestingly, such constraints appear to be very stable for a given protein in evolutionary time (Simon et al. 2002).

How does "positive" selection relate to such functional constraints? It is helpful to recall the basic definitions inherent in the dN/dS approach to measuring narrow sense selection (see chap. 9). For any given region, the rate of silent nucleotide substitutions sets a kind of "baseline" rate of substitution. Any rate of amino acid substitution significantly less than that expected based on the silent substitution rate (dN/dS < 1) indicates functional/selective constraint, with the strength of the constraint (stringency of selection) greater the lower the ratio. Likewise, in this framework, "positive selection" occurs when amino acid substitutions occur more frequently than expected based on the silent substitution rate (dN/dS > 1), with the strength of this selection greater the higher the ratio. Two other terms that come up in the context are *relaxation of selection*, which refers to the situation where functional constraint weakens (the limit being complete neutrality), and *diversifying selection*, which refers to a situation where there is positive selection that varies across individuals.

In a recent paper on incorporating recombination into estimates of selection (Wilson and McVean 2006), such "diversifying selection" was contrasted with "purifying selection" as follows: "An excess of nonsynonymous relative to synonymous polymorphism is a clear signal of diversifying selection, whereas a lack of nonsynonymous relative to synonymous polymorphism is indicative of purifying selection imposed by functional constraint" (1411). What is of interest here is that selection appears in two rather different roles—on the one hand, as "purifying selection" or "functional constraint" preventing change in amino acid composition, and on the other, as "diversifying selection" promoting it. But clearly "functional constraints" must be maintained during diversifying selection, at least to the extent that nonfunctional changes won't be tolerated. It is merely the case that purifying selection has been relaxed or reversed for some individuals with respect to at least one alternative amino acid, so that the predominant amino acid is no longer the one under functional constraint not to change. Thus, both "positive" and "diversifying" selection may be viewed simply as a shifting constraint.

By contrast with such functional or selective constraints on protein evolution, the meaning and interpretation of developmental and other constraints at the gross phenotypic level is much less consistent (for some recent perspectives, see Schwenk and Wagner 2004; Brakefield 2006; Salazar-Ciudad 2006). Historically, the idea of such constraints appears to have developed in the context of the revival of evo-devo and the critique of the neo-Darwinian paradigm by Stephen J. Gould and his colleagues, though the historical timing suggests that the "constraint" terminology itself may have migrated from the field of protein evolution. Gould and Lewontin's famous 1979 paper on "The Spandrels of San Marco and the Panglossian Paradigm," already discussed, is the first place I am aware of that the term saw prominent use. Their abstract began dramatically as follows: "An adaptationist programme has dominated evolutionary thought in England and the United States during the past 40 years. It is based on faith in the power of natural selection as an optimizing agent. It proceeds by breaking an organism into unitary 'traits' and proposing an adaptive story for each considered separately. Trade-offs among competing selective demands exert the only brake upon perfection; non-optimality is thereby rendered as a result of adaptation as well" (581). Having explained what it is they are attacking, they go on to state that they are promoting an alternative vision, one that they note has typically held sway in continental Europe. This vision "holds instead that the basic body plans of organisms are so integrated and so replete with constraints upon adaptation . . . that conventional styles of selective arguments can explain little of interest about them. It does not deny that change, when it occurs, may be mediated by natural selection, but it holds that constraints restrict possible paths and modes of change so strongly that the constraints themselves become much the most interesting aspect of evolution" (594).

Gould and Lewontin describe two categories of constraints. The first is *phyletic constraints*, exemplified by the observation that "humans are not optimally designed for upright posture because so much of our *Bauplan* evolved for quadrupedal life" (594). They also invoke phyletic constraint to explain why no mollusks fly in air and no insects are as large as elephants. *Developmental constraints* are for them a subcategory of phyletic constraints. Second, and distinct from such phyletic and developmental constraints, are *architectural* ("*bautechnischer*," Seilacher 1970) *constraints*: "These arise not from former adaptations retained in a new ecological setting (phyletic constraints as usually understood), but as architectural restrictions that never were adaptations, but rather the necessary consequences of materials and designs selected to build basic *Baupläne*" (595). Such architectural constraints are of course the famous "spandrels" of the paper's title. Their primary example comes from the work of Seilacher (1972) involving divaricate ribs and patterns on bivalve shells. These ribs or patterns are in most cases nonfunctional, but in occasional cases they may secondarily acquire a function (e.g., in burrowing, in mimetic resemblance, or in hosting symbiotic algae). Their larger point—that it is a mistake to atomize organisms and explain each part as the solution to a problem raised by the environment—is quite congenial to the view I have been promoting here. However, their preferred alternative of a "pluralistic view" also atomizes organisms; it merely tries to explain each part as either a direct adaptation *or* as a product of some other factor or factors. This hardly clarifies matters.

What are such "architectural constraints" or "spandrels" from the viewpoint of the conditions for existence? The phenomenon of bivaricate ribbing or coloration, as Gould and Lewontin note, appears to arise from the process by which the bivalve mantle lays down the shell. That it may sometimes have a function is of interest, as are their observations that when it has an apparent function, it is much less variable than otherwise. All of this is consistent with the principle of the conditions for existence of characters in populations; clearly if such patterns can continue to exist, they must be satisfying their conditions for existence, most likely because they are intimately connected with the process by which shells form (which process itself is clearly functional and thus maintained by narrow sense natural selection). A related example might be the mammalian navel. If one looked to the adaptive or functional significance of the adult navel to explain its existence, one would clearly be ignoring the fact that the navel is a remnant of an earlier structure that was indeed functional. There is hardly anything surprising here—other than the tendency of some other biologists to go too far in expecting every "character" or "trait" we can distinguish in an organism to have a function.

What is of interest to us in this first attempt to classify the role of constraints in evolution is that for Gould and Lewontin, such constraints are not constraints on *evolution*, but constraints on *adaptation by natural selection*. Look at the language they use: "the *constraints* themselves become more interesting and more important in delimiting *pathways of change* than the *selective force* that may mediate change when it occurs" (581, my emphasis). The alternatives in play are the adaptationist program, which views each trait as an optimal solution to a problem set by the environment, with natural selection serving to drive the evolutionary process toward that solution, and the constraints perspective, which focuses on other factors that may serve to limit the changes that can occur. Gould and Lewontin's caricature of the adaptationist program is a teleological one, in which "current utility" is equated with "reasons for origin" (in Gould's colorful language). In fact, it is precisely the deterministic teleological mode of explanation I have been criticizing as well. But their solution is to argue that there may be other reasons for origin aside from natural selection, in the form of "phyletic and architectural constraints." These constraints are thus conceived of as constraints on a fundamentally teleological process of selection, in which there is a desire to achieve some end, but constraints either prevent one from getting there or influence the form by which the end is achieved.

The problem with the adaptationist program is not that it considers each trait to represent a solution to an environmental problem, which is merely utilizing the Darwin-Wallace principle under certain simplifying assumptions about genetics. The problem is instead with the teleological step of evaluating the causes of the evolutionary process with respect to particular real or hypothetical end states, when these end states are projected as implicit goals of the evolutionary process. Merely arguing under some circumstances that a trait we identify as complex was not produced by natural selection but by some "constraints" on natural selection does not get at the root of the problem, because the teleological perspective with respect to selection remains.

Developmental constraints have now become standard fare in evolutionary biology. However, their meaning and interpretation remain somewhat open (Maynard Smith et al. 1985; Gould 1989; Antonovics and van Tienderen 1991; Schwenk 1995; Raff 1996; Arthur 2002; Richardson and Chipman 2003; Schwenk and Wagner 2004; Brakefield 2006). My own graduate advisor, Pere Alberch, was one of the first to utilize Gould and Lewontin's concept of developmental constraints; he proposed that developmental constraints can be investigated empirically as a way to understand the paths that evolution takes (e.g., Alberch 1980, 1982). In carrying out this program he was able to show, for example, that the pattern of digit loss that occurs evolutionarily in frogs and salamanders reflects innate differences in the underlying developmental system, and that the patterns are easily replicated by experimental reduction of limb-bud cell number in the lab (Alberch and Gale 1983, 1985).

Alberch's argument was that the highly structured epigenetic interactions involved in development of multicellular organisms impose a limit on the sorts of novel morphologies that can emerge. He pointed out that the question of what range of phenotypic forms is generated by mutation is a different one than considerations of the survival of those forms after they are generated. For example, in illustrating several "forbidden morphologies" indicated by a model of limb development, he noted that "regardless of how adaptive the morphologies illustrated . . . could be, they cannot be selected for since the necessary variation will not be produced" (1982, 327).

From one point of view, such developmental constraints are irrelevant to the main topic of this book. If we consider a fertilized egg over the time from fertilization to hatching, the egg is itself a functional system that generates the hatchling. This functional system must (in a conditional teleological sense) generate a hatchling, which in turn must be able to satisfy its conditions for continued existence. The developmental system itself must satisfy certain conditions for existence of the individual and, at the level of the population, conditions for its own existence. This functional requirement, however, says absolutely nothing about the range of divergent morphologies that can be generated. By analogy with the conditions for *existence* of forms, we might also postulate conditions for the *origin* of forms (Müller and Newman 2003). These conditions would of course include certain functional characteristics of the developmental system. From this point of view, phylogenetic/developmental constraints (conditions for origin) and functional/selective constraints (conditions for continued existence) are complementary, not competitive. They correspond to the "generative" and "selective" constraints, respectively, of Richardson and Chipman (2003; cf. also Amundson 1994; Schwenk 1995).

However, from another point of view, it does seem that generative constraints and functional constraints can serve as alternative explanations. If I am trying to explain why there are no six-legged tetrapods, it is indeed quite a different matter if in five hundred million years there never have been any produced by the developmental systems of tetrapods, or if they have been produced constantly but have always died off at a higher rate than the four-legged members of the population. In such a situation, it is only reasonable to investigate *both* the developmental system and the functional context, the first to try to understand the limits on the range of morphologies that can be generated, the second to try to understand why the morphologies that do occur are retained. These are simply different issues.

As recently pointed out by Salazar-Ciudad (2006), the real question is not "Whether development constrains evolution or not?" but rather "How different kinds of developmental functioning influence evolutionary dynamics (through the morphological variation they can produce)?" and "Which proportion of developmentally possible variation is filtered out by selection?" (113). His own work on the mechanisms involved in generating the patterns of tooth cusps in mammals (Salazar-Ciudad and Jernvall 2002) provides an elegant example of such an approach (see also the recent work of Prusinkiewicz et al. 2007 on plant inflorescences).

THE CONDITIONS FOR EXISTENCE IN
MACROEVOLUTIONARY EXPLANATION:
THE ORIGIN OF BIRD FLIGHT

Background

In the previous chapter, I discussed the relation of the *state* of adaptedness and the *process* of adaptation to a changed environment. My general point was that during any process of evolutionary change, overall adaptedness is necessarily maintained, so that the process of adaptation can only mean a change in the *mode* of adaptedness. At the level of individual traits, such a change might mean either an improved performance of the derived trait (compared with the ancestral) at some task critical in the new environment, or an improved effect on overall survival. Let's look at a simple example of macroevolution and consider the question of how we should represent the process of evolutionary adaptation.

In particular, how should we represent the role of narrow sense natural selection and of the conditions for existence in this process? The example I have chosen to look at is the evolution of bird flight, a classic example of a major morphological and functional transition.

There are more than ten thousand species of living birds. Over the last twenty years or so, a wealth of new fossil discoveries have strongly supported the view that birds are a subgroup of theropod dinosaurs, as Huxley argued years ago (Chiappe 2007). Since most other theropods don't fly, clearly flight originated somewhere along the lineage leading to modern birds. How? Why? What can we reasonably hope answers to these questions will look like?

I would argue that the *how* of the evolutionary process can be studied in at least three different ways:

1. One can use phylogenies to reconstruct (with some degree of error) ancestral character states, and thus the path evolution took through character space. One usually does so under a minimum evolution (gradualism, parsimony) or likelihood criterion.
2. One can examine such reconstructed paths of evolutionary change and posit major functional role(s) for the characters under study at every time along the path, with the implicit understanding that these roles were responsible for their evolutionary maintenance.
3. One can examine such paths as reconstructed alterations to developmental/ genetic systems and interpret the particular path followed as one that was generated by the limited range of possible (not just viable) alterations in such physicochemical systems.

However, to say *why* this particular path was chosen in anything other than a con- ditional teleological sense requires a causal account of the transformations involved. This is exactly where teleological determinism enters. As I have noted repeatedly, when we say that the path was chosen because of a derived state or function, we are implicitly using the end state achieved as the goal of the process of selection (unless we have independent evidence of such causes). This is precisely what we need to be able to avoid.

Keeping in mind this general classification of the descriptions and associated explanations we can attach to evolutionary change, let's consider some of the views that have been held through the years on the origin of bird flight. The origin of flight of course has been quite closely connected with the question of the origin of birds as a group, or, in modern terms, the question of the phylogenetic relations of birds, and with the origin and early function of feathers. I will try to focus on flight itself, however, rather than these related issues.

The discovery of *Archaeopteryx* in 1861 was a key turning point in the study of the origin of flight, because it provided a clear intermediate stage between "reptiles" and birds—one in which, for example, the tail was still quite long. It was partly based on this fossil that Huxley supported a dinosaurian origin of birds.

The modern debate began in 1927, with the publication in English of Heilmann's *The Origin of Birds*. Heilmann attacked the dinosaurian view, instead supporting a pseudosuchian, or basal archosaur, ancestry of birds. From this starting point, he proceeded to construct a scenario for the origin of birds, based on a hypothetical "Proavis" intermediate between reptiles and birds. The functional morphology of the transition was integral to this scenario. Following Abel (1911), who had argued that the backwardly directed large toe in the hind foot of both birds and theropods indicated an arboreal ancestry, Heilmann suggested a path of functional transformation from an arboreal climber, to glider, to powered flier. The evidence and arguments advanced on the nature of the "Proavis" finds Heilmann (1927) taking us on

> a mental excursion to the remote period when the transformation of the reptile into a bird was in the process of realization, and the bird-class, so to speak, had its first vernal bloom. At that far away time the blood of the small reptile-like creatures, hopping from branch to branch in the trees of the Lower Trias, must have been fermenting with new yearnings and longings. Impelled by an unconscious desire, they tried to jump farther and farther; there was a queer and stimulating sensation of pleasure in straining the efforts to the utmost, almost beyond the bounds of possibility, in challenging the chance of missing, the risk of a dangerous fall; there was an immense discovery in feeling the first supporting effects of the air-current along the tips of their scales.
>
> Then it was that these commenced to lengthen wherever the pressure of air was strongest: on the back margins of the limbs, on the flanks, and along the sides of the tail. And the animal leaping, the fore-limbs spreading more than the hind-limbs, the lobed elongations of the scales grew stronger along the posterior edge of the arms, so that gradually a new shape of parachute was formed, and winged flight came into existence. (201)

This is a rather remarkable vision of the elongation of scales under the mechanical stimulation of a rush of air, combined with the role of "unconscious desire" driving the process by making the animals take increasingly death-defying leaps. I need hardly point out that this is a rather teleological way of looking at the situation. This teleology is borne out by his overall vision of the process aiming toward the production of a bird: "We must not think of the entraînement from reptile to bird as taking place along a single line only. . . . No doubt many futile attempts have been made in attaining the character of the bird, but all of these came only a part of the way; certain defects or weak points have made these stocks succumb in their competition with the one that was best equipped, the only one that finally succeeded in becoming a bird" (201–202).

But while we may smile today at Heilmann's way of expressing his evolutionary vision, and its implicit teleology, there is no doubt that his interpretation, in which small arboreal, lizardlike animals (figure 11.1) adopted a gliding lifestyle, and

FIGURE 11.1. "Restoration of the hypothetic Proavian" by Heilmann (1927, fig. 142). Reprinted from *The Origin of Birds* by Gerhard Heilmann, © 1927 by D. Appleton & Co., renewed © 1955 by Gerhard Heilmann. Used by permission of Dutton, a division of Penguin Group (USA) Inc.

subsequent powered flight, is a testable hypothesis concerning the path evolution took. Moreover, his emphasis on the role of functional constraint on evolutionary transitions is a key theme of all evolutionary biology since Darwin (1859) and one that we must acknowledge, however inadequate our understanding of what such functional constraints actually are. What is plainly absurd in his vision is merely the mechanism of transformation, not the transformation itself. It is nevertheless easy to see why the founders of the modern synthesis reacted so strongly to the evolutionary biology of the 1920s, of which Heilmann is by no means an unusual example.

The Modern Synthesis

How did the theoreticians of the modern synthesis do with this same case? Discussion of the origin of flight occurs sporadically in the early synthesis (Simpson 1946, 1953; de Beer 1954). As a representative—if somewhat iconoclastic—view developed in the context of the synthesis, I'd like to examine Walter Bock's (1965) seminal discussion of "The Role of Adaptive Mechanisms in the Origin of Higher Levels of Organization." For Bock, the problem at issue is "whether or not the appearance of new major forms of organization is adaptive in its entirety, and whether or not it can be explained in terms of known mechanisms of biological adaptation" (273). His general argument is that "if the sequence of events involved in the origin of new groups of organisms is outlined properly, the known mechanisms of biological adaptation will provide a completely adequate explanation for the adaptive nature of macroevolutionary changes" (274).

Bock begins rather nicely, by defining what he means by adaptation (see also Bock and von Wahlert 1965). In particular, he distinguishes three senses of adaptation. The first is what he calls "universal adaptation":

> All living organisms exist in an environment; life cannot be isolated from its inorganic environment. The term *universal adaptation* best describes the essential interaction between living organisms and their environment. Universal adaptation is an absolute property of life that can never be lost; otherwise the organism would not be alive. Universal adaptation, being an absolute property of life, is a cause of evolution. It is not a result of evolution, and, moreover, it does not change during evolution. Because organisms must maintain universal adaptation, they must change as their external environment changes—*ergo*, they evolve. If they do not change, they become extinct. There is no other alternative. (274)

Clearly, what Bock is calling "universal adaptation" is merely satisfaction of the conditions for existence (compare this passage to the chapter epigraph by E. S. Russell). Unfortunately, this is the last we hear of it. Because it is universal, it is apparently of no interest.

Bock's other two categories of adaptation are evolutionary and physiological; it is only the former that interests him. *Evolutionary adaptation* is defined as "the relationship of a species to a specific set of environmental conditions," with the caveat that "this term designates both a process and a state of being." Bock connects "adaptation" as a state of being with features he calls "adaptations." These are defined in terms of their interaction with the environment to create a form-function complex that may have one or more "synergs," which are individual form-function couplings (Bock and von Wahlert 1965).

Bock's perspective provides one of the most sophisticated attempts to solve the problem of adaptation to arise out of the synthetic theory, and there is much to like about it. Nevertheless, Bock does not escape a teleological view of the action of selection. This enters—rather ironically—when he is trying to limit its power:

> Not all aspects of the origin of evolutionary novelties can be explained in terms of adaptive mechanisms. The development of all features seems to be adaptive, i.e., under the control of selection forces; however, the particular feature that appears to *fill a certain selective demand* is based on chance. This paradox of accident or design in evolution has been elegantly analyzed by Mayr (1962). One of the consequences of this chance aspect in the origin of new features is the principle of multiple evolutionary pathways, which accounts for the different but equally adaptive answers that appeared in response to the same selection force. (1965, 276, my emphasis)

Here we see new features evolving to "fill a demand," but the existence of the demand is of course inferred from the evolution of the feature (for more instances of such a teleological view, see the 1962 article by Mayr that Bock refers to).

How does all of this work out for Bock when he comes to analyze the origin of birds and their flight? Bock contrasts his view that birds evolved from reptiles along a series of functionally and morphologically intermediate stages with the view that they instead evolved in one step. The difference for him is that

> in the single-step evolution only one set of *selection forces* is involved, while in the series of several steps a sequence of different sets of *selection forces* is involved. In the evolution of flying birds from terrestrial reptiles, the steps would include walking on two legs, climbing, leaping, parachuting, gliding, and finally actively flying. The different features of the ancestral group would have evolved at different times and at different rates *in response to the selection forces* associated with each step; these changes would occur in a very definite sequence, depending on the order of the stages. (277, my emphasis)

This picture of correlated morphological, functional, and ecological change, in which different body parts change at different rates, does not seem highly controversial, and the alternative proposal he considers—the single-step evolution of birds from terrestrial reptiles—is unlikely, to say the least. What is the real target of Bock's argument? This becomes clear a little later in the paper, when he discusses Simpson's (1953) views on intermediate "adaptive zones" (related to Simpson's modified version of Wright's adaptive landscapes). Bock is fundamentally arguing against the teleological perspective involved in viewing an evolved lifestyle/morphology relationship ("adaptive zone") as more stable or significant than the "transitional zones" that organisms passed through in their evolution—that is, against the idea that organisms will pass through a "poorly adapted" phase as they go through such

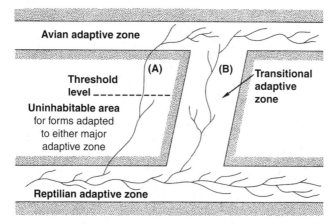

FIGURE 11.2. Bock's (1965, fig. 2) model for the evolution of flight. The legend reads: "Schematic models contrasting two different methods of crossing from one major adaptive zone to another. In A, the organism must cross an ecologically unstable region, which may be considered a threshold level. In the second model, B, the organism crosses from one major adaptive zone to another through a transitional adaptive zone. In this model the regions between the major adaptive zones are regarded as uninhabitable for organisms adapted to the major adaptive zone on either side." Reprinted by permission of the Society of Systematic Biologists.

an intermediate "adaptive zone" (figure 11.2). He does so on the reasonable basis that there is no evidence supporting such a view.

He ends, then, rather satisfied with the model he has proposed for avian origins. It is, he says,

> in complete agreement with the known fossil record and with the mechanisms of adaptive evolutionary change advocated in the synthetic theory. If these mechanisms can explain adequately the evolutionary change within one form of organization, they can explain change from one form to another. No new mechanisms need be proposed. All changes occur by natural selection acting on the genetic variation generated within populations of interbreeding individuals; a major evolutionary change is nothing more than a cumulative series of small steps. The change from one form of organization to another is adaptive in that all intermediate groups are fully adapted to their environments, and all changes are under the control of known mechanisms of adaptive evolutionary change. (286)

The Modern Phase

The next major phase in the debate over the origins of birds and their flight came in the 1970s, with the publication of Ostrom's (1973, 1974, 1976) work supporting a

FIGURE 11.3. Ostrom's (1979, fig. 10) reconstruction of "a pre-*Archaeopteryx* stage *(top)* and *Archaeopteryx (bottom)*" in accordance with his "insect net" hypothesis for the initial function of wing feathers. Reprinted by permission of *American Scientist*.

theropod dinosaur ancestry for birds. Ostrom pointed out that in spite of much positive evidence for a bird-dinosaur relationship, Heilmann's views had reigned supreme for fifty years. The work of such stalwarts of the modern synthesis as G. G. Simpson (1946) and Gavin de Beer (1954) supported Heilmann's views with the assumption that all the saurischian features of birds, especially *Archaeopteryx*, were due to "parallelisms and convergences" (Simpson 1946, 94).

Interestingly, along with his marshaling of the anatomical evidence linking birds, particularly *Archaeopteryx*, and theropod dinosaurs, Ostrom published a new theory of the early functional significance of wings and wing feathers—the insect net hypothesis (Ostrom 1974; see figure 11.3):

> From the size of the head and mouth, and the shape of the teeth in *Archaeopteryx,* we may reason that the usual prey of *Archaeopteryx* consisted of relatively small animals, most probably large insects, and perhaps small lizards and mammals. Enlargement of the primitive contour feathers on the forelimb could have increased the efficiency of the forelimb as a prey-catching appendage by gradually converting it into a large snare, both arms becoming a trapping device or "net" with which to corral or surround small prey so they could be more easily grasped in the mouth or hand. (34)

Ostrom's scenario continues:

> If vigorous flapping of the feathered forelimbs played a part at any stage in the business of catching prey, the increased surface area of the enlarged contour feathers would undoubtedly have produced some lift during such assaults. From this point, it is a small evolutionary step for *selection to improve* those features that were important for

flapping, leaping attacks on prey—perhaps to "fly up" after escaping insects: e.g., enlargement of the primaries and secondaries and their firm, rigid attachment to the forelimb skeleton; elongation and specialization of the bones of the forelimb and hand; retention of the theropod-like scapula and coracoid; enlargement of the pectoral adductor muscles; and stabilization of the shoulder joints by fusion of the clavicles. We can speculate further that predation in this manner would succeed only in proportion to the precision of the leaping attack. Thus, *selection would tend to improve* not only the "flight power," but also the "flight controls"—the associated sensory and motor neural components. (35, my emphasis)

Note that the role of selection here is "improvement" of features with respect to a particular functional role. Again, the evidence for such selection is that evolution of that functional role occurred.

In short, Ostrom's 1974 review provides an excellent example of the sort of reasoning then especially popular in evolutionary morphology, derived from the synthetic tradition as represented by Simpson and Bock. Evidence in favor of a particular functional series is marshaled, and then selection is brought in to explain the transition from one functional state to the next by some "advantage" of the transition. It is this tradition that Croizat (1962) reacted so strongly to. Consider Ostrom's (1974) summary of the three scenarios he discussed:

The arboreal theory assumes that the incipient wing or patagium, fixed or not, had some *selective value* as an aerodynamic or lifting surface right from the very beginning, and that *selection favored* those individuals with larger and larger airfoils. The cursorial theory postulates the incipient wings as "propellers," pushing backward against the air and assisting the hindlegs in building up ground speed, until *selection of* larger and larger "propellers" resulted in minimum flight velocity. (45, my emphasis)

The *Ornitholestes*-like osteology of the forelimb and shoulder of *Archaeopteryx* indicates that enlargement of the contour feathers of the forelimbs might well have been related to a predatory function—with *selection favoring* those individuals that possessed improved prey-catching ability. It is postulated that the incipient "wings" of avian ancestors were perfected first as snares to trap insects or other small animals against the ground, or to knock them down from escape, until they could be grasped in the hands or teeth. The "wings" subsequently were modified to stronger, flapping appendages, perhaps *to subdue larger prey*, and then coincidentally aided in leaping attacks on that prey. (46, my emphasis)

What is clear here is that the real issue under debate is in the functional series postulated. Bringing in "selection" to aid in moving evolution along any one of the trajectories postulated is gratuitous.

Current Thinking

A lot has happened in the more than thirty years since Ostrom revived the dinosaurian theory of bird origins. In particular, as already noted, a wealth of new fossil data has supported his proposal of a theropod origin of birds (Paul 2002; Chiappe 2007;

but see also Feduccia 1999). What do current views of the reconstructed morphological/functional path from theropods to *Archaeopteryx* look like? How is selection incorporated into these scenarios?

In writing this book, it has been difficult not to worry that I am quibbling about minor details—that in spite of rather strange locution at times, there are no fundamental errors in thinking in evolutionary biology. It has been rewarding, then, to occasionally find rather blatant examples of teleological thinking in all its glory. Greg Paul's engaging and stimulating semipopular book on *Dinosaurs of the Air* (2002), which contains extensive discussion of the problem of the origin of flight, provides a fine example of such thinking. Paul is not a professional paleontologist, but rather a dinosaur artist and amateur "dinosaurologist" who has published extensively in both peer-reviewed and popular venues. In spite of its wealth of factual information, Paul's book can read almost like a caricature of selectionist thinking. It is extremely useful to us for just that reason.

To analyze this bounty of data on the use of natural selection in macroevolutionary explanation, I searched an online version of the book for examples of the use of the terms *selective* and *selection*. I attempted to translate these into nonteleological language, then sorted the samples by category. Some judgment was involved in how best to translate the statements, but nevertheless I felt that for most I could achieve a translation with no loss of content. Examples of each category are given in table 11.2.

By far the most frequent use of *selective pressure, selective force*, and *selective advantage* in Paul's book was as a synonym for *function* or *functional significance*, in the context of a change in structure and function that did occur or might occur in evolution. An advantage of the change in structure and function is inferred, and selection is assumed responsible. A second category of use involved *functional constraints on pathways of evolutionary change*; the presence of such a constraint was interpreted as selection for the end state constrained to occur, and the removal of such a constraint as the unleashing of the selective forces at hand. A third category of use is much less common; it is when *evolutionary results are used to explain themselves*. For example, in the statement that there is "intense selective pressure" to improve a new capability, such as flying, the "intense selective pressure" is first being inferred from the rapid change in wing but slower change in leg morphology, but then used to explain this very change. A fourth category involved *evolutionary predictions* made from ideas about the sorts of functional changes that might be likely (or unlikely) to occur. Finally, a fifth category involved entirely gratuitous use of the term *selection*, when *evolution* was clearly meant.

Paul's perspective is striking for its reliance on selection pressures, forces, and advantages for everything that occurs in the evolution of flight. However, in looking more broadly at the current literature, what is soon apparent is that most authors tend to be much more restrained in their invocation of selection than Ostrom was thirty years ago or Paul is today. Two factors are probably responsible. First, the

FUNCTION, FUNCTIONAL SIGNIFICANCE, OR FUNCTIONAL CONTEXT

Paul's Passage	Explication
"However, the *selective forces* that would encourage runners to evolve gliding flight are obscure: speed would fall off rapidly after take-off and any energy savings over walking or running the same distance would be at best modest." (p. 115)	However, the *functional significance* of gliding flight in runners is obscure: speed would fall off rapidly after take-off, and any energy savings over walking or running the same distance would be at best modest.
"Countless earthbound predators have gnashed their teeth in frustration as their avian targets simply flew beyond their reach. This *selective advantage* is so powerful that some birds—grouse and turkeys, for example—have developed their ability to take off at a steep angle at the expense of long-range flight performance." (p. 149)	Countless earthbound predators have gnashed their teeth in frustration as their avian targets simply flew beyond their reach. This *function of flying* can be critical for survival, and some birds—grouse and turkeys, for example—have developed their ability to take off at a steep angle at the expense of long-range flight performance.
"The nonflight *selective forces* that may have *promoted* the evolution of the theropod furcula and straplike scapula blade are explored in this chapter." (p. 228)	The changes in nonflight *functions* that may have been *correlated with* the evolution of the theropod furcula and straplike scapula blade are explored in this chapter.
"This explanation is even truer if the elevation and control system is better developed than that of *Archaeopteryx*, in which case predation as the *selective agent* for such an advanced complex becomes implausible." (p. 229)	This explanation is even truer if the elevation and control system is better developed than that of *Archaeopteryx*, in which case defense from predation as the main *function* of such an advanced complex becomes implausible.
"If *selective forces* have not been strong enough to lead to the retention of folding arms bearing large arrays of feathers for brooding in flightless birds, then the possibility that such complexes initially *evolved* primarily *for* brooding must be ranked as low." (p. 231)	If folding arms bearing large arrays of feathers used for brooding have not been retained in flightless birds, then the possibility that such complexes initially *functioned* primarily in brooding must be ranked as low.

FUNCTIONAL CONSTRAINTS

Paul's Passage	Explication
"Why . . . would arboreal gliders that can climb and then travel from tree to tree aerially be under *selective pressure* to evolve powered flight?" (p. 112)	Is there any way in which a *transition* from an arboreal glider that can climb and then travel from tree to tree aerially to one that has powered flight could occur?
"Because it is difficult for small forest fliers to soar, the *selective pressure* to increase horizontal range by switching to flapping should be considerable." (p. 121)	Because small forest fliers can't soar, an increase in horizontal range *can only occur* by switching to flapping.
"This scenario starts with the presumption that *selective pressures* to increase horizontal range while keeping the wings short can be met only by developing powered flight." (pp. 121–122)	This scenario starts with the presumption that an increase in horizontal range while keeping the wings short *requires* powered flight.

(*continued*)

TABLE 11.2 (*continued*)

FUNCTIONAL CONSTRAINTS

Paul's Passage	Explication
"Once flight is abandoned as a major mode of movement, the *selective pressures* to reduce and eliminate the costs of growing and maintaining the flight apparatus are very strong." (p. 152)	Once flight is abandoned as a major mode of movement, the *functional constraints* preventing the reduction in size of the flight apparatus are lost.
"In reality, there was no *inescapable evolutionary pressure* for the first avian fliers to retain flight. As Gould explains, they were free to lose flight if *selective pressures directed them to* do so, and we should be surprised if they did *not* do so." (p. 249)	In reality, there was no *constraint* on the first avian fliers requiring them to retain flight. As Gould explains, they were free to lose flight, and we should be surprised if they did *not* do so.

EVOLUTIONARY RESULTS (EXPLAINED BY THEMSELVES)

Paul's Passage	Explication
"There often is intense *selective pressure* to improve a new capability, and this pressure continues until its ultimate limits are reached." (p. 131)	There often is *rapid improvement of* a new capability, and this improvement continues until its ultimate limits are reached.
"Likewise, saber-toothed marsupial thylaco-smilids, nimravids, and felids are similar in form because they evolved under similar *selective pressures.*" (p. 218)	Likewise, saber-toothed marsupial thylaco-smilids, nimravids, and felids are similar in form because they evolved similar *lifestyles.*

EVOLUTIONARY PREDICTIONS

Paul's Passage	Explication
"Priede (1985) notes that if energy efficiency really has been the *primary goal* of *natural selection,* then 'there should be a progressive reduction in energy expenditure by animals together with a general increase in food intake.'" (p. 397)	Priede (1985) notes that if energy efficiency really is *maximized* in evolution, then "there should be a progressive reduction in energy expenditure by animals together with a general increase in food intake."
"Considering that this classic dino-theropod hunting mode worked well during the entire span of the dinosaur reign, *selective pressures* to veer away from this pattern should not have been strong as long as size remained at least modest." (p. 233)	Considering that this classic dino-theropod hunting mode worked well during the entire span of the dinosaur reign, *they would not have veered away from this pattern* as long as size remained at least modest.

GRATUITOUS USE OF SELECTION

Paul's Passage	Explication
"Whatever the reason, *selective evolution* has resulted in an arrangement in which dinosaurs best adapted for operating in well-lit conditions mainly fly during the day, and mammals well-suited for flying in the dark take over after sunset." (p. 303)	Whatever the reason, *evolution* has resulted in an arrangement in which dinosaurs best adapted for operating in well-lit conditions mainly fly during the day, and mammals well suited for flying in the dark take over after sunset.

critique of Gould and Lewontin (1979) has had its effect, and workers are properly chastened about constructing untestable adaptive scenarios for the origin of features. Second, the rise of cladistic methodology for studying phylogeny and character evolution has promoted a focus on testable hypotheses of character evolution rather than adaptive scenarios driven by selective advantage, with cladistics from the beginning being frequently tied to a rejection of "evolutionary systematics." As stressed by Padian (2001a), "Cladistics took the focus off direct ancestry and put it on shared ancestry, to be assessed strictly on the possession of synapomorphies. It also removed hypotheses of process and function from assessments of relationship, which were now based solely on the analysis of pattern" (599). This general attitude has certainly had a salutary effect on the discussion.

In most current analyses of the functional transitions in the origin of flight (Padian and Chiappe 1998; Garner, Taylor, and Thomas 1999; Padian 2001a, 2001b; Dial 2003; Zhou 2004; Gatesy and Baier 2005; Clarke and Middleton 2006; Longrich 2006; Chiappe 2007), selection is barely mentioned or not mentioned at all. For example, in his proposal of wing-assisted incline running (WAIR) as a possible intermediate functional stage on the path to avian flight, Dial (2003) completely avoids it, focusing on the construction of an evolutionary series consistent both with the fossil evidence and functional considerations. Likewise, in a recent review of the origin of bird flight (Zhou 2004) selection never explicitly enters the picture. A quote should make the general tenor of Zhou's approach clear: "The basal phalanges of the hand are relatively shorter than the distal phalanges in the hands of *Archaeopteryx*, *Jeholornis*, and *Confuciusornis*, which probably indicate that the hands of these early birds were used in grasping tree trunks or branches. Thus, there existed an evolutionary transition from a less arboreal foot with a climbing hand to a perching foot with a reduced hand. Therefore, a typical perching foot was perfected gradually in the early evolution of birds" (464). Even in the popular book by Chiappe (2007), the discussion is quite refreshingly sober. His general attitude is conveyed by the statements that "understanding the functional advantage of an evolutionary novelty is far more complex than understanding its evolutionary origin" (79), and "the origin of flight is a problem for which we have no direct data from the fossil record" (85).

In short, most current authors talk of functions and adaptation, but not selection. The demands they make on evolutionary theory are thus entirely consistent with my interpretation here in terms of the principle of the conditions for existence as equivalent to the functional constraints on evolution. In other words, in spite of their lack of explicit mention of selection, underlying all of the functional scenarios considered is the assumption of a close connection between form and function, which implicitly depends both on the principle of the conditions for existence of organisms, and on the principle of the conditions for existence of characters in populations (the Darwin-Wallace principle). Meanwhile, the mechanism of natural selection, as a cause of evolutionary change, is nowhere in evidence.

By largely eliminating selection from their explanations (except implicitly, as a factor maintaining adaptedness and thus accounting for the form/function relationship), current workers studying the origin of flight have generally come to lose the teleological overtones so apparent in earlier work. However, this change is not due to a principled assessment of the relationship between selection, adaptedness, form, and function, but rather to the confluence of the effects of the "spandrels" view and the cladistic critique of adaptationism. Unlike Greg Paul, most paleontologists today have some hesitance in proposing selective advantages to explain the evolutionary path chosen, rather than simply tracing the functional path itself. If they are selectionists, they are selectionists with a bad conscience and prone to hide their selectionism between the lines of their papers. What we are left with is a rigorous study of macroevolution that has no explicit theoretical connection to the microevolutionary process we can observe in populations living today—hardly a desirable state of affairs.

. . .

In this chapter, I have tried to show that in historical contexts, it is the conception of natural selection as a force or agent that leads to teleological determinism; when understood as the conditions for existence, natural selection is better conceived of as the range of *boundary conditions* within which evolution occurs. In tracing the origin of flight in birds or any other macroevolutionary transformation, we implicitly use the functional integrity of the organism and its lifestyle as a constraint on the transformations we are willing to allow. Beyond this point we simply can't go. However, this does not require us to sever all connection between micro- and macroevolution but merely to recognize that in macroevolutionary explanation, the proper role of the microevolutionary process of narrow sense natural selection is as a conservator of adaptedness, not as a force for change.

In my introduction to the problem of bird flight earlier in the chapter, I suggested three ways one might hope to study and understand the path of macroevolution. The first was to use phylogenies to reconstruct ancestral character states, while the second was to give a functional interpretation of such ancestral character states. We have already seen extensive use of these approaches. The third approach involves the examination of such paths as involving specific alterations to developmental/genetic systems, with the hope that one can interpret the particular path followed as one that was generated by the limited range of possible (not just viable) alterations in such physicochemical systems. Although flight itself is difficult to approach this way, this path to understanding the origin of feathers is now being pursued with vigor, and much recent work has been done on the evo-devo of the scale-to-feather transformation (Prum 1999, 2005; Prum and Brush 2002; Harris et al. 2005; see also Xu 2006). Though we will never know *why* the feather evolved, such studies provide significant insight into exactly *how* the feather evolved and why this particular path was available; they reveal the mechanism of evolution at its most basic. This is nothing to sneeze at.

Beyond simply understanding *how* the transformation occurred, we certainly can hope to know as well *what* all those feathered dinosaurs were doing with their feathers. We can be sure that whatever it was, it enabled the dinosaurs to satisfy their conditions for existence (Cuvier's principle), and the feathers to satisfy their conditions for existence as well (the Darwin-Wallace principle). In fact, we can be almost certain that feathers were an integral part of the lifestyle of all the animals along the entire series connecting present-day birds with their Jurassic ancestors. Thus, both birds and their feathers have come down to us today across more than 150 million years, as parts of systems in which their effects on the world actively support their own continued existence.

12

The Conditions for Existence as a Unifying Concept in Evolutionary Biology

Natural History nevertheless has a rational principle that is exclusive to it and which it employs to great advantage on many occasions; it is the conditions for existence.

—GEORGES CUVIER, 1817, *LE RÈGNE ANIMAL*

Most of this book has been rather backward looking. I began with the presumption that there is something slightly twisted about the way we evolutionary biologists typically view the relation of natural selection to adaptation. To see what this might be, I have sought to follow Fisher's injunction that we should "attempt to understand the thoughts of the great masters of the past, to see in what circumstances or intellectual *milieu* their ideas were formed, where they took the wrong turning or stopped short on the right track" (1959 [1974], 449). In this chapter, I would like to look not only back but forward. In particular, I would like to examine some of the areas in which I believe recognition of the principle of the conditions for existence can highlight interconnections that are not often made, both within evolutionary biology and between evolutionary biology and other fields. The areas I look at briefly are quantitative genetics, the levels of selection, evo-devo, the ecological niche, physiology, and conservation biology.

QUANTITATIVE GENETICS AND THE CONDITIONS FOR EXISTENCE

An uneasy relationship currently exists between Mendelian/molecular population genetics and quantitative genetics. Population genetics considers Mendelian genes, represented as sequences of DNA that have certain known biochemical properties and are replicated and repaired in specific ways by specific enzymes. Quantitative genetics, in contrast, considers phenotypic features of individual organisms. The key question, of course, is how the inheritance of phenotypic features relates to the inheritance of DNA—or, to put this another way, how the continued existence of characters in populations relates to the continued existence of genetic (and higher-level) lineages.

Fisher (1918, 1922a, 1930a, 1941) sought to solve the problem by constructing his additive model of quantitative inheritance, in which individual Mendelian genes have certain "average effects" on the quantitative measure considered, and he showed that the observed phenotypic correlation among relatives could be accounted for on this basis. However, as Fisher was well aware, quantitative genetics is essentially a *local* theory (cf. Feldman and Lewontin 1975). The average effect of a gene is not just a property of the gene itself but of the gene and its environment, so all quantitative genetic measures are relative to the constancy of the environment (including the genetic environment).

The other major problem with the application of quantitative genetics to evolution comes from attempting to determine causes from observed correlations. In particular, when one attempts to assign observed differences in "fitness" to observed differences in traits, it is always possible one is dealing not with cause but merely correlation. In the popular multivariate approach of Lande and Arnold (1983), one can statistically eliminate all but one observed feature as the cause of observed differences in reproductive rate (realized fitness) among individuals, but there is no way to account for the possibility that hidden correlated variables are the true cause of the observed differences (Lande and Arnold 1983, 1214). This fundamental problem with the Lande-Arnold approach has recently been critiqued in great detail by Pigliucci and Kaplan (2006), who see little hope for a predictive role of quantitative genetics in evolutionary biology.

Population genetics, however, also has its problems, which come from assigning certain fixed "fitnesses" to particular alleles (genes) in populations, as we have seen in chapters 7 and 8. Fitness, or expected rate of increase, is obviously not a fixed property of an allele, but a property of an allele in a particular genetic, organismic, and extrinsic environment, and it can't be assumed constant over evolutionary time. Another way of putting this is to note that the fitness of an allele, let alone an organism, is itself a quantitative trait and thus calls out for a quantitative genetic, rather than Mendelian approach.

In the beginning, Fisher (1930a) realized all of this quite clearly. In considering Fisher's views, it pays to remember that he was the originator not merely of quantitative and population genetics but of much of modern statistics as well. As a statistician, he is well known as a "frequentist," yet unlike most frequentists, his conception was that probability expresses not the limit of an infinite series of trials, but rather the frequency of an outcome (from a reference set of possibilities) across an infinite population of trials (Fisher 1958b; Savage 1976, 461). Or, as he explained it in 1925:

> The idea of an infinite hypothetical population is, I believe, implicit in all statements involving mathematical probability. If, in a Mendelian experiment, we say that the probability is one half that a mouse born of a certain mating shall be white, we must conceive of our mouse as one of an infinite population of mice which might have been produced by that mating. . . . Being infinite the population is clearly hypothetical, for not only must the actual number produced by any parents be finite, but we might wish to consider the possibility that the probability should depend on the age of the parents, or their nutritional conditions. We can, however, imagine an unlimited number of mice produced upon the conditions of our experiment, that is, by similar parents, of the same age, in the same environment. The proportion of white mice in this imaginary population appears to be the actual meaning to be assigned to our statement of probability. Briefly, the hypothetical population is the conceptual resultant of the conditions which we are studying. The probability, like other statistical parameters, is a numerical characteristic of that population. (700)

As we have seen, Fisher attempted to solve the problem of the relation of quantitative traits to Mendelian genes by considering each gene to have a certain "average effect" on a quantitative measurement, defined as the partial regression of the measurement on the presence or absence of that gene, *with all other genes held constant*. In accordance with his general statistical views, the "population" to which this concept of average effect applies is not the actual population, in the sense of the organisms actually (presently) out there, but instead an infinite population of populations in the same "nature and circumstances," of which the actual population is only a sample (1930a, 23). Thus, when Fisher defined average excess in association with his discussion of "the genetic element in variance," he was quite clear that

> this definition will appear the more appropriate if, as is necessary for precision, the population used to determine its value comprises, not merely the whole of a species in any one generation attaining maturity, but is conceived to contain all the genetic combinations possible, with frequencies appropriate to their actual probability of occurrence and survival, whatever these may be, and if the average is based upon the [measures] attained by all these genotypes in all possible environmental circumstances, with frequencies appropriate to the actual probabilities of encountering those circumstances. (1930a, 30–31)

I have already discussed the issues involved in defining just what is meant by considering populations to be in "the same nature and circumstances." If the population we are interested in is "the conceptual resultant of the conditions we are studying," then the specification of these conditions is critical. In fact, this problem relates directly to what Fisher (1922b) himself called the problem of "specification": the problem of determining the form of distribution in the hypothetical infinite population from which the observed sample is drawn (see Spanos 2006). What, then, is the form that the distribution of average excess and average effect takes in this infinite hypothetical population of which the observed value is a sample? Is it a normal distribution? If we assume so, can this be justified on independent grounds? And even assuming a normal distribution, how do we determine the variance of the distribution? Unfortunately, Fisher is entirely unclear on such issues.

Modern approaches to evolutionary quantitative genetics tend to be even less clear on the resolution of such issues than was Fisher, often not even acknowledging the question of the hypothetical population of which the observed data represent a random sample (e.g., Lynch and Walsh 1998), although rightly paying attention to the assumptions of the model on which such parameters as "heritabilities" are considered to be genetically determined.

Problems with the additive genetic model of quantitative inheritance have been acknowledged, but largely ignored in practice, in the extensive literature on evolutionary quantitative genetics developed on the basis of the Lande and Arnold (1983) approach. A valuable critique by Heywood (2005) compares the traditional approach based on the "breeder's equation" and narrow sense heritability to the "Price equation" (Price 1970, 1972a), which has recently seen increased popularity in both the philosophical (e.g., Okasha 2006) and biological (e.g., Frank 1998) literature. His explicit decomposition of evolutionary change based on the Price equation shows that the breeder's equation leaves out many other possible sources of change, other than differences in fitness or narrow sense heritability. He classifies these as "constitutive transmission bias," "induced transmission bias," and "spurious response to selection." Such difficulties are at the root of many observations that contradict the assumptions of the simple linear model inherent in the breeder's equation, such as the results of Kruuk et al. (2002) showing that antler size in red deer is positively correlated with realized fitness and is heritable, but not evolving. They explain this result by environmentally induced covariance of antler size and realized fitness—well-fed deer have both big antlers and lots of offspring.

In the spirit of positive contribution to the debate, I'd like to make three points. First, my search for data on the standardized variance in rate of increase (opportunity for selection) to use in chapter 8 made it clear that studies of natural selection conducted in a quantitative genetic framework have been so focused on selection differentials, selection gradients, and heritability that authors often neglect to report their data on the mean and variance of rate of increase (realized fitness) within the population (see similar comments in Houle 1992; Hersch and Phillips 2004). It is unfortunate that much of the demanding fieldwork that has gone into quantitative genetic studies of evolution in natural populations has not contributed to our knowledge of this basic natural history information, which is directly relevant to the rate of evolution of these populations.

Second, as just noted, a key issue for quantitative genetic studies is the assumption that the environment is constant from one time period to the next. The very notion of heritability relies, in fact, on the constancy of additive effects, which in essence means that the genic and larger environment is constant. This not only makes no sense in a population evolving genetically, in which the genic environment is changing, but also fails to take into account the ecological variation from one time period to the next, associated with temporal variation in the weather and the populations of other organisms. It is only by minimizing such genic and environmental effects that heritability has been a useful concept in plant and animal breeding (Feldman and Lewontin 1975; Jacquard 1983; Kempthorne 1997; Heywood 2005). Here I might make the simple suggestion that quantitative genetic parameters having to do with genetic causation over a particular interval be distinguished from those having to do with transmission of traits from one interval to another (see my earlier comments on the study of heritability of fitness in red deer by Kruuk et al. 2000).

Finally, I might make this more general comment. Fisher's project in 1918, 1922, and 1930 was to recast Darwinism as a scientific theory (Gayon 1998). In this theory, evolution by natural selection became a statistical law, as embodied in the fundamental theorem of natural selection. It is no accident that Fisher compared the fundamental theorem to the second law of thermodynamics, as a law of statistical regularity in populations. The rate of increase in mean fitness is proportional to the variance in fitness at a given time—this *is* the law (statistical regularity) of evolution by natural selection. At the same time, a consequence of this definition is that the actual change in fitness (rate of increase) can never be predicted: "I do not question that the selective intensities acting instantaneously may well be equivalent to those derivable from such a [potential] function, but I think it should be emphasized that both changes in time, that is in the environmental *milieu* and in the gene ratios themselves, that is the heritable constitution of the organism, will change this virtual function in a way that cannot be specified in terms of the quantities used in formulating the fundamental theorem" (Fisher, 3 May 1956, letter to Motoo Kimura, quoted in Bennett 1983, 229).

As stressed earlier, this perspective comes down to saying that the fundamental theorem is exact and rigorous, but completely useless for understanding evolutionary change on its own, because it neglects the other side of the equation, the deterioration of the environment. Even for Fisher, to understand evolution, we must understand not just what natural selection does to increase fitness (expected rate of increase), but how the environment changes so as to reduce it. This involves both genetics and ecology. In his vision, natural selection maintains adaptedness in the face of constantly deteriorating environment. This at least makes sense.

But my discussion so far has focused mainly on rate of increase (realized fitness) itself. When we come to consider organismal *traits*, the situation becomes even more complex. Fisher's quantitative genetics was developed for simple metric traits such as stature; and when he came to consider natural selection, he dealt with fitness as a trait, but not with the effect of other traits on fitness. In fact, as we have seen, he was quite skeptical of the possibility that the appropriate data to examine natural selection in action could ever be obtained for a natural population. By contrast, traits play two rather distinct roles in modern evolutionary quantitative genetics theory (e.g., Lande and Arnold 1983). On the one hand, they are merely phenotypic features of organisms, and differences in rates of increase among organisms with distinct forms of a particular trait, together with heritability of trait differences along lineages, are used to explain evolutionary change in a population (for complications, see Heywood 2005; Okasha 2006; Godfrey-Smith 2007). On the other hand, it is commonly assumed that the trait differences among the organisms considered are *causing* the differences in rate of increase, though nothing in either classical quantitative genetics theory or the Price approach requires such a connection. This causal relation is implicit in most modern reconstructions of natural selection (see Godfrey-Smith 2007).

However, quantitative genetic studies of natural (rather than experimental) populations are poorly suited to examining causes, because they are by nature observational studies. Correlation is not causation, even if when one finds a correlation, one certainly always hopes to be able to determine the underlying causation. Realistically, we can at best hope to elucidate *some* of the major causes of any particular evolutionary change. For example, the Grants' elegant work on Darwin's finches discussed in chapter 10 is certainly quite suggestive of some of the major causes in play.

If we give up on Fisher's exact but inapplicable definitions to develop a more practical version of quantitative genetics applicable to particular metric traits, as workers such as Heywood (2005) have been trying to do, we must recognize that such a theory is only an approximation to what is really happening in the population. From the perspective of the conditions for existence, the most important question is whether the trait of interest is in fact changing over the time period studied (is or is not satisfying its conditions for existence). If it is, we must attempt to explain the change as due to some combination of mutation, differential survival and reproduction, migration, heredity, and environmental change. If it is not, we must attempt to explain the lack of change as due to some combination of mutation, differential survival and reproduction, migration, heredity, and environmental change. If the goal of evolutionary prediction of change in phenotype is not rigorously realizable, what we can hope for is a retrospective analysis that will explain the necessary conditions for whatever changes (or lack thereof) occurred over the time period considered.

Looking back over the last 150 years, we see that for Darwin and Wallace, the principle of natural selection (Darwin-Wallace principle) had to do with the conditions for the continued existence of traits in populations. The central condition they both hypothesized was that the organisms with a particular trait or of a particular type would tend to survive and reproduce at a greater rate than those with alternative traits or of other types. When these conditions were no longer fulfilled— namely, when the ancestral form was no longer the one with the greatest chance of surviving and reproducing—evolution would occur. But where Wallace largely ignored heredity, Darwin realized that the mechanics of heredity are central to understanding the continued existence of traits in populations, and he tried (if unsuccessfully) to solve the conundrum (Gayon 1998). After the rediscovery of Mendel, Fisher sought to rescue Darwin by pointing out that Mendelism was compatible with the data on correlations among relatives and that with Mendelian inheritance, variation was conserved within the population, allowing for long-continued evolutionary change, even with no new mutation. Nevertheless, Fisher's consideration of particular *traits* was largely along the lines laid down by Darwin himself. For traits, he assumed either a simple Mendelian basis or some general sort of heritability no different from Darwin's. Thus, for Fisher, natural selection was indeed a general statistical law of populations, but to apply it to particular traits in real populations involved an approximation of the real structure of heredity.

As it develops further, quantitative genetics theory seems well positioned to take on a central role in elucidating the necessary conditions for the continued existence of particular traits in populations. This is likely to show some surprises, as compared with the traditional approach. I have already noted Kruuk et al.'s (2002) work showing that antler size in red deer is stable, in spite of heritability and associated variation in fitness. Heywood (2005, following Price et al. 1988) likewise suggested that breeding date in birds can be stable in spite of variation in rate of increase associated with different breeding dates, and heritability of breeding date (see the review by Merilä, Sheldon, and Kruuk 2001). It appears that quantitative genetics can achieve a closer approximation of the true structure of the inheritance of quantitative traits, but it can do so only by rejecting the universal validity of the strict *hypothesis* of evolution by narrow sense natural selection—that hereditary differences among organisms in their ability to survive and reproduce is the only factor necessary to account for the continued existence of traits in populations.

LEVELS OF SELECTION AND THE CONDITIONS FOR EXISTENCE

The debate over the units and levels of selection has been a long and contentious one in the history of evolutionary biology. To ground my discussion of the issues, I would like to briefly review this history. The modern debate is commonly held to have begun in the 1960s, when George C. Williams published his classic, hugely influential work: *Adaptation and Natural Selection: A Critique of Some Current*

Evolutionary Thought (1966). Williams's book was written as a critique of group selectionist explanations for features of organisms, particularly the argument of Wynne-Edwards (1962) that animals refrain from reproducing as much as they are able to when population density gets high, thus limiting population growth "for the good of the species." It is perhaps worth pointing out that Williams's book came out just a few years after the revised edition of Fisher's *Genetical Theory* (Fisher 1958a), with its newly added attack on just such thinking.

The next major sally in the units/levels of selection debate came a few years later, when Richard Lewontin (1970) published his review of "The Units of Selection." Lewontin examined selection at a variety of possible levels, including molecules, organelles, cells, gametes, individuals, kin, and populations. He admits selection at all of these levels, though noting that lower-level selection is generally more effective than higher-level. In particular, he notes that "the rate of evolution is limited by the variation in fitness of the units being selected" (8)—that is, the standardized variance (opportunity for selection) at each level.

In 1976, a book burst on the scene that greatly stimulated the units/levels of selection debate: *The Selfish Gene* by Richard Dawkins, an animal behaviorist who had studied with Niko Tinbergen (Dawkins 2006; Grafen and Ridley 2006). Dawkins's goal was to expound the theory of genic selection as applied to animal (including human) behavior for a popular audience, countering such popular works as Ardrey's *The Social Contract* (1970), Lorenz's *On Aggression* (1966), and Eibl-Eibesfeldt's *Love and Hate* (1971). All of these authors, according to Dawkins, "got it totally and utterly wrong" because they "misunderstood how evolution works." In particular, "They made the erroneous assumption that the important thing in evolution is the good of the *species* (or the group) rather than the good of the individual (or the gene)" (Dawkins 2006, 2). Dawkins set out to "examine the biology of selfishness and altruism" (1) without this erroneous assumption. As predecessors in this project, Dawkins cites not only Williams but also J. Maynard Smith, W. D. Hamilton, and R. L. Trivers.

Although Dawkins's discussion of the "selfish gene" may have been largely in the context of the genetic basis for behavior, he did note parenthetically that the large quantity of noncoding DNA might be an indication of genes that were selfish in a more extreme sense—namely, in conferring no benefit at all at the phenotypic level. "The true 'purpose' of the DNA is to survive, no more and no less. The simplest way to explain the surplus DNA is to suppose that it is a parasite, or at best a harmless but useless passenger, hitching a ride in the survival machines created by the other DNA" (45). This thought was developed further in a pair of papers that appeared in the 17 April 1980 issue of *Nature*, one by Orgel and Crick entitled "Selfish DNA: The Ultimate Parasite" and one by Doolittle and Sapienza entitled "Selfish Genes, the Phenotype Paradigm, and Genome Evolution."

Doolittle and Sapienza took the most aggressive tack. Citing Gould and Lewontin's paper on "The Spandrels of San Marco," which had appeared the previous year, they accuse their fellow scientists of being addicted to a "phenotype paradigm" in which genome structure has to be understood with reference to some function at the phenotypic level, and "adaptive stories" are invented for what such a function might be. But "the only selection pressure which DNAs experience directly is the pressure to survive within cells." Because of this, "if there are ways in which mutation can increase the probability of survival within cells without effect on organismal phenotype, then sequences whose only 'function' is self-preservation will inevitably arise and be maintained by what we call 'non-phenotypic selection.' Furthermore, if it can be shown that a given gene (region of DNA) or class of genes (regions) has evolved a strategy which increases its probability of survival within cells, then no additional (phenotypic) explanation for its origin or continued existence is required" (601). Doolittle and Sapienza focused in particular on the behavior of transposable elements. Their contrast of the phenotypic selection and non-phenotypic selection views is instructive in highlighting the basic issues of the levels of selection debate.

The storm of debate that met these papers (see, e.g., the 26 June 1980 issue of *Nature*) has not yet died down. The recent appearance of an entire book devoted to *The Biology of Selfish Genetic Elements* (Burt and Trivers 2006) attests both to the diversity and prevalence of the phenomena to be explained, and the diversity of possible modes of differential survival and reproduction occurring at something lower than the organismal level (see also chap. 9).

At about the same time the debate within the scientific community was initiated, philosophers of science became interested in the issues involved in the levels/units of selection debate. A key event was the appearance in 1980 of a paper by David Hull, arguing that there were really two different issues involved: (1) the selection of *replicators* and (2) the selection of *interactors*. Hull's "replicators" and "interactors" were generalizations from the model of organismal selection, with genes being the replicators and organisms the interactors. The term *replicators* came from Dawkins (1976), while *interactors* corresponds approximately to Dawkins's "vehicles."

Hull's paper, coming on the heels of Dawkins's book, enshrined the replicator versus the interactor as the two fundamentally different faces of natural selection. It was part of a flurry of papers that appeared in the early 1980s, many of which were collected in a volume edited by Brandon and Burian in 1984. Arnold and Fristrup (1982) argued rather persuasively that selection can occur at many levels, and developed a generalization of Price's covariance approach (Price 1970, 1972a) to examine this selection mathematically. Gould and Vrba's promotion of "species selection" (e.g., Vrba and Gould 1986) supported their perspective, as did increasing evidence that "selfish DNA" was real. Books by D. S. Wilson (1980) and E. Sober (1984) contributed to the debate.

Since then, there has certainly been no end to interest in the subject. Just to mention a few of the more prominent contributions, G. C. Williams (1992) returned to the issues in his *Natural Selection: Domains, Levels, and Challenges*, Keller (1999) edited a volume entitled *Levels of Selection*, and Gould's magnum opus on *The Structure of Evolutionary Theory* (2002) reiterated his endorsement of a hierarchical model of evolution. More recently, an excellent and thorough book-length treatment by Okasha (2006) on *Evolution and the Levels of Selection* has appeared. Related issues, such as the role of multilevel selection in major evolutionary transitions, have also been extensively discussed (e.g., Maynard Smith and Szathmáry 1995; Michod 1999).

This brief historical review barely hints at how vast this literature has become. I could not hope to digest, let alone significantly add to, such a concentration of thought and effort. Nevertheless, some comments on the current state of the field may be of some use in relating the debate to the perspective I have developed here. From the viewpoint of the conditions for existence, the key question in evolution is what continues to exist through evolutionary time (and conversely, what doesn't). As noted in chapter 7, only lineages (e.g., genetic lineages, cell lineages, species lineages), not genes, not organisms, continue to exist in evolutionary time. It is lineages that evolve (Hull 1980). In the simple formalism developed there, we always need to deal with a defined number of lineages (a population) of which at least some continue to exist over a specific period of time. We can allow new members of the population to be added by splitting and/or recombining of existing lineages. Finally, we need to have an explicit understanding of the ancestor-descendant relations between the initial and final lineages belonging to the population. Given this, we can calculate absolute and relative rates of increase for the lineages belonging to the initial population, and thus determine the degree to which they are satisfying their conditions for continued existence over the period examined.

The first point I'd like to make is that our perception of differential survival and reproduction necessarily depends both on the lineages we are looking at and on the way we break them into classes. Even in the simple asexual case, for example, where genetic lineages, cell lineages, and organismic lineages are isomorphic, with genetic lineages nested within cell lineages nested within organismic lineages—and even with a constant number of genetic lineages per cell—as soon as we look at *classes* of entities, our description of what is happening can differ at the various levels. In fact, as I have noted in my discussion of the opportunity for selection (chap. 8), this is true even at a single level. For example, if one is looking at a population of genotypes at a particular locus over a given time period, one will get a certain variance in rate of increase between them. If one looks at the genotypes at a neighboring locus, one might get a different variance—merely because one is dividing up the individual rates of increase (which are the same in both cases) in different ways. This is, of course, what we mean when we talk of different types of (medium sense) selection at different loci in the same population.

The same thing can happen if one looks at different levels within a single population. In our asexual population, let's consider the cell lineages as well as the genotypes at a single locus. Unless we class the cell lineages by their genotypes at this particular locus, any differential survival and reproduction we observe among them will be different from that calculated for the genotypes. Nevertheless, the standardized variance ("opportunity for selection") is identical at the higher and the lower level—provided that two conditions are met: (1) each higher-level lineage must contain the same *number* of lower-level lineages, and (2) no differential survival/reproduction among lower-level lineages *within* higher-level lineages can occur.

Now consider relaxing one of our assumptions. Assume that there is in fact a different number of lower-level lineages within each higher-level lineage. To give a simple example, consider a situation in which a varying number of repeat sequences exist among a series of aphid clones. Obviously, we will get a different picture of what is going on if we focus on individual repeat lineages or on individual organismal lineages, merely because the populations do not have a consistent ratio of lineages. We are dealing merely with a matter of perspective, or how we define groups and individuals, not with any real difference in what is happening at the two levels.

Given these instances of what is *not* a process of multilevel selection, it is clear that there are several distinct ways for a true process of multilevel selection (differential survival and reproduction) to occur among *lineages*. To begin with, we need to consider whether there are any necessary connections between the levels. It turns out that there are. For individual lineages, in the situation of complete nesting of lineages, it is apparent that the continued existence of lower-level lineages is completely dependent on the continued existence of the higher-level lineage that contains them. If mitochondrial lineages can't escape from the cell lineage that contains them, when the cell dies, so do the mitochondria.

Higher-level lineages may have varying levels of dependence on the continued existence of lower-level lineages. For example, some genetic lineages may be critical for continued existence of the cell, others less so; nevertheless, a cell lineage could not continue to exist without any genes at all. Likewise, some plastid lineages in plants are not necessary for organismal survival, and if we think back to the initial formation of eukaryotes, it is clear that the ancestors of mitochondria, chloroplasts, and so forth, could not initially have been necessary for cell survival (or there would have been no cells to invade). Survival of at least some organisms is necessary for species survival, since a species is made up of organisms. The general point here is that even with different criteria for survival and reproduction at different levels (i.e., different things we count, like genes and organisms), complete independence between levels is never going to obtain, if there is any reason to be talking about nested levels at all. Nevertheless, partial independence in the differential survival and reproduction at different levels can certainly occur. It is such partial independence that is generally meant by multilevel selection.

Considered at the level of individual lineages, multilevel differential survival and reproduction occurs if there is a change in the *ratio* of a lower-level lineage to a higher-level lineage. For example, consider a transposon lineage that increases its number within a single cell lineage. The absolute rate of increase of the transposon lineage is the multiple of the rate of increase of the transposon lineage with respect to the cell lineage, and the rate of increase of the cell lineage itself. If we want to consider the relative rate of increase of the transposon lineage, we must of course think of the entire population of transposon lineages within the population of cell lineages.

But although differential survival and reproduction at different levels among individual lineages means that survival and reproduction of lower-level lineages is *partially* decoupled from survival and reproduction of higher-level lineages, it is important to remember that if the survival of lower-level lineages is dependent on that of higher-level lineages, or vice versa, then there is necessarily a relationship between them. If the conditions for continued existence of lower-level lineages include the continued existence of the higher-level lineage that contains them, then they have an "interest" in that continued existence. Such common interests have been key to understanding the origin and stability of the major levels of organization themselves (Maynard Smith and Szathmáry 1995; Michod 1999). In this general context, the functions and adaptedness of individual features (of genes, organisms, etc.) can be defined with reference to the lineages at more than one level (e.g., the replication of DNA has a function both for the genetic lineages themselves and for the cells and organisms that contain them).

Multilevel selection as differential survival and reproduction at two or more different levels is not the only possible meaning of multilevel selection, however. As just noted, if one considers not merely individual lineages but particular classes of these lineages distinguished by certain traits, then even with a fixed ratio between lower- and higher-level lineages (e.g., genes in cells) it is possible to see different phenomena at the two levels.

For example, in Williams's (1966) primary model of group versus individual selection, group selection occurred when variation in the mean value of a trait such as body size *across* groups and variation in the individual values of the same trait *within* groups had opposite associations with survival and reproduction. In the situation he considered (see especially his fig. 2), the rate of increase of organismic (and their contained genetic) lineages is higher for smaller organisms within groups, but the group mean rate of increase is higher for groups with larger mean size. In one sense this is not multilevel selection at all, in that there is really no need to pay attention to the relation with the groups: variation in absolute rate of increase among individuals, or in relative rate of increase of individuals in the total population of individuals, entirely accounts for the total variance (i.e., the among group variance is just part of the total variance). To consider this "group selection" depends entirely on how we define our groups.

That said, in another sense this can be multilevel selection. Clearly, if the groups are discrete, not exchanging individuals (e.g., species), and lineages characterized by a particular trait have different absolute rates of increase in different groups (group-dependent rates of increase), then it does make some sense to talk of more than one level of survival and reproduction.

But why would individual lineages characterized by the same trait have different rates of increase depending on the group they're found in? To answer this question, we must look not only at differential survival but at the *causes* of such differential survival. This is the realm of Hull's "interactors." If we want to talk about narrow sense natural selection occurring at different levels, we need to think about interactors, the causal interactions they are involved in, and how survival depends on these interactions. In this context, too, it can make sense to talk of functions (and adapted features) at different levels. For example, if groups survive at a higher rate the greater their mean size, but smaller organisms survive at a higher rate within groups, then one could say that large mean body size is better adapted with respect to group survival, and small individual body size with respect to individual survival. Still, group selection in this sense fundamentally means that the group an organism finds itself in is an important aspect of the environment determining its rate of increase.

Thus, when we speak of multilevel differential survival and reproduction, we can distinguish multilevel processes involving change in the ratio of lower- to higher-level lineages, from those that merely involve traits that have group-dependent effects on rate of increase. The issues in the levels of selection debate depend on *both* the conditions for the continued existence of individual lineages (Cuvier's principle) and those for the continued existence of traits in populations of lineages (the Darwin-Wallace principle).

For this latter meaning, a major problem, and one I have not really dealt with here, is the role of heredity, because heredity is central to understanding the effects of differential lineage survival at different levels on trait survival at different levels. For example, conservation of genetic information is not necessarily equivalent to conservation of nucleotide sequence, because this does not allow for the possibility of various conservative changes in nucleotide sequence that have little or no effect on protein function. Likewise, morphological characters can persist even while their developmental genetic basis changes. The Price approach (Price 1970, 1972a, 1995), with its dual emphasis on covariance between particular traits and rate of increase, and "transmission bias" in passing traits on to the next generation, is one attempt at a general solution to the problem (cf. Arnold and Fristrup 1982; Okasha 2006), but it is a solution only at the level of medium sense selection. In other words, the

Price approach deals with differential survival and reproduction as a cause of evolution, but not the cause of such differential survival and reproduction itself. In the previous section, we have seen how quantitative geneticists are striving for a more accurate model of heredity in natural populations and presumably something like this is needed in a multilevel analysis, too.

Fortunately, the problem of heredity at multiple levels is well beyond the ostensible subject of this book. What is more critical for us here is how a multilevel perspective affects our view of function. Because it is lineages that continue to exist in evolutionary time, functions are necessarily relativized to the continued existence of some lineage or other. We might distinguish *direct functions* (or adaptedness) of traits that can be ascribed to any level at which differential survival and reproduction of individual lineages occurs due to these traits, from *indirect functions* (or adaptedness) relative to any other level lineage at which the traits are a condition for its continued existence. Conflicting functions can occur both among different levels of lineages and among different temporal segments within a single lineage (e.g., larvae and adults, haploid and diploid stages). Even at the organismal level, the traditional locus of the study of function and adaptedness, some features (such as those related to reproduction) only have functions relative to the survival of the genetic and cellular lineages carried by the organism, not the organism itself.

Yet while functions are of the greatest interest here, in the end heredity cannot be ignored if we are interested in the continued existence of traits in populations. And although direct function can occur at various levels, heredity occurs primarily at the level of nucleic acids; it is the specificity of replication of nucleic acids that provides the specificity and stability necessary for the continued existence of life as we know it. A more explicit understanding of their hereditary basis is obviously critical for any conclusions about the conditions for the continued existence of traits in populations, at any level.

EVO-DEVO AND THE CONDITIONS FOR EXISTENCE

The interrelation of development and evolution has been a central theme in evolutionary theory at least from the time of Chambers (1844), who tied evolutionary advance to modifications in development. In the late 1800s, Haeckel made ontogeny central to the study of evolution with his dual proposals that "phylogeny is the mechanical cause of ontogeny" and "ontogeny is the brief and rapid recapitulation of phylogeny." But since the publication of Gould's work on *Ontogeny and Phylogeny* in 1977, the interrelationship of development and evolution has received renewed interest, greatly invigorated by the rise of evolutionary developmental genetics (Amundson 2005; Laubichler and Maienschein 2007).

In chapter 11, I discussed some conceptual aspects of evo-devo (evolutionary developmental biology) in association with the concept of developmental constraints.

The main point emerging from that discussion was that to think in terms of developmental constraints is often to think of evolution—or more commonly, of evolution by natural selection—as a teleological process, a process that wants to get somewhere but is constrained by the available variation. I suggested that true developmental, or generative, constraints are better viewed as the *process* (modification of developmental mechanisms) that produces phenotypic variation, rather than as constraints on some other process. The modification of developmental mechanisms is the phenotypic level analogue of mutation at the genetic level (and of course the two are interconnected). As new forms are generated by the modification of development, they are subject to the dual constraint of the conditions for existence of individuals (Cuvier's principle) and of the conditions for the continued existence of traits in populations (the Darwin-Wallace principle). This view has been central to the field since its beginning. Van Valen (1973) famously commented that "evolution is the control of development by ecology." Alberch (1980) emphasized that "development plays the crucial role in this interaction, since it defines the realm of the possible" (664; see also Alberch 1989).

Where this leaves us, then, is with the question of the relation of genotype to phenotype, which lies at the heart of evo-devo, as much as it does quantitative genetics. Given an existing developmental system, we can ask: what genetic perturbations of the system are possible? And what functional requirements (selective constraints) limit the viability of the resulting phenotypes? This contrast of "internalist" or "structuralist" properties of the developmental system constrained by "externalist" or "functionalist" demands for viability has been a common theme in evolutionary developmental biology (e.g., Alberch 1980, 1982; Wake 1991; Shubin and Wake 1996; cf. Amundson 2005). The main point I want to emphasize here is that appeals to such properties of the developmental system as "wholeness" and "integration" are really appeals to functional constraints, not variational properties of the developmental system itself, unless we can demonstrate such properties through experimental analysis and/or modeling of developmental processes.

How does this perspective compare to that current in the field? In a recent issue of *Evolution & Development*, Hendrikse, Parsons, and Hallgrimsson (2007) argue that evo-devo—for all its progress toward respectability as a scientific discipline—has not really succeeded in articulating its central question. This they claim to be true in spite of the fact that such conceptual issues as modularity, integration, canalization, biased variation, heterochrony, and allometry have been extensively discussed (reviewed by Müller 2007). Instead, they argue that much work that purports to be evo-devo is an examination of either how developmental mechanisms have changed in evolutionary time or how changes in development have produced new morphologies. Yet, "By itself, elucidating the developmental mechanisms by which an evolutionary sequence occurred adds nothing to evolutionary explanation other than inserting a developmental 'how' clause" (394).

What do Hendrikse et al. propose should be the central organizing question of evo-devo? It is the question of *evolvability*, or the capacity of developmental systems to evolve (see also Dawkins 1989; Raff 1996; Wagner and Altenberg 1996; Wagner 2005). For them, evolvability is a question of the way that developmental systems do or don't allow evolution to occur; as such, evolvability depends both on biases in the variation that occurs and limits on the amount of variation. Development is critical for understanding the evolvability of organisms, because it is critical for understanding the type and amount of variation that is generated. Such a perspective is indeed similar to the one I have offered here.

However, for Hendrikse et al. (2007) development matters only because "it structures the ways in which variation is presented to natural selection" (400). In other words, in spite of their focus on the autonomy and explanatory potential of the field of evo-devo, they conceive of it merely as a sort of helpmate to the selectionist accounts of the modern synthesis, remedying the lack of a theory for biases in the production of phenotypic variation. Natural selection is still the creative force in evolution; it just works with a more limited range of materials than the synthetic theory allows.

As an example of Hendrikse et al.'s viewpoint, consider their take on the wonderful work of Abzhanov et al. (2004) on developmental control of beak size in Darwin's finches. Abzhanov et al. showed that expression levels of *Bmp4* (which codes for a developmental signaling molecule) in the upper-beak primordium are significantly higher during development of finches with large beaks, and that large beaks can be induced in chicken embryos by infecting the upper-beak mesenchyme with retroviruses expressing *Bmp4*. Nevertheless, Hendrikse et al. aren't satisfied that this work at the descriptive level counts as evo-devo, because "all evolutionary transformation must have a developmental basis and uncovering it in and of itself contributes nothing to explaining the diversity of life." Instead, they suggest that the work of Abzhanov et al. counts as an example of evo-devo because "by suggesting a relatively simple developmental basis for integrated evolutionary transformations in a functionally important character—beak length and width—they provide a developmental explanation for how this character could evolve as rapidly as has been shown in previous work . . . and why it varies so remarkably among a closely related group of species. Developmental biology here contributes essential information to understanding the *evolvability* of this system" (397).

This view does not go far enough toward giving the developmental mechanisms underlying beak shape a role in the evolutionary process. In particular, it tends toward the teleological perspective critiqued previously, in which developmental constraints and biases are constraints on the power of natural selection. It would be better to say, rather, that we now know that beak size in Darwin's finches (and other birds) is susceptible to large variation in size due to (among other things) the influence of varying levels of BMP4 during embryogenesis, but that each species is functionally constrained by diet to a small part of the range of beak sizes available. When food item availability shifts, and the finch diet changes, this is equivalent to a shift in the functional constraints on the system, allowing previously unrealized morphologies to emerge at the population level. In such a view, it is the developmental genetic system that is creative, because it generates phenotypic diversity, and selection (both broad sense and narrow sense) is the limiting factor, containing that variation within strict boundaries.

The view of Hendrikse et al. does at least recognize some importance for development in structuring phenotypic evolution. By contrast, a number of workers seem to think that evo-devo has nothing to teach us about phenotypic evolution beyond understanding the way that developmental mechanisms evolve, under the influence of natural selection for the phenotypic diversity that we see. For example, Sean Carroll, one of the most prominent evo-devo researchers, offers a thoroughly selectionist account of the role of development in evolution in his popular book *Endless Forms Most Beautiful* (2005). For Carroll, the main lesson we have learned from the advances in our understanding of developmental genetics is that the genetic tool kit for development is conserved across all animals. Nevertheless, understanding this requires no change in the synthetic view of evolution driven by natural selection for small changes; it only informs our understanding of how those changes occur: "Evo Devo reveals that macroevolution is the product of microevolution writ large" (Carroll 2005, 291). Even Carroll's views are not orthodox enough for workers like Hoekstra and Coyne (2007), who criticize him for emphasizing the role of *cis*-regulatory changes (changes in promoters and enhancers) in morphological evolution—in their view, most morphological evolution can instead be explained by point mutations in structural genes.

As I have already noted, a worker such as Salazar-Ciudad (2006) has a very different view of the relevance of development to evolution (see also Müller and Newman 2003; Müller 2007). For him, "the role of development is not to bias the morphological variation that would be produced otherwise. Development is simply the phenomenon responsible for the production of morphological variation. It determines what is possible and among it what is likely" (113). This view is fully in accordance with the vision of my own graduate adviser, Pere Alberch, who argued, "If development is to play any role at all in evolutionary biology, its contribution will be to provide an understanding of the possible morphological transformations" (Alberch 1982, 327). This role of development in explaining evolution is complementary to the role of the conditions for existence, or natural selection (in both broad and narrow senses). Evolution requires both a mechanism that generates variation and one that restricts the variations that are allowed to persist to those that are functional in the context of the organism's life cycle. Neither can safely be ignored.

But the study of the role of development in generating phenotypic variation is only half of the story. The other half is the study of the evolution of evolvability itself, a prominent theme of evo-devo not touched on by Hendrikse et al. (2007). Richard Dawkins (1989), the archetypal gene selectionist, was one of the first to suggest that there might be a kind of "higher-level selection" for evolvability that occurs during evolution. Interestingly, he arrived at this view from his experience playing with a program that he created to generate "biomorphs" for the book *The Blind Watchmaker* (1986). The program explicitly incorporated both a genotype and developmental rules by which the genotype determined a phenotype, and Dawkins found that the range of developmental rules he allowed was a key factor in controlling the diversity of forms he could generate.

Since then, selection for such evolvability has become associated with concepts of the evolutionary role of modularity, robustness, and other such properties of developmental systems (e.g., Wagner and Altenberg 1996; Kirschner and Gerhart 1998; Schlosser and Wagner 2004; Wagner 2005; Müller 2007). When one considers the differential survival and multiplication of lineages from a multilevel perspective, it is clear that the ability to evolve is a critical condition for the continued existence of all lineages. The environment is constantly changing, and only those organisms that are able to change in ways compatible with survival continue to exist. To put this another way, it is clear that all of the forms of life around us (with the possible exception of "living fossils") have evolved extensively over millions of years; thus, from a conditional teleological perspective, they must have been able to. What are the properties of developmental systems that allow such continuing evolution to occur, while constantly maintaining functionality, and how have these properties themselves evolved? And conversely—as Hendrikse et al. (2007) ask—how do these properties feed back on the evolutionary process to determine the types of change that occur? It is the task of answering both of these questions, not merely the second, that evolutionary developmental biology, or evo-devo, has now set itself. It appears quite possible that explicit answers will be forthcoming.

THE ECOLOGICAL NICHE AND THE CONDITIONS FOR EXISTENCE

As discussed in chapter 6, the ecological niche, conceived of as a "place in the economy of nature" existing apart from the organism that occupies it, and to which natural selection adapts the species, was central to Darwin's teleological conception of natural selection. In Darwin's view, environmental change is the driving force of evolution, because environmental change makes old niches disappear and new niches appear, and natural selection then adapts the organisms to the new niches (or they go extinct). This basic scenario still occurs in most textbook accounts of natural selection.

But the concept of a niche has itself undergone significant evolution since Darwin and since its more explicit formulation by Grinnell (1917) and Elton (1927). An excellent overview of niche theory is provided by Chase and Leibold in their book *Ecological Niches: Linking Classical and Contemporary Approaches* (2003). Here they defend niche theory against the proposal by Hubbell (2001) of his *Unified Neutral Theory of Species Abundance and Diversity* and, in so doing, greatly clarify the concept of the niche itself.

Let's follow Chase and Leibold's historical reconstruction of the development of the theory of the niche. They begin with Grinnell and Elton. In their view, the theory of the niche can be seen as having followed a dialectical process. At its roots, the theory encompassed two divergent viewpoints: Grinnell's view that the niche fundamentally involves the *requirements* of the species, and Elton's that it has to do with its *role* in the ecosystem (i.e., its effects on the biotic and abiotic environment). Subsequent workers largely in Grinnell's tradition include Gause (1964) and Hutchinson (1957, 1959, 1965, 1978). As will prove especially important for us, Hutchinson's "n-dimensional hypervolume" definition of a niche was also associated with the distinction between the fundamental and realized niche of a species.

In Chase and Leibold's view, a thesis formed of these two divergent traditions occurred in the "niche theory" developed by MacArthur (1969), Levins (1968), and May (1973), among others, in which the effects of species on their environment and of the environment on the species were modeled by coupled equations. But niche theory called forth a strong critique in the form of Simberloff's (1978, 1981) charges that it was unfalsifiable, because it did not have a specified null model to test results against. This critique led in the 1980s and 1990s to the rise of the use of random null models and experimental systems, which allow manipulation of key parameters, and to a neglect of general concepts like the niche in favor of narrower concepts appropriate to the model systems examined. Finally, in 2001, Hubbell launched the latest assault on the niche by proposing his "unified neutral theory of abundance and diversity" in which all species were treated as ecological equivalents. In spite of the fact that this makes no sense when considered a reflection of reality, this hypothesis nevertheless led to some success in predicting the statistical patterns seen (and Hubbell had some ideas as to how such a situation might arise).

Against this background, Chase and Leibold propose to resuscitate the concept of the niche by providing a new definition, one that draws on the tradition of Grinnell and Hutchinson, in which the niche is defined as a species' requirements, and on that of Elton, in which the niche is defined as a species' impact on its environment. They offer several versions of the definition, but the following one is clearest: the niche of a species is "the joint description of the environmental conditions that allow a species to satisfy its minimum requirements so that the birth rate of a local population is equal to or greater than its death rate along with the set of per capita effects of that species on these environmental conditions"

(Chase and Leibold 2003, 15). In other words, they simply place the two aspects, the requirements and impacts of a species, side by side. It is worth noting, however, that requirements pertain to the fundamental niche of a species, whereas impacts pertain to its realized niche.

In the terminology of this book, the ecological requirements of a species are merely its ecological conditions for existence. Closely connected with the concept of the niche is the concept of a limiting factor or condition. In fact, the language used by Chase and Leibold is quite reminiscent of the sort of language I have been using here: "the environmental conditions that allow a species to satisfy its minimum requirements" does not differ in any significant way from "the environmental conditions for existence of a species."

Interestingly, this type of language has been used throughout the history of this side of the concept. Thus, Grinnell (1917) concluded his seminal paper by noting that the range of the California thrasher "is *determined* by a narrow phase of *conditions* obtaining in the Chaparral association, within the California fauna, and within the Upper Sonoran life-zone" (433, my emphasis). Likewise, Hutchinson (1957) defined the niche as an "*n*-dimensional hypervolume" "every point in which corresponds to a state of the environment which would *permit* the species S_1 to *exist indefinitely*" (416, my emphasis). He considers this hypervolume to correspond to what he calls the "fundamental niche" of a species. By contrast, a species may not occupy its entire fundamental niche, due largely to interactions with other species (competition), resulting in a realized niche smaller than the fundamental niche. Hutchinson viewed the realized niche as variable in space, so that different parts of the fundamental niche might be occupied across the range of a species.

Hutchinson returned to the concept of a niche in 1965, in his essays in a book with the wonderful title *The Ecological Theatre and the Evolutionary Play* and in 1978, in his textbook introduction to population ecology. Here he once again defined the fundamental niche as a hypervolume constituting "a set of values of the variables *permitting the organism to exist*" (1965, 32), or "the *conditions for the existence* of a species" (1978, 158, my emphasis). As in 1957, he contrasted this fundamental or "preinteractive" niche with one that occurs in the situation of niche overlap with a second species, the realized or "postinteractive" niche, in which one or the other species is excluded from the area of overlap, or they somehow divide the niche space between them. At least one aspect of the evolutionary relation Hutchinson envisioned between fundamental and realized niches is structured by his focus on competition as the key factor restricting species to only portions of their fundamental niche: "Interactions are, however, likely always to occur and one might expect there to be continual pressure to take over marginal parts of the realized niche of one species by a second species whose fundamental niche happens to trespass on the realized niche of the first species" (1978, 161). This "continual pressure" he explicitly credits to natural selection, which functions during the competitive interaction implicit in overlap of fundamental niches.

In the two-part niche envisioned by Chase and Leibold (2003), the "requirement" part of the niche corresponds directly to the "n-dimensional hypervolume" of Hutchinson, as they are quite aware. However, Chase and Liebold allow a variety of factors to enter into this part of the niche, including factors defined as resources, predators, and stressors, where Hutchinson had focused largely on environmental variables such as elevation and nutrient concentration.

Interestingly, for Chase and Leibold, the fundamental and realized niches are also defined after Hutchinson: "The fundamental niche represents all of the conditions in which a species could *potentially* exist, whereas the realized niche represents those conditions in which it *does* exist in the presence of interacting species" (2003, 53). What is worth noting in this formulation is the qualification "in the presence of interacting species," which harkens back to the focus on competition by Hutchinson (1957, 1978). This seems an unnecessary limitation. The realized niche is no doubt a subset of the fundamental niche, for the simple reason that all of the conditions compatible with survival of the species are unlikely to be available. However, the effect of competitors is only one of a number of possible causes for the lack of availability of parts of the fundamental niche (as will be discussed later).

Of course, the focus on competition is understandable if we recall the origin of quantitative niche theory in Gause's work with microorganisms. When there is some limiting resource, such as a single food supply, then effects of another species on the availability of that resource are sure to play an important role in restricting the availability of parts of the fundamental niche of a species. More generally, two species can interact by sharing not only resources but also predators or stressors. What especially interests Chase and Leibold (along with other ecologists) is the *conditions for coexistence* of two species in such a situation. These conditions involve not only the requirements of the two species, but the impacts of the species on each other's requirements and the environmental supply of these requirements. There is no need here to follow them in the details of this approach; I merely want to point out that this way of thinking is quite congenial to the viewpoint of the conditions for existence.

What is particularly relevant to us here is how Chase and Leibold conceive of the evolutionary relations of the niche. The final chapter of their book is entitled "The Evolutionary Niche." Rather surprisingly, however, they conceive of the relevance of the niche to evolution in narrow terms—namely, in terms of the effect of natural selection on species traits that define niches, and conversely, in terms of the effects of community processes in defining the (realized) niches of species, and thus the conditions under which trait evolution occurs. They use their niche-based approach to examine such evolutionary phenomena as character displacement, the development of resistant versus tolerant phenotypes with respect to predation or herbivory, and phenotypic plasticity. All of this is fine, but what seems to be missing from their account is a view of how the fundamental and realized niches themselves evolve over time, and how in particular evolutionary shifts to previously unutilized parts of the fundamental niche may play a role in evolution.

I noted that in Hutchinson's (1957, 1965, 1978) original definitions, the explanation considered for the occupancy of only a restricted part of the fundamental niche—the realized niche—was largely in terms of species interactions. But clearly, other reasons can occur for a species not satisfying its conditions for existence in a manner that it is capable of doing. The simplest of these is that part of the fundamental niche may not be physically available, either because of dispersal barriers or because it does not yet exist anywhere on the face of the planet. For example, consider an invasive species that has been transported to a new area by humans (e.g., rats in New Zealand). Assuming that the species can indeed exist there, it will be occupying a different part of its fundamental niche than previously, merely because the resources, threats, and other variables are different. Or consider the effect of Europeans moving into North America and setting out bird feeders in their yards. Numerous species of birds were suddenly able to add bird feeders to their realized niche, although utilizing this resource was clearly part of their fundamental niche for untold millennia previously.

Finally, consider the ancestral Darwin's finch arriving on the Galápagos. I have already had occasion to critique Darwin's conception that there were empty niches, or places in the economy of nature, just waiting to be filled.* But clearly the realized niche of the immigrant birds on the Galápagos was immediately different from that of the ancestral population—they had shifted to a new part of their fundamental niche. If this shift couldn't occur, the birds could never have established themselves. Yet once this shift had occurred, and a population existed, the new realized niche set the stage for all the further evolution that occurred. It is the interplay between environmental changes and ecological shifts to unutilized areas of species' fundamental niches that permits the evolutionary process to continue at all, and that helps solve the problem of "maladaptation" posed by Lyell for Darwin. In one sense, this does involve vacant niches, because the new realized niche was indeed vacant until the species shifted into it. But such a vacant niche existed as an unutilized property of an existing species, not as a future evolutionary possibility toward which the evolutionary process was to be driven by natural selection; by definition, any realized niche that occurs without genetic change was already part of the species' fundamental niche.

---

* Of course, there could be unutilized *resources* that an immigrant species might exploit, such as a nut too tough for any native to crack.

Consideration of the relation between realized and fundamental niches brings up an issue that has not been at the forefront of my discussion so far in this book—the *mode* or *way* in which the conditions for existence are satisfied. In general, I have focused on the conditions for existence of organisms as the boundary conditions within which evolution must occur, and the conditions for existence of traits in populations (based on the Darwin-Wallace principle) as a further constraint on the evolutionary process. However, within these boundaries more than one lifestyle or mode of existence is certainly possible. The range of lifestyles, or realized niches, that occur both within and between populations is a topic clearly worth pursuing further, as are the mechanisms that adjudicate between them. Behavioral mechanisms may certainly exist that promote diversity of realized niches even among individuals within populations. As noted, genotype-environment covariance is a likely effect of such mechanisms. And narrow sense natural selection may certainly be involved in regulating such behavioral mechanisms. Such a view is entailed in another recent development at the intersection of evolution and ecology, the theory of "niche construction" (Odling-Smee, Laland, and Feldman 2003).

Niche construction had its origin in the work of Richard Lewontin in the early 1980s, who suggested that it makes no sense to consider organisms only the object of evolutionary forces, because the activity of organisms itself affects their environment and thus feeds back on their evolution (Lewontin 1983, 2000). This rather obvious point is surprising only because it conflicts with the "organism as artifact" model fundamental to Darwin's teleological conception of natural selection. It can be followed out to some interesting effects, which have been pursued at length by Odling-Smee et al. (2003). They define niche construction as a process that occurs "when an organism modifies the feature-factor relationship between itself and its environment by actively changing one or more of the factors in its environment, either by physically perturbing factors at its current location in space and time, or by relocating to a different space-time address, thereby exposing itself to different factors" (41). In their view, this is a process that occurs because of the activity of individuals, but it feeds back on either the evolution of the population itself or the evolution of other populations. It does so partly by a phenomenon they call "ecological inheritance," or the persistence of environmental modifications beyond the life span of individuals.

As many have noted (e.g., Dawkins 2004; Brodie 2005; Erwin 2005; Griffiths 2005), niche construction, at least thus defined, is a very broad and all-encompassing concept, involving virtually any activity of an organism that has some persistent effect on the environment or on the environment it experiences. It is of some interest to our project to try to express the concepts of niche construction in terms of Chase and Leibold's (2003) definition of the niche.*

---

* Odling-Smee et al. (2003) prefer a selection-based definition that seems much less tractable empirically: "the sum of all the natural selection pressures to which the population is exposed" (40).

If we think in Chase and Leibold's terms, niche construction involves a situation where the *impacts* of individual organisms in a population alter the environment (*supply* of resources, etc.) either for themselves or their descendants, or for some other organism. This of course is merely the impact side of the Chase and Leibold niche. The *requirement* side of the niche is represented in niche construction theory partly as selection for niche construction activities, which ties in with feedback on individual organisms (or lineages) that can improve their own chance of survival by altering their environment and partly by environmental changes driven by other organisms, which act as exogenous constraints on lineage survival.

To try to clarify this further, let's consider an earthworm (any one of the more than five thousand living species). Earthworms are one of Odling-Smee et al.'s (2003) favorite examples of niche construction, with a pedigree going back to Darwin (1881). For them, an earthworm is a paradox, because it has a physiology suited for living in a freshwater environment, not on land. It can live in the soil only because the soil *is* soil, not sand or clay, and it is soil because of the activities of countless generations of previous earthworms: "In this case it is the soil that does the changing, rather than the worm, to meet the demands of the worm's freshwater physiology. So what is adapting to what?" For them, this case demonstrates that "standard evolutionary theory short-changes the active role of organisms in constructing their environments. From the standard evolutionary perspective, which is externalist, the niche-constructing traits of earthworms can only be described as adaptations by earthworms to their soil environments in response to natural selection."

While I must admit to some sympathy for their critique of selectionist explanations, it is unclear to me that they have really understood the problem with such explanations. The alternative reconstruction they offer is as follows: "Either organisms can change to suit their environments, or environments can be changed by the organisms to suit them, and probably their descendants. In the earthworm case, there is no denying that in contemporary populations the match between earthworms and their soil environment is brought about at least in part by the second route, that is, through earthworm-induced changes in the soil" (375). So the argument for niche construction is based on the present environment, in which earthworm modified soil provides the necessary conditions for the existence of the worms.

Before considering the niche construction hypothesis in detail, let's take a look at the phylogenetic background to earthworm terrestriality. In a recent analysis, Erséus and Källersjö (2004) found support for an aquatic origin of the Clitellata (which includes leeches, bloodworms, and earthworms) and for a sister group relationship between earthworms and enchytraeids, or potworms. Potworms include both terrestrial and aquatic animals, as in fact do earthworms (Crassiclitellata) themselves. The partially aquatic genera of earthworms, including *Spargonophilus*, *Criodrilus*, and *Lutodrilus*, formed a clade named the Aquamegadrili by Jamieson (1988), in a phylogenetic analysis based on morphological data. A more recent analysis based on molecular data (Jamieson et al. 2002) found the genus *Komarekiona*, which includes riparian and terrestrial forms, to be nested within this group, and the common earthworms, *Lumbricus*, to be quite closely related. Based on their analysis, Jamieson et al. (2002) suggested that the aquatic habit of these worms is probably primitive: "It is not unlikely that the aquamegadrile families . . . have always had an aquatic or amphibious existence" (710). The broad-scale picture is thus of earthworms as part of a large radiation of freshwater oligochaetes, within which they form part of a clade exploring the boundary between aquatic and terrestrial environments.

Let's imagine, then, an ancestral freshwater annelid, which burrows in the mud at the bottom of a pond, feeding on interstitial inhabitants. Let's further imagine that over time its descendants are able to venture farther and farther from the pond, because the activity of their ancestors has modified soil chemistry and material properties such that it is now a more hospitable environment for them. Furthermore, so that this will be a pure process of niche construction, imagine that the worms themselves retain the exact same genes, physiology, morphology, and so on, as their ancestors but that the modifications of the environment allow them to survive (satisfy their conditions for existence) farther and farther from the pond. In this scenario, there has been no evolution at all, merely a change in the environment due to the activity of worms. This change in environment, however, must be based on the behavior of individual worms, which were able to expand their territory into zones farther from the pond, exploring an unrealized part of their fundamental niche (this *is* niche construction in one of their senses).

However, to follow the niche construction hypothesis, this movement of individual worms exploring their fundamental niche has produced environmental change (niche construction in another of their senses). This change has permitted a further shift in the realized niche of the worms, in which they have moved to a new part of their fundamental niche, one that was not available anywhere prior to the environmental modifications due to their activity. Thus, niche construction can be glossed as follows: niche construction occurs when a species' activities make parts of its fundamental niche available to it that would otherwise not exist. This can clearly be a process in time, but it is an ecological, not an evolutionary process.

If we add evolution into this scenario, the process looks a little different. Once the worms have moved to a new part of their fundamental niche, they now are satisfying their conditions for existence in a new way. This will feed back on the traits of the worms, such that traits that were stably maintained in the old realized niche now may not be and may be replaced by others (the Darwin-Wallace principle). Further shifts in the realized niche will permit other traits to arise, and as the worms progressively modify the environment farther and farther from the pond, a continual interplay will ensue among the exploratory behavior of individual worms, the environmental change due to the activity of the worms, the realized niche experienced by the worms, and the stability of trait values. What makes the whole process work is the combination of worm effects on the environment and the previous existence of a fundamental niche that includes at least occasional forays into the unmodified environment (i.e., "preadaptation").

I picked the earthworm case to examine in some detail because it does clearly provide some features that make the whole process work. In particular, there is an ecological feedback ("ecological inheritance") from the individual worm lineages on their own future success. In the scenario as I have traced it, the by-products of earthworm activity that promote environmental changes conducive to their survival have been associated with the expansion of those lineages having this effect. Thus, earthworms have facilitated their own expansion into new habitats. The key event here, however, is the "choice" by individual earthworms to explore beyond the confines of their ancestral niche.

Other cases of niche construction do not fit this pattern. For example, "ecosystem engineering" refers to all the effects (impacts) of a species, either on itself or on other species: "An ecosystem engineer is an organism whose presence or activity alters its physical surroundings or changes the flow of resources, thereby creating or modifying habitats and influencing all associated species" (Crain and Bertness 2006, 211). Yet Odling-Smee et al. (2003) consider such effects on other species to also be part of niche construction; they give the example of desert lichen-eating snails that are important for soil formation, with no suggestion that it matters to the snails themselves whether there is soil or not. The lumping together of individual organisms behaviorally "choosing" to experience distinct habitats and those modifying the environment in a way that will affect the current or future survival and reproduction of their own lineage also seems an unfortunate thing to do. We have already seen that a simple scenario for increasing terrestriality in earthworms requires each of these processes to play a distinct role. Thus, I must agree with those (e.g., Dawkins 2004; Brodie 2005; Erwin 2005; Griffiths 2005) who see "niche construction" as a collection of heterogeneous phenomena. In particular, Dawkins (2004) is quite right to emphasize the distinction between "niche-constructing" activities that can feed back on the success of a lineage, and "niche-changing" activities that merely change the environment in a way that affects other lineages.

However, if we restrict our attention to more narrowly defined niche-constructing activities, in which organisms change the environment experienced by themselves or their lineage—as Laland and Sterelny (2006) seem willing to do—we are still left with a variety of interesting phenomena. Odling-Smee and his colleagues are right that such evolutionary factors as the role of individual behavioral choice, the interplay between the realized niche, the fundamental niche, and environmental change, and the effects of organisms on their own and other organisms' environments have not been central to much of recent evolutionary thought. Perhaps lumping all of these diverse phenomena together under the common banner of "niche construction" will indeed encourage people to pay attention to these phenomena (Laland and Sterelny 2006).

In doing so, however, I believe that the descriptive and empirical framework provided by Chase and Leibold (2003) will prove to be the more general and useful conception. In this view, each species has a particular realized niche (the way it satisfies its conditions for existence), which not only is a subset of the fundamental niche (conditions for existence) but also includes the impact of the species on the availability of those (and other) conditions (see McGill et al. 2006). Thus, the ecological niche entails both the ecological boundary conditions for the evolutionary process to continue, an evolving property of the organismal lineage itself, and the specific way of life achieved, which results from the interaction of the evolving organisms and their environment.

PHYSIOLOGY AND THE CONDITIONS FOR EXISTENCE

Traditionally, organismic (as opposed to genetic) physiology has had little direct relation to evolutionary biology, and vice versa. In particular, since the rise of the modern synthesis, it has not often been thought necessary to take account of physiology in understanding the fundamental mechanisms of evolution. Within physiology itself, evolution has generally appeared as a kind of background condition, being explicitly invoked only to explain instances of evolutionary adaptation to new and, especially, extreme environments (e.g., Schmidt-Nielsen 1979; Hochachka and Somero 2002). For example, in the textbook of animal physiology by Hill, Wyse, and Anderson (2004), the authors define evolutionary physiology as the study of "the evolutionary origins of physiological traits," but for them this means essentially the origins of adaptations to new environments. Thus, evolutionary physiology deals with evolutionary adaptation: "polar bears are well suited to deal with cold, and dromedary camels, with heat. Evolution by natural selection is believed by modern biologists to be the primary process that has produced this match between species and the environments they inhabit" (22–23).

What seems to be missing from this view is any role of physiology itself in the evolutionary process. Yet physiology lies at the heart of biology. Organisms don't just exist passively, they exist actively—this is the essence of life, which is a nonequilibrium thermodynamic system that maintains internal order by external entropy production (Prigogine and Stengers 1984). And physiology is the science of how organisms manage to do this—that is, how they satisfy their conditions for continued existence as organized physical systems (see Wouters 2005b). Where ecology looks at the external functional possibility of organisms, relative to their external properties ("niche relations"), physiology is concerned with their internal functional possibility, given certain external conditions. As Cuvier noted, "the different parts of each being must be coordinated in such a way as to render possible the existence of the being as a whole" (1817, 6). This functional constraint is bound to be a key one in evolution, yet it has received very little attention through the years. Instead, physiologists have mostly worked out the basic mechanisms of life independent of evolutionary concerns, and they have focused their evolutionary interests on environmental adaptation under the assumed influence of past selection.

Against this background, the recent work by J. Scott Turner (2000, 2007) stands out. Turner is a comparative physiologist who became dissatisfied with the orthodox view of physiology's role in evolution while investigating the physiology of termite mounds, as structures that regulate the physiological state of the nests. He published a work in 2000 on the physiology of such animal-built structures, and he has continued his explorations recently in a book entitled *The Tinkerer's Accomplice: How Design Emerges from Life Itself* (2007). As might be expected from the title, his argument is well worth examining in the present context.

Turner's perspective grows out of a dissatisfaction similar to the one I have been pursuing throughout this book. For Turner, like me, there is "a difficulty with our conventional understanding of biological design" (2007, 6). But for Turner, the problem with the conventional view—that apparent design emerges only by natural selection—is that it does not sufficiently account for the "designed" appearance of life. His first chapter, "Cleanthes' Dilemma," tells the story of the debate constructed by Hume between Philo and Demea, with Cleanthes as adjudicator. Turner takes Hume's perspective to be unresolved among the three, with the following relevance to our present situation:

> Remarkably, after all the revolutionary developments in biology during the twentieth century, we are today no closer to a resolution of the disagreement than was Hume more than two centuries ago. On the one hand is Philo—the evolutionary biologist, the molecular biologist, who sees the world as being shaped by the "tinkerer" of natural selection, and who excludes on principle any possibility that a designing force or purposefulness can shape the natural world. And on the other hand is Demea—the naturalist, the creationist, the deep ecologist—for whom the natural world virtually shouts that it is designed, some way, somehow.

But Turner's take is that "perhaps Cleanthes was right after all: that there *is* a common ground between Philo and Demea and that a satisfactory explanation for the phenomenon of biological design rests there" (12). Although I—as a true partisan of Philo—find it difficult to be open to this point of view, let's reserve judgment until we see just what Turner means by this.

For Turner, "at the heart of the problem" is the definition of adaptation. On the one hand, adaptation can mean evolutionary adaptation by natural selection: "a phenomenon of genetic evolution, a progressively 'good fit' over many generations between assemblages of genes and the environment in which they live." On the other hand, it can mean immediate physiological adaptation, "the work of thermodynamic machines that maintain the ephemeral and orderly stream of matter and energy we call an organism" (13). This physiological adaptation is clearly synonymous with the physiological mechanisms that are involved in satisfaction of the internal (physiological) conditions for existence of organisms. What does this mean for Turner? "My thesis is simple: organisms exhibit their marvelous harmony of structure and function—an attribute I call designedness—not because natural selection of genes has made them that way, but because agents of homeostasis build them that way. Key to this is what I believe to be a universal phenomenon of life: the inexorable partitioning and creating of environments upon which homeostasis can be imposed. Out of this relentless busyness comes the exuberant diversity of well-functioning— well-designed—life" (14).

Turner calls living organisms "Bernard machines" after Claude Bernard, the founder of modern physiology, and in particular the founder of the doctrine of the constancy of the *milieu intérieur* (recall that Bernard's conception of the organism descended from Cuvier's; see chap. 5). As Turner notes, Bernard's doctrine was later transformed by Walter Cannon into the concept of *homeostasis*, which Turner defined in an earlier paper (Turner 2004) as the "persistence of function." As he noted there, "Physiological function requires an orderly environment that specifies particular pathways for flows of mass and energy. Persistent cellular metabolism, for example, requires an elaborately constructed *milieu* of catalytic shapes within the cell, embodied in proteins encoded by genes. The second law of thermodynamics asserts that this orderly environment inexorably degrades, and along with it the physiological function it specifies. Function persists only if work is done to restore this orderliness as rapidly as it degrades. This is homeostasis at its most fundamental level" (332–333). Homeostasis is thus for Turner fundamentally the mechanisms that maintain function, in the face of changing environmental relations—that is, the mechanisms that help satisfy the internal conditions for existence of organisms under a variable environment. His basic point is that although we wish to rule teleology out of science, organisms are fundamentally teleological machines. In the terminology I introduced in chapter 2, they are "purposive systems" that contain an internal representation of the properties of the world, of the effect of these properties on their own state, and of their own preferred future state, allowing them to respond appropriately (in terms of their continued survival) to many external challenges. Some, such as we humans, are even intentional systems.

In his book, Turner examines the behavior of a variety of Bernard machines, including termites in termite mounds, fibroblasts in bodies, endothelial cells in blood vessels, osteoblasts and osteoclasts in bone remodeling, the nervous system and antler growth. The fundamental point he is making with all these examples is that the feedback-regulated behavior of lower-level systems can produce adaptive behavior at the whole-system level, thus fibroblasts automatically lay down collagen fibers along lines of stress, and termites automatically build their mounds by following simple rules about CO_2 levels. Such behavior is the essence of "life itself." The adaptedness of organisms is not just encoded in their genes; it is constructed by them in their own lifetime by their adaptive response to environmental demands. This is what organisms are.

Moreover, whole organisms are also "Bernard machines." For example, consider a starfish, such as *Pisaster*, which typically preys on mussels. Turner asks,

> How does the starfish know there is a mussel nearby? How does it navigate there to capture it? By what mechanisms does the starfish pry open the shell? How does the starfish come to be hungry? How does the starfish come to *want* a mussel? There is also a kind of mentality here, but a very different kind from the ethereal mentality of eidos [i.e., the Platonic idea]: the starfish itself must somehow know what it wants, and be aware of the opportunities available for satisfying its desire. And there is a different kind of intentionality at work, too: the starfish itself must somehow formulate and carry out actions to satisfy the desire. (2007, 145)

Turner rightly connects this conception of striving with Aristotle's concept of *physis*, or nature (see chap. 3), in opposition to Plato's *eidos*. For him, this intrinsic intentionality (or at least apparent intentionality) is again the essence of "life itself." But according to him, this poses a huge problem for evolutionary biology, because "the *fact* of evolution itself cannot be rationally explained *with* intentionality, but the means whereby evolution works cannot be explained rationally *without* it." So what is to be done?

Turner's answer is "to bring intentionality firmly into the Darwinian fold" (148). He does so by arguing that intentionality is itself a form of homeostasis. Consider the eye. The production of a retinal image, for all its exquisite functionality, is the easy part to understand. The hard thing to understand is how the organism can use this image to construct an internal representation of the world that can guide its actions. Turner points out that the development of a sensory map of the world depends in an important way on sensory feedback during construction of the system, which involves homeostatic mechanisms at the cellular level.

Where all of this previous discussion comes to a head is in the final chapter of the book, in which Turner endeavors to break down "Biology's Bright Lines." One of these is the heredity/function distinction, as embodied in the central dogma of molecular biology. Turner proposes to introduce another type of entity—one that is process based, rather than thing based—into evolutionary theory. These entities are what he calls "persistors," which he defines as "persistent environments created and managed by systems of Bernard machines" (219). These include environments within organisms, which function as persistors for the contained cellular lineages, and external environments that persist beyond the lifetime of individual organisms, such as the *heuweltjies* of the South African Karoo: self-maintaining ecological patches involving both termites and vegetation. Thus, for Turner, evolution is produced by the interplay between Bernard machines, or persistors, and Darwin machines, or replicators.

Yet in spite of all Turner's eloquence and well-explained examples in support of the vital importance of physiology, especially homeostasis, to evolution, precisely on what basis the reconciliation of Darwinism and physiology is to be achieved remains rather unclear, at least to me. In the final paragraph of his book, Turner only hints at what the synthesis he is looking for might be. The big problem for him is

> resolving Cleanthes' dilemma, to understand how the tinkerer—the Darwin machine— and his purposeful accomplice—the Bernard machine, work together to make life and evolution happen. And resolving Cleanthes' dilemma means making evolution a biological phenomenon once more: not a pale atomistic imitation of life, as a Neo-Darwinist Philo would have it, and not the inscrutable thought of an intelligent designer that a modern Demea would favor, but a living phenomenon replete with the purposefulness and intentionality that is the fundamental attribute of life itself. (227)

What are we to make of Turner's perspective? Turner is certainly right to stress the evolutionary importance of the nature of organisms as physiological systems and in particular as self-regulating, purposive systems. Yet, from my perspective, the relevance of this to evolution is that we need to incorporate purposiveness and intentionality not as the driving force for evolution, but rather as a necessary condition for evolution.

The concept of homeostasis is a way of talking about how the internal and external systems conditions for life are maintained, and the maintenance of these conditions is of course necessary for life to continue. To understand the origin of such mechanisms, it is merely necessary to think about the conditions for organismal existence. Work on "robustness" and "evolvability" has already highlighted what some of the systems requirements for continued evolution are. It is clear from the characteristics of the life we see around us that homeostatic mechanisms are likewise conditions for the continued existence of all life forms. In fact, if the ecological niche is (at least partly) the state space defining the external conditions for existence of a species, one might similarly define an internal, or physiological, state space giving the internal conditions for the existence of organisms. In this view, homeostatic mechanisms are those that keep the internal environment within this state space.

However, as cautioned in chapter 2, when discussing such mechanisms, one must always be wary of bringing in purposiveness and intentionality as first principles. If biological systems are self-regulating and even forward looking, then this is not an intrinsic property of such systems but instead a property maintained only to the extent it is a condition for their continued existence. This is Philo's enduring contribution to the debate, and it can't be denied without returning to a semimystical, vitalist perspective. As Hume (1947) had Philo say: "It is in vain, therefore, to insist upon the uses of the parts in animals or vegetables, and their curious adjustment to each other. I would fain know how an animal could subsist, unless its parts were so adjusted?" (185). In spite of his clear lack of any such intention, Turner flirts with the danger of introducing obscurantism into his formulation, much as Russell (1946) did in his insistence that "directive activity" is a primary characteristic of life.

Due to no doubt partly to the clarity of his own vision, and partly to the constraints imposed by the format of a semipopular book, Turner does not treat some other relevant views on how to relate the phenomenon of homeostasis to evolution. One of the most prominent is the emphasis by L. L. Whyte (1965) on the role of *Internal Factors in Evolution*, in particular on the evolutionary importance of "internal selection" within the highly ordered, functional homeostatic system of an organism (see also Arthur 1997, 2004). Whyte states that the goal of his approach is the discovery of the "coordinative conditions" of life, defined as "the general algebraic *conditions* expressing the biological spatio-temporal coordination, the rules of ordering *which must be satisfied* (to within a threshold) by the internal parts and processes of any cellular organism capable of developing and surviving in some environment" (Whyte 1965, 35, my emphasis).

Homeostatic regulation is one of the mechanisms involved in the satisfaction of these conditions. Clearly, if an organism is to exist, it is necessary that it be able to deal with environmental fluctuations in an "adaptive" way—if I went into convulsions every time a new situation arose, I wouldn't last long. Mutations must satisfy not only the external sieve of ecological functionality but also the internal sieve of physiological functionality. As Turner insists, organisms are hierarchically organized, functionally integrated systems, and such systems may have their own evolutionary dynamics.

More recently, Günter Wagner and Kurt Schwenk (2000) developed the concept of "evolutionarily stable configurations," drawing on not only Whyte's vision of "internal selection" but also related concepts such as that of "functional integration" (Bock and von Wahlert 1965; Wake, Roth, and Wake 1983). They defined these as "character complexes that interact in their contribution to fitness in a way that leads to evolutionary self-stabilization" (156). Variations in characters that form part of such a complex are "selected" for their effects on system properties, not for their direct interaction with the environment; it is instead the entire system that interacts with the environment. The primary example they use is the lingual feeding mechanism of iguanian lizards. This has been retained through a huge radiation of lizards of varying diet and ecology. Nevertheless, one lineage of lizards, the scleroglossans (including monitors and snakes), has "escaped" from the constraint of this lingual feeding mode, evolving a mode of jaw-mediated feeding. This development allowed the tongue to become specialized for chemoreceptive function as a means of transporting odorants to the vomeronasal organ. Wagner and Schwenk view the evolutionary process as one in which such evolutionarily stable configurations of features are stabilized by their functional interrelations.

Interestingly, this vision is rather close to that of Cuvier some two hundred years earlier, in a nonevolutionary context. Recall that for Cuvier, his four *embranchements* represented distinct modes of functional organization. Within each *embranchement*, functional differences characterized the lesser groups. Likewise, Wagner and Schwenk consider the hypothesis that body plans *(Baupläne)* are conserved in animals because of such functional constraints. They conclude on the contrary that the stability of body plans across hundreds of millions of years is due both to developmental constraints and to functional constraints.

As already noted, however, developmental constraints can mean two rather different things: functional constraints on the developmental system, and true "generative constraints" due to limited variability of the developmental system. If we ascribe all limitation of variation in body plan to the limitations imposed by functional constraints during development and during independent existence, then we have indeed a thoroughgoing "selectionist" or "conditions for existence" account for such stability. Nevertheless, as Wagner and Schwenk acknowledge, the practical difficulties in distinguishing among the potential causes of such phenotypic stability are substantial, if not necessarily insurmountable.

Another relevant view on the relation of physiology to evolution comes with the concept of phenotypic or developmental plasticity, most notably developed by Mary Jane West-Eberhard (2003). This concept of plasticity is in fact closely related to Turner's point that organisms themselves are adaptive systems, which respond appropriately to environmental changes, stimuli, and other variables. West-Eberhard begins with the phenotype as the locus of her evolutionary interest. For her, "The piece that is missing from a synthesis of development and neo-Darwinism is an adequate theory of phenotype development that incorporates the influence of the environment" (19). The emphasis on phenotype development is characteristic of the evo-devo school in general, but the inclusion of the environment as a central factor distinguishes West-Eberhard's approach.

West-Eberhard is quite aware of her conceptual predecessors, discussing such evolutionary classics as the Baldwin effect (Baldwin 1896, 1902), in which behavioral plasticity is the motor of evolutionary change. In the terms used earlier, such behavioral plasticity involves a shift of the realized niche to a new part of the fundamental niche. This sort of effect, driven by the purposive, even intentional nature of individual organisms, is exactly the sort of thing that Turner has in mind in his critique of the conventional wisdom. West-Eberhard sees homeostatic mechanisms in exactly the same light: "Usually, these workaday processes are associated with maintenance of the status quo. But it does not take much imagination to realize that they could also function to accommodate new conditions and structures as well as unusual ones even though they are not especially evolved to do so [please read: 'have not functioned that way in the past']" (45). In this way, homeostatic mechanisms facilitate evolution by allowing continued existence under changed environmental conditions.

Unlike Turner, West-Eberhard stresses the Darwinian basis of the adaptive behavior of organisms. It is quite interesting to compare her discussion of bone and muscle plasticity, for example, with Turner's. Her emphasis is on the plasticity of bone features, and she notes that "natural selection may mold the responsive phenotypic structure to make it more likely to produce certain response" (40). I would prefer to say that "the structure of phenotypic response is itself a functional feature of the developmental system," although the underlying point is essentially the same. But we need Turner to remind us that the adaptive functional response of the developmental system is due to the action of cells that themselves behave as "Bernard machines," regulating certain features of their cellular environment.

If internal functional constraints are not well incorporated into current evolutionary theory, this is perhaps because of the emphasis in traditional selectionist models on variation within populations as the source of all evolutionary change. What tends to be left out of the equation, then, is variation that is not compatible with life in any available environment and thus defines the boundaries of possible existence. Yet as Whyte (1965) emphasized, this is the first sieve through which all variation must pass. And as the explosive growth of interest in "systems biology" is teaching us (e.g., Kaneko 2006; Konopka 2006; Palsson 2006), the internal conditions for the existence of organisms are now becoming amenable to theoretical and experimental analysis.

CONSERVATION BIOLOGY, GENETIC LOAD, AND THE CONDITIONS FOR EXISTENCE

The field of conservation biology is a relatively new one, though its historical roots are deep. Its birth can be timed rather precisely: in 1980, Michael Soulé and Bruce Wilcox published an edited collection of papers under the title *Conservation Biology*; a textbook appeared the next year (Frankel and Soulé 1981), followed in the late 1980s by the birth of a scientific society and journal. From the start, the field of conservation biology has been an applied one; the passage of the Endangered Species Act (ESA) in the United States in 1973, in particular, created the need for a science of conservation. With its focus on protection of individual species from extinction, the ESA highlighted the importance of understanding the necessary conditions for the continued existence of individual species. The overall goals of conservation biology were clearly spelled out in a foreword to the 1980 volume by Thomas Lovejoy, director of the World Wildlife Fund: "The demands on science are large. New questions arise when many populations and entire biomes are being fragmented and reduced on such a scale and at such rates. These questions are a great challenge to the ingenuity of biologists and are the central focus of this volume. Unless we solve them, we will end up with less than we intend, struggling in our ignorance to protect genetically eroding populations and decaying ecosystems" (Soulé and Wilcox 1980, ix). Lovejoy's portrait of "genetically eroding populations and decaying ecosystems" pointed the way toward the twin focus that defines conservation biology: a focus on understanding both the *genetic* and the *ecological* conditions for continued species survival, so that enlightened managers can make appropriate, science-based decisions. The focus on genetics and ecology is understandable, because these are the factors that we can manipulate.

On the ecological side, the 1980 volume included a body of work looking at habitat associations in tropical forests, which was directed at understanding how one might best design reserves for preservation of the maximum number of species. Wilcox (1980) and Terborgh and Winter (1980) examined the causes of extinction in isolated populations, focusing on the size of protected areas as the primary determinant of number of species retained (based on the theory of island biogeography developed by MacArthur and Wilson), and pointing out that "faunal collapse" occurred when islands got too small (Wilcox) and the species most vulnerable to extinction were those that were initially rare (Terborgh and Winter).

Even more interesting for us to examine is the genetic side of the equation, in view of the emphasis placed in earlier chapters on the distinction between absolute and relative rate of increase (fitness), and the confusion between them that is common in the field (see chap. 7). Conservation biologists can't afford to be confused between absolute and relative rates of increase: they are necessarily concerned with absolute rate of increase, because this *is* population growth, which is the measure of success. In fact, one might say that an absolute rate of increase of 1.0 or greater is *the* goal of conservation; this is equivalent to satisfaction of the ecological conditions for existence. Given this focus on absolute realized fitness, conservation biologists have been particularly interested in individual variation in overall genetic "quality," with the assumption being that poor-quality individuals are detrimental to a population; they have also been concerned about depletion of genetic diversity as a cause of vulnerability to environmental change. From the start, then, concerns about genetic drift (Franklin 1980) and inbreeding and outbreeding depression (Senner 1980; Soulé 1980) have been central to the field.

Soulé's contribution to the founding volume had the interesting title "Thresholds for Survival: Maintaining Fitness and Evolutionary Potential." For Soulé, there are "three survival problems or issues: "(1) the short-term issue is immediate fitness— the maintenance of vigor and fecundity during an interim holding operation, usually in an artificial environment; (2) the long-term issue is adaptation—the persistence of the vigor and evolutionary adaptation of a population in the face of a changing natural environment; (3) the third issue is evolution in the broadest sense—the continuing creation of evolutionary novelty during and by the process of speciation" (Soulé 1980, 151).

The problem of "immediate fitness," which he identifies with "vigor and fecundity," has to do with the ability of organisms to continue to exist in small, usually captive populations; it is thus closely related to the question of the satisfaction of the physiological or internal conditions for existence of the organisms, under permissive environmental circumstances. Conventional population genetic theory suggests that survival can be affected by "deleterious" recessives that appear in the homozygous state due to inbreeding, and by genetic drift in small populations, allowing "deleterious" alleles to increase in frequency. Soulé notes that the quantitative genetics literature provides abundant data on the relation between inbreeding and fitness, generally showing that inbreeding is bad for fitness (absolute rate of increase). In fact, he notes that in domesticated animals, intense inbreeding is generally associated with extinction of most or all lineages in a short number of generations. His preferred approach, however, is to examine heterozygosity levels of allozymes in individuals in natural populations as predictors of fitness. In a survey of thirteen species for which relevant data had been reported, eleven showed increased heterozygosity to be positively correlated with such fitness measures as survival to a particular size or age, growth rate, and developmental stability, and negatively correlated with morphological variability and aggressiveness.

His conclusion is cautious: "Enough is known about inbreeding to justify concern that any increase in homozygosity reduces absolute vigor and fecundity, but so far we have no way of deciding how much is tolerable. We can not even pinpoint with any certainty the cause of inbreeding depression, although we suspect that this loss of fitness is mostly attributable to a load of deleterious recessive genes and to altered gene interactions rather than to a loss of single gene heterosis concomitant with decreasing heterozygosity" (160).

Unfortunately, as Soulé notes, population genetic theory is generally of no help in modeling this situation, because the parameters of the models are unknown for the populations concerned. Instead, Soulé suggests that conservation biologists adopt a rule of thumb developed on the basis of the experience of animal breeders: that inbreeding be restricted to no more than 1 percent per generation, or an effective population size of at least fifty individuals.

How have these conclusions held up, more that twenty-five years later? An interesting place to check in is an exchange that occurred in a recent issue of *Animal Conservation*. Reed, Nicholas, and Stratton (2007a) contributed a "featured paper" entitled "Genetic Quality of Individuals Impacts Population Dynamics," building on earlier work (Reed and Frankham 2003; Reed 2005). Their goal was to assess "whether the mean fitness (genetic quality) of individuals impacts the realized population growth rate despite density-dependent population dynamics" (276). Clearly, here "mean fitness" means something other than "population growth rate," or there would be no need to assess the relationship between them. What do Reed et al. (2007a) mean by "genetic quality" or "mean fitness"?

In earlier work by Reed and his colleagues (e.g., Reed and Frankham 2003; Reed 2005), "mean fitness" is mean absolute realized fitness as measured by some component of fitness (e.g., seed set). Thus, reduced fecundity, survivorship, and so forth, would be considered to be reduced genetic quality. The environment in which the genetic quality is measured can be either the laboratory or the field. However, the term *genetic quality* also has another, overlapping use, in the context of mate choice and sexual selection. In this context, Hunt et al. (2004) define *genetic quality* in a quantitative genetic framework as "the breeding value of an individual for total fitness." They argue, "We can therefore learn little about genetic quality from measures of only a few fitness components" (329). Likewise, in a recent review, Neff and Pitcher (2005) define *genetic quality* in a way that seems to be a synonym of *total fitness*: "We define genetic quality based on the contribution a gene variant (allele) or genotype (alleles) makes to an individual's fitness; an individual is of higher genetic quality when it possesses an allele or genotype that increases its fitness relative to that of an individual with a different allele or genotype. Fitness, in turn, can be defined by the individual's lifetime reproductive success (LRS)" (20). What is especially interesting for us about Reed et al.'s approach is that they separate "genetic quality" from lifetime realized fitness, instead considering it best estimated by a particular component of fitness, one presumably less likely to be influenced by environmental capacities.

In their featured study, Reed et al. (2007a; see also Reed 2007a, 2007b; Reed et al. 2007b) examined seven populations of wolf spiders of two different species, over a period of three years. They used "genetic variation" and "population size" as "surrogates" for the average genetic quality of individuals in the population. Genetic variation in one species was assessed using variation at fifteen microsatellite loci, that in the other using broad sense heritabilities estimated using the correlation of phenotypic values among siblings. Population size was assessed by standard mark-recapture methods. These spiders are an excellent model system, because adults are easily surveyed and marked, and they are semelparous, synchronous breeders in which the fecundity of females can be assessed by counting hatchlings emerging from egg cases.

An important variable included by Reed et al. (2007a) was prey capture rate. When spiders were observed with prey in their web, they were scored as positive; lacking such prey, they were scored as negative. Prey capture rate thus gave Reed et al. some measure of environmental "quality." They note that there is reason to believe that prey availability is a limiting factor for these spider populations. Unfortunately, as in most of the studies examined in chapter 8 with respect to opportunity for selection, Reed et al. (2007a) don't report their data for population size and growth rate on a yearly basis (but see Reed et al. 2007b, where mean and standard error of fecundity are reported). Nevertheless, in their most compelling data, they show that if the effect of prey capture availability is removed (by taking the residuals of a regression of population growth rate on prey capture availability), population size still explains much of the variation in population growth rate. This effect, however, is found only during "high stress" years of high population levels and/or decreasing prey capture. For them, this is tantamount to evidence that increased heterozygosity (which correlates with population size) causes higher population growth rates in times of stress: "We have shown that the genetic quality (mean fitness) of individuals in populations of two species of wolf spider directly impact the dynamics of those populations" (Reed et al. 2007a, 280).

In the accompanying "Commentaries," McGowan, Wright, and Hunt (2007), Brodie (2007), and Keller, Biebach, and Hoeck (2007), aren't quite so confident about what Reed and his colleagues have shown. McGowan et al. (2007) accept the authors' claim to "successfully demonstrate an overall impact of genetic quality on population growth rate, over and above the effects of food limitation and density dependence" (284). But they point out that "none of the populations studied by Reed *et al.* (2007) actually went extinct. It would seem essential to know the levels of inbreeding and the type and magnitude of environmental stress populations need to experience before extinction is the only outcome" (285).

The other two commentaries are even more critical. Brodie (2007) entitled his pointedly "Population Size Is Not Genetic Quality." He notes that the approach of Reed et al. sweeps a number of distinct factors together under the heading of "genetic quality," including inbreeding effects such as homozygosity of recessive alleles and genetic drift of deleterious alleles to higher frequency due to decreased population size, resulting in lowered genetic diversity. Nevertheless, Brodie's fundamental criticism is that the authors have demonstrated correlation but not causation: the close association of small populations, low heterozygosity, and low diversity makes it impossible to ascribe variation in survival to any one of these factors.

Finally, the commentary by Keller et al. (2007) comes the closest to themes emphasized elsewhere in this book when dealing with the relationship between fitness and population growth. They note that "inbreeding depression in individual fitness components is . . . of limited importance to conservation biology unless these reductions in individual fitness translate into reduced population growth rates" (286). In this context, they emphasize in particular the distinction made by Wallace (1970, 1975) between "hard" and "soft" selection. Under density-dependent population regulation, of course, a change in mean fecundity may not affect population size at all. For them, the empirical problems involved in understanding the relationship between fitness components like fecundity, which show clear effects of inbreeding, and overall population growth, remain daunting: "Defining the conditions under which population growth rates are depressed by inbreeding will remain one of the major challenges for conservation genetics today" (Keller et al. 2007, 287).

This window into the current debate within conservation biology over genetic effects on population growth (mean absolute fitness) is illuminating. Clearly, the theoretical issues I have been documenting in this book are far from being merely "academic" in their import. In these times of declining biodiversity, with numerous species worldwide unable to satisfy their conditions for continued existence and succumbing to extinction, it makes the science of these conditions—the science of conservation biology—more critical than ever. If I have argued for anything in this book, it is for escape from assumptive approaches in which natural selection, genetic drift, ecological opportunity, or any other factors are invoked without an explicit definition of what is meant, and in which verbal models involving such terms are substituted for explicit discussion of the variables that are actually measured. As Thomas Lovejoy noted at the founding of the field, to understand the conditions for the continued existence of species, we clearly need to understand both their ecological (external) and genetic/functional (internal) conditions. We must also understand how these interact. Assumptions here can only get in the way of understanding.

As a parting shot, let's take another look at the spiders examined by Reed et al. (2007a, 2007b) from the point of view I have been developing in this book. These authors clearly showed that smaller populations of these spiders had lower genetic diversity, heterozygosity, fecundity, and population growth rates, and that these differences in population growth rate could not be explained merely by differences in prey capture rates. Moreover, years with reductions in prey capture rates were associated with reductions in population size and fecundity. If we accept for a moment the hypothesis that "genetic quality" of individuals was lower in the smaller populations, accounting for reduced fecundity, what does this mean in real terms? It means that individual spiders are less likely to be able to satisfy their conditions for existence. But given an assumed constant environment, this must mean that the fundamental niche of these "low genetic quality" individuals is smaller than that of "high genetic quality" individuals, or at least has less overlap with the circumstances that the individuals manage to find. Thus, they are less likely to successfully meet their requirements, to fall within the boundaries of their fundamental niche. But given that the environment is here held constant, we must be talking about the *internal* physiological abilities, the coadaptedness of the parts, of the individual organisms—that is, their ability to satisfy their conditions for existence as a characteristic that is largely environment *independent*. Even in a hard-nosed, empirical science like conservation biology, both internal and external factors necessarily come into consideration when considering the conditions for existence of a species.

When Muller (1950) talks about "genetic load" or Reed et al. (2007a) talk about "genetic quality," they are implicitly pointing to the internal coadaptedness of the organism as a constraint on evolution. I have insisted that we must never assume that absolute and relative rates of increase are measuring the same thing (see chaps. 7 and 8), because confusion between these quantities has led to a disconnected, disembodied notion of "fitness." Nevertheless, it has been clear from the early days of genetics that most mutations are unconditionally bad for survival, that these are usually recessive in their effects, and that inbreeding depression in outbred populations is a real phenomenon. I remember hearing Robert Vrijenhoek give a talk in the mid-1980s on some Mexican guppies; he spoke of one population as being "a bunch of crappy little inbred fish." This makes sense. If we can't believe that all selection is "hard," we can't believe that it's all "soft," either. The key is to recognize that the question of the relation of selection to population growth is still entirely unresolved; many more empirical data, data collected with an open mind, are needed. As stressed by Sacherri and Hanski (2006), "it is unwise to assume that genetic variation and selection are irrelevant to population dynamics" (346). But equally, as I have emphasized in chapter 8, it is unwise to assume that population dynamics are irrelevant to genetic variation and selection.

· · ·

We need an evolutionary theory that integrates empirical data on population dynamics and evolutionary dynamics, free of assumptions about the effects of genetic variation on rates of increase. Only then can we hope to understand the interrelation of the conditions for the continued existence of traits in populations (under natural selection *and* other evolutionary processes) and the conditions for the continued existence of species, the fundamental concern of conservation biology. Achieving such understanding is certain to be a condition for the survival of many species, including perhaps our own.

Conclusion

It is amazing how the strictures of the old teleologies infect our observation, causal thinking warped by hope.

—JOHN STEINBECK AND ED RICKETTS, 1941, THE SEA OF CORTEZ

The story I've traced in this book, though long and convoluted, is at root a simple one. It is the story of the legacy of the conflict between the perspective of the atomists, Leucippus, Democritus, Epicurus, and Lucretius, who insisted that all order in the universe existed only because it was a condition for existence, and that of Socrates and his follower Plato (not to mention the Stoics and Christians), who argued for intelligent design as an explanation for the order and adaptedness of the world. Putting Darwin in this picture has always been a bit problematic: did Darwin do away with design, or did he naturalize it?

The main historical point of this book has been that Darwin, as a creature of his time and place, accepted the basic premise of the argument to design: that adaptedness is a problem crying out for explanation and, further, that the only possible explanation for "apparent design" is a historical one, a story of creation *for* the functional role played. These preconceptions, combined with his focus on environmentally driven evolution, produced the theory of natural selection as it has come down to us today, as a fundamentally teleological theory in which the achieved result of the evolutionary process is seen as a previously vacant place in the economy of nature, a vacant place that became the cause of its own realization.

Why do we need to "retire Darwin's watchmaker"? Because we don't need him, and we never did. If we accept a naturalistic world view, then we must accept that all organisms we see existing on the planet today are doing so only because they are satisfying the conditions for their existence. Moreover, no matter what we think about how wonderfully "adapted" they are, if their population isn't growing or shrinking, their mean overall adaptedness is exactly 1.0, no more, no less. Thus, organisms are not any better adapted than they need to be to survive (or they'd be surviving at a higher rate). This is the argument of the atomists, and it remains logically compelling today. As Cuvier saw, the fact that the existence of organisms is explained by their satisfaction of the conditions for their existence can be taken as a first principle on which to found a scientific view of life.

So whence, then, design? Given the existence of a straightforward and compelling argument to the contrary, why have so many through the ages thought to see design in nature? First is, of course, the problem of origins. While the Epicurean hypothesis may be logically unassailable, its story of origins is intuitively implausible, especially for organisms. The understanding still wants to know where all these complex, organized bodies *came from*. The hypothesis of evolution by common descent, with relatively gradual modification, essentially solves that problem (aside from the origin of life itself).

Second is the problem of *adaptedness* of the forms around us. Even granted that in the existing world such "monsters" as envisaged by Crombie (1829) and others could not survive, how is it that the world as a whole has such a harmonious appearance, such a high level of apparent "adaptedness"? And how is it that Aristotle, the great empiricist, was led to conclude that nature "does nothing in vain"? This is where Darwin and Wallace come in. The principle of natural selection is a principle for the continued existence of traits in populations. Not only do individual organisms need to satisfy their own conditions for existence, but the *traits* of these organisms need to satisfy *their* conditions for continued existence, or they'll disappear from the population. At a first approximation, the condition for the continued existence of a trait is that organisms with that trait survive at a higher rate than those without. We can thus follow Aristotle this far in saying that nature "does nothing in vain": to the extent that variation occurs, is heritable, and so forth, then nonfunctional traits will not persist in populations. Nevertheless, we know that in fact nature does many things "in vain," for a variety of reasons. Overdominance, with persistence in the population of disfavored homozygotes, is a simple example of how this adage can break down in sexual populations. Linkage disequilibrium, mutation pressure, and phenotypic plasticity are further complications, and many more have been detailed in this book.

By retiring the watchmaker, we can finally throw off the nineteenth-century chains that have locked us into the contrast of drift and selection, of mutations of large and small effect, of internal constraints versus external causes. We can be free to pursue the study of evolution under the dual view of mechanisms that generate variation and the conditions for the continued existence of that variation (functional/selective constraints). Darwinians have always been leery of admitting any evolutionary forces other than natural selection as an explanation for adaptation, or even evolution, because such explanations smacked of teleological mysticism; how could undirected forces produce such exquisite adaptedness? But when we realize that whatever adaptedness we perceive exists only by virtue of the fact that certain lineages are satisfying their conditions for existence, we are freed to admit any evolutionary process into the tent, secure in the knowledge that if it destroys adaptedness, then that evolving lineage will end—and other lineages will survive.

Epilogue
Evolutionary Biology and Intelligent Design

When I planned this book, I had thought to leave "intelligent design" out of it. After all, this is a book about science, not about politics, and the intelligent design debate is a political debate. Yet the issues touched on in this book are indeed central to the debate over intelligent design, and some explicit comments are in order, if only so there should be no chance of misunderstanding my position. Rather than attempting to debunk the arguments used in favor of intelligent design, a simple and yet necessary exercise already well executed by numerous authors (Forrest and Gross 2003; Perakh 2004; Shanks 2004; Scott 2005; Brown and Alton 2007; Isaak 2007), here I'd like to comment on some aspects of the debate that look somewhat different from my perspective.

As noted in the introduction, Darwinians, by accepting the premise of the argument to design (i.e., the premise that "apparent design" must have some historical explanation), have left the door wide open for the intelligent design enthusiasts. Darwin's intellectual ancestor is Paley, not Lucretius, and from this perspective it does become a somewhat tricky matter distinguishing the Darwinian view from that of design. Yet it was Darwin himself who noted, "If it could be proved that any part of the structure of any one species had been formed for the exclusive good of another species, it would annihilate my theory, for such could not have been produced through natural selection" (1859, 201). As has been discussed by many authors, what makes the theory of natural selection (in the broadest sense) useful as a scientific hypothesis, and the theory of "intelligent design" not, is that the theory of natural selection tells us what sorts of organisms and what sorts of features we should expect to find in the world; namely, we should expect to find organisms and features that can satisfy their conditions for existence. We have some well-established theories, such as theories of sex-ratio evolution, that do a pretty good job predicting just what these features will be (e.g., Herre, Machado, and West 2001).

It is to the great credit of evolutionists that a phenomenon like sex, which is prevalent and seems so familiar as to be quite "natural," presents a real problem for evolutionary theory. Clearly, sex satisfies its conditions for existence, but precisely how it does so remains something of a mystery. But most organismal features are not so puzzling; we don't expect to find a wild tree that makes delicious fruits with no seeds, no effect on the growth of the parent tree or related lineages, and so forth, but that do provide food for monkeys. That such phenomena are generally not known is a strong argument against the additional assumption of design, as something superadded to mere existence. And if the designer merely intended organisms to exist, it is hard to know how we could distinguish the hypothesis of design from mere existence: the design hypothesis explains nothing that is not already explained by existence itself.

The Epicurean philosophy is relevant in another way as well. In the ongoing debates over origins, equal time for "intelligent design theory" in the classroom, and the like, it is not often remarked that another, alternative, unquestionably scientific theory for the origin of the universe and of life is available, one that can justifiably be contrasted with that of evolution by diverging lines of descent with modification. This is the Epicurean hypothesis, in which the world popped into existence one day as the self-sustaining result of random rearrangements of atoms. This may not be very plausible, but it is a scientific hypothesis. In fact, the phylogenetic relations we can establish among organisms, the fossil record showing the gradual and continuous change through the history of the Earth, and astronomical evidence for evolutionary origins of the solar system and universe, all give us a good reason to accept the evolutionary hypothesis and reject the Epicurean one.

A third feature lending emotional support to the design hypothesis through the years is the fact that organisms themselves are purposive, even intentional systems. It is this intrinsic purposiveness that was at the heart of Aristotle's philosophy. But though this purposiveness is indeed an extremely important aspect of what organisms *are*, and it has important implications for the *character* of the evolutionary process, this feature of organisms speaks to "design" no more than do all the other incredible examples of "exquisite adaptations."

In short, when we recognize that organisms exist only by virtue of the fact that they are satisfying their conditions for existence, and that features of organisms likewise exist because they are satisfying *their* conditions for existence, we are freed to consider evolution in a dispassionate manner. Life is not designed, or at least it shows no evidence of design for anything other than continued existence, which needs no designer. In other words, Paley's watch not only winds itself and makes other watches (which can differ from their parent), but this is *all* it does. To truly retire the watchmaker, we must retire not only Paley's watchmaker but also Darwin's, inherited from Paley. We must admit that there is not only not design but indeed not even "apparent design" in the biological world, in the sense of entities doing any more than they need to do to continue to exist. This is the true message of Epicurus, and there is no reason that a Darwinian should think any differently. Evolution is literally the *evolution* or unfolding of life's potential, constrained by the functionality of the cell, the organism, the population, and the ecosystem that is the condition for its continued existence. This is a position with worthy ancient roots. It is also a strong position from which to confront the revival of the argument of design by the advocates of intelligent design.

GLOSSARY

Adaptation (Evolutionary)	A genetically based change in the mode of overall adaptedness (lifestyle) of organisms (or genes, cells, etc.). This often involves the evolution of features that boost rate of increase (realized fitness) when compared with the ancestral state of the feature (adapted features), and these features often have improved performance at some function necessary for the derived lifestyle.
Adaptedness	*Overall adaptedness:* the degree to which individual organisms (or genes, cells, etc.) are adjusted to their environment. Measured by rate of increase (survival and reproduction).
	Adapted features: features of organisms (or genes, cells, etc.) that have the effect of boosting the rate of increase of that organism (or gene, cell, etc.) relative to other features in a specified phenotype set.
	Performance: the efficiency with which a feature carries out a particular biological role (Bock and von Wahlert 1965) or function.
Argument of Design (Design Argument)	The argument that the structure of the universe is the result of a teleological design process—namely, a process aiming for the results achieved. Has two aspects:
	Argument to design: the argument from the structure of the universe *to* the existence and characteristics (e.g., power, wisdom, and goodness) of a designer.

Argument from design: the argument *from* the assumed design purpose of organisms or their parts to an explanation of the characteristics of the organism or its part. Because the eye was intended for seeing, the cornea was made clear, and so forth.

Average Effect For a quantitative variable in a population of organisms, the difference between the value of its partial linear regression on the gene dosage at a locus, holding all other genes constant, and the population mean value (Fisher 1930a, 1941, 1958a).

Average Excess For a quantitative variable in a population of organisms, the difference between the mean value of individuals carrying a gene and the population mean value (Fisher 1930a, 1941, 1958a).

Conditions for Existence (of Life) The requirements for life; the boundary conditions within which life can occur. These conditions are a property of individuals (genes, cells, organisms, lineages, populations, etc.).

Conditions for Existence (Principle of) *General form:* anything that exists must be satisfying its conditions for existence.

Cuvier's principle: any organism that exists must be satisfying its conditions for existence.

Darwin-Wallace principle: any trait (or type) that continues to exist in a population with variation in traits (or types) must be satisfying its conditions for existence.

Evolutionarily Closed Population A population of organisms (or other type of lineage) over a defined time period, in which all the ancestors of the final population were present in the initial population (no immigration).

Fitness *Realized fitness:* overall adaptedness, as measured by rate of increase.

Parametric fitness: the expected rate of increase in a specified environment.

Function (Natural) An effect of a part (organ, behavior, etc.) that is a condition for the continued existence of the system to which it belongs. A type of conditional function.

Function (Other Types) *Conditional function:* an effect of a part on the containing system that is necessary for occurrence of some state or event that we have defined as the "end."

Design function: an (expected) effect of a part that is the reason why the designer included it in the system.

Use function: an (expected) effect of a part that is the reason why the user is using it.

Purposive function: an effect of a part that aids in achieving an intrinsic goal of the purposive system containing it.

Lineage	An ancestor and all of its descendants.
Natural Selection	*Broad sense:* satisfaction (or not) of the conditions for existence (by an organism, lineage, character, population, etc.).

Medium sense: variation in rate of increase among classes of lineages within a population.

Narrow sense: variation in rate of increase among classes of lineages within a population, caused by the distinguishing features of the classes. Any features responsible are (by definition) better adapted than their alternatives.

Evolution by (narrow sense): variation in rate of increase among genetically distinct classes of lineages within a population, caused by the genetic differences between them. In quantitative genetic terms, it equals the additive genetic variance in rate of increase.

Hypothesis of (evolution by narrow sense): a trait will continue to exist in a population of organisms (or other lineages) if it is inherited, and causes those organisms characterized by it have a greater rate of increase than those that aren't. Conversely, a trait will disappear from a population if it causes those organisms characterized by it have a reduced rate of increase. Note that this involves inheritance as well as natural selection (see Gayon 1998).

Principle of (evolution by narrow sense): if there is variation in average ability to survive and reproduce among genetically distinct classes of lineages within a population, caused by the genetic differences between them, then there will be genetic improvement in average ability to survive and reproduce (relative to existing circumstances). In quantitative genetic terms, this is Fisher's (1930a) *fundamental theorem of natural selection*: at any moment, average adaptedness as measured by rate of increase (population growth rate) is tending to increase due to evolution by narrow sense natural selection (additive genetic variance in rate of increase), at a rate equal to additive genetic variance in rate of increase.

Rate of Increase	*Absolute:* the number of descendants of a lineage per time period considered.

Relative: the number of descendants of a lineage per time period considered, divided by the mean number of descendants across all lineages in the population.

Standardized Variance in Rate of Increase (= Opportunity for Selection)	The total among-lineage variance in relative rate of increase over a period.

REFERENCES

Aagaard, J. E., X. Yi, M. J. MacCoss, and W. J. Swanson

2006 Rapidly evolving zona pellucida domain proteins are a major component of the vitelline envelope of abalone eggs. *Proceedings of the National Academy of Sciences of the USA* 103:17302–17307.

Abel, O.

1911 Die Vorfahren der Vögel und ihre Lebensweise. *Verhandlungen, Zoologisch-Botanische Gesellschaft in Wien* 61:144–191.

Abzhanov, A., M. Protas, B. R. Grant, P. R. Grant, and C. J. Tabin

2004 *Bmp4* and morphological variation of beaks of Darwin's finches. *Science (Washington, DC)* 305:1462–1465.

Alberch, P.

1980 Ontogenesis and morphological diversification. *American Zoologist* 20:653–667.

1982 Developmental constraints in evolutionary processes. In J. T. Bonner (ed.), *Evolution and Development* (pp. 313–332). Berlin: Springer.

1989 The logic of monsters: Evidence for internal constraint in development and evolution. *Geobios (Paris)* 12:21–57.

Alberch, P., and E. A. Gale

1983 Size dependence during the development of the amphibian foot: Colchicine-induced digital loss and reduction. *Journal of Embryology and Experimental Morphology* 76:177–197.

1985 A developmental analysis of an evolutionary trend: Digital reduction in amphibians. *Evolution* 39:8–23.

Albon, S. D., T. H. Clutton-Brock, and F. E. Guinness

1987   Early development and population dynamics in red deer. II. Density-independent effects and cohort variation. *Journal of Animal Ecology* 56:69–81.

Albon, S. D., T. H. Clutton-Brock, and R. Langvatn

1991   Cohort variation in reproduction and survival: Implications for population demography. In R. D. Brown (ed.), *The Biology of Deer* (pp. 15–21). New York: Springer.

Alexander, H. G.

1956   *The Leibniz-Clarke Correspondence*. Manchester: Manchester University Press.

Alfonso-Sanchez, M. A., R. Calderón, and J. A. Peña

2004   Opportunity for natural selection in a Basque population and its secular trend: Evolutionary implications of epidemic mortality. *Human Biology* 76:361–381.

Amundson, R.

1994   Two concepts of constraint: Adaptationism and the challenge from developmental biology. *Philosophy of Science* 61:556–578.

1996   Historical development of the concept of adaptation. In M. R. Rose and G. V. Lauder (eds.), *Adaptation* (pp. 11–53). San Diego, CA: Academic Press.

2000   Against normal function. *Studies in History and Philosophy of Biological and Biomedical Sciences* 31:33–53.

2005   *The Changing Role of the Embryo in Evolutionary Thought: Roots of Evo-Devo*. Cambridge: Cambridge University Press.

Amundson, R., and G. Lauder

1994   Function without purpose: The uses of causal role function in evolutionary biology. *Biology and Philosophy* 9:443–469.

Andersen-Nissen, E., K. D. Smith, K. L. Strobe, S. L. Rassoulian Barrett, B. T. Cookson, S. M. Logan, and A. Aderem

2005   Evasion of toll-like receptor 5 by flagellated bacteria. *Proceedings of the National Academy of Sciences of the USA* 102:9247–9252.

Andersson, J. O.

2005   Lateral gene transfer in eukaryotes. *Cellular and Molecular Life Sciences* 62: 1182–1197.

Antonovics, J., and P. H. van Tienderen

1991   Ontoecogenophyloconstraints? The chaos of constraint terminology. *Trends in Ecology & Evolution* 6:166–168.

Appel, T. A.

1980   Henri de Blainville and the animal series: A nineteenth-century chain of being. *Journal of the History of Biology* 13:291–319.

1987   *The Cuvier-Geoffroy Debate: French Biology in the Decades before Darwin*. New York: Oxford University Press.

Ardrey, R.

1970   *The Social Contract*. London: Collins.

Aristotle

1952   *Physica*. In W. D. Ross (ed.), *The Works of Aristotle* (vol. 2, 184a–267b). Trans. R. P. Hardie and R. K. Gaye. Oxford: Oxford University Press.

Arnold, A. J., and K. Fristrup

1982   The theory of evolution by natural selection: A hierarchical expansion. *Paleobiology* 8:113–129.

Arnold, S. J.

2003   Performance surfaces and adaptive landscapes. *Integrative and Comparative Biology* 43:367–375.

Arnold, S. J., M. E. Pfrender, and A. G. Jones

2001   The adaptive landscape as a conceptual bridge between micro- and macroevolution. *Genetica (Dordrecht)* 112–113:9–32.

Arnold, S. J., and M. J. Wade

1984a   On the measurement of natural and sexual selection: Applications. *Evolution* 38:720–734.

1984b   On the measurement of natural and sexual selection: Theory. *Evolution* 38:709–719.

Arthur, W.

1997   *The Origin of Animal Body Plans: A Study in Evolutionary Developmental Biology.* Cambridge: Cambridge University Press.

2002   The emerging conceptual framework of evolutionary developmental biology. *Nature (London)* 415:757–764.

2004   *Biased Embryos and Evolution.* Cambridge: Cambridge University Press.

Asma, S. T.

1996   *Following Form and Function: A Philosophical Archaeology of Life Science.* Evanston, IL: Northwestern University Press.

Ayala, F. J.

1999   Adaptation and novelty: Teleological explanations in evolutionary biology. *History and Philosophy of the Life Sciences* 21:3–33.

Bach, T.

1994   Kielmeyer als "Vater der Naturphilosophie"? Anmerkungen zu seiner Rezeption im deutschen Idealismus. In K. T. Kanz (ed.), *Philosophie des Organischen in der Goethezeit: Studien zu Werk und Wirkung des Naturforschers Carl Friedrich Kielmeyer (1765–1844)* (pp. 232–251). Stuttgart: Steiner.

Bailey, C.

1964   *The Greek Atomists and Epicurus: A Study.* New York: Russell & Russell.

Bakewell, R.

1833   *An Introduction to Geology.* New Haven, CT: Howe.

Baldwin, J. M.

1896   A new factor in evolution. *American Naturalist* 30:441–451, 536–553.

1902   *Development and Evolution.* New York: Macmillan.

Balme, D. M.

1972   *Aristotle's* De Partibus Animalium I *and* De Generatione Animalium I *(with passages from II. 1–3).* Trans. D. M. Balme. Oxford: Clarendon.

Barlow, N., ed.

1958   *The Autobiography of Charles Darwin, 1809–1882.* New York: Norton.

Barrett, P. H., P. J. Gautrey, S. Herbert, D. Kohn, and S. Smith, eds.

1987   *Charles Darwin's Notebooks, 1836–1844.* Ithaca, NY: Cornell University Press.

Barrett, P. H., D. J. Weinshank, and T. T. Gottleber

1981   *A Concordance to Darwin's* Origin of Species, *First Edition.* Ithaca, NY: Cornell University Press.

Barrow, J. D., and F. J. Tipler

1986   *The Anthropic Cosmological Principle.* Oxford: Oxford University Press.

Barrowclough, G. F., and R. F. Rockwell

1993   Variance of lifetime reproductive success: Estimation based on demographic data. *American Naturalist* 141:281–295.

Baum, D. A., and A. Larson

1991   Adaptation reviewed: A phylogenetic methodology for studying character macroevolution. *Systematic Zoology* 40:1–18.

Bayle, P.

1734   *Mr. Bayle's Historical and Critical Dictionary.* 2nd ed. London: Knapton.

Bazin, E., S. Glémin, and N. Galtier

2006   Population size does not influence mitochondrial genetic diversity in animals. *Science (Washington, DC)* 312:570–572.

Beatty, J.

1992   Random drift. In E. F. Keller and E. A. Lloyd (eds.), *Keywords in Evolutionary Biology* (pp. 273–281). Cambridge, MA: Harvard University Press.

Beck, L. W.

1969   *Early German Philosophy: Kant and His Predecessors.* Cambridge, MA: Harvard University Press.

Beckner, M.

1967   Teleology. In P. Edwards (ed.), *The Encyclopedia of Philosophy* (vol. 8, pp. 88–91). New York: Macmillan.

Begon, M., C. A. Townsend, and J. L. Harper

2006   *Ecology: From Individuals to Ecosystems.* 4th ed. Malden, MA: Blackwell.

Bennett, J. H.

1983   *Natural Selection, Heredity, and Eugenics: Including Selected Correspondence of R. A. Fisher with Leonard Darwin and Others.* Oxford: Oxford University Press.

1999   Foreword. In R. A. Fisher, *The Genetical Theory of Natural Selection* (pp. vi–xxii). Oxford: Oxford University Press.

Bernard, C.

1927   *An Introduction to the Study of Experimental Medicine.* Trans. H. C. Greene. New York: Macmillan. Reprinted 1957, New York: Dover.

Bigelow, J., and R. Pargetter

1987   Functions. *Journal of Philosophy* 84:181–196.

Bird, A. P.

1980   DNA methylation and the frequency of CpG in animal DNA. *Nucleic Acids Research* 8:1499–1504.

Birdsell, J. A.

2002   Integrating genomics, bioinformatics, and classical genetics to study the effects of recombination on genome evolution. *Molecular Biology and Evolution* 19:1181–1197.

Bishop, J. A.

1972   An experimental study of the cline of industrial melanism in *Biston betularia* (L.) (Lepidoptera) between urban Liverpool and rural North Wales. *Journal of Animal Ecology* 41:209–243.

Blackledge, T. A., J. A. Coddington, and R. G. Gillespie

2003   Are three-dimensional spider webs defensive adaptations? *Ecology Letters* 6:13–18.

Bobzien, S.

1998   *Determinism and Freedom in Stoic Philosophy*. Oxford: Clarendon.

Bock, W. J.

1965   The role of adaptive mechanisms in the origin of higher levels of organization. *Systematic Zoology* 14:272–287.

Bock, W. J., and G. von Wahlert

1965   Adaptation and the form-function complex. *Evolution* 19:269–299.

Boorse, C.

1976   Wright on functions. *Philosophical Review* 85:70–86.

2002   A rebuttal on functions. In A. Ariew, R. Cummins, and M. Perlman (eds.), *Functions: New Essays in the Philosophy of Psychology and Biology* (pp. 63–112). Oxford: Oxford University Press.

Bostrom, N.

2002   *Anthropic Bias: Observation Selection Effects in Science and Philosophy*. New York: Routledge.

Bowler, P. J.

1983   *The Eclipse of Darwinism: Anti-Darwinian Evolution Theories in the Decades around 1900*. Baltimore: Johns Hopkins University Press.

2003   *Evolution: The History of an Idea*. 3rd ed. Berkeley: University of California Press.

Bowmaker, J. K., and D. M. Hunt

2006   Evolution of vertebrate visual pigments. *Current Biology* 16:R484–R489.

Boylan, M.

1986   Monadic and systemic teleology. In N. Rescher (ed.), *Current Issues in Teleology* (pp. 15–25). Lanham, MD: University Press of America.

Boyle, R.

1772   *The Works of the Honourable Robert Boyle*. London: Johnston.

Brakefield, P. M.

2006   Evo-devo and constraints on selection. *Trends in Ecology & Evolution* 21: 362–368.

Brandon, R. N., and R. M. Burian, eds.

1984   *Genes, Organisms, Populations: Controversies over the Units of Selection*. Cambridge, MA: MIT Press.

Brodie, E. D., III

2005   Caution: Niche construction ahead. *Evolution* 59:249–251.

2007   Population size is not genetic quality. *Animal Conservation* 10:288–290.

Brodie, E. D., III, A. J. Moore, and F. J. Janzen

1995   Visualizing and quantifying natural selection. *Trends in Ecology & Evolution* 10:313–318.

Brooke, J. H.

1989   Scientific thought and its meaning for religion: The impact of French science on British natural theology, 1827–1859. *Revue de Synthèse* 4:33–59.

Brown, B., and J. P. Alton

2007   *Flock of Dodos: Behind Modern Creationism, Intelligent Design and the Easter Bunny*. New York: Cambridge House Press.

Brown, D.

1988   Components of lifetime reproductive success. In T. H. Clutton-Brock (ed.), *Reproductive Success: Studies of Individual Variation in Contrasting Breeding Systems* (pp. 439–453). Chicago: University of Chicago Press.

Browne, J.

1995   *Charles Darwin: Voyaging, a Biography.* Princeton, NJ: Princeton University Press.

2002   *Charles Darwin: The Power of Place.* New York: Knopf.

Brunet, F. G., H. R. Crollius, M. Paris, J.-M. Aury, P. Gibert, O. Jaillon, V. Laudet, and M. Robinson-Rechavi

2006   Gene loss and evolutionary rates following whole-genome duplication in teleost fishes. *Molecular Biology and Evolution* 23:1808–1816.

Buckland, W.

1823   *Reliquiæ Diluvianæ: Or, Observations on the Organic Remains Contained in Caves, Fissures, and Diluvial Gravel and on Other Geological Phenomena Attesting the Action of a Universal Deluge.* London: Murray.

1841   *Geology and Mineralogy considered with reference to Natural Theology.* New ed. Philadelphia: Lea & Blanchard.

Buffon, G. L. L., Count de

1791   *Natural History, General and Particular.* Trans. W. Smellie. London: Strahan.

Bulmer, M.

2005   The theory of natural selection of Alfred Russel Wallace FRS. *Notes and Records of the Royal Society* 59:125–136.

Burt, A.

1995   The evolution of fitness. *Evolution* 49:1–8.

Burt, A., and R. Trivers

2006   *Genes in Conflict: The Biology of Selfish Genetic Elements.* Cambridge, MA: Harvard University Press.

Burtt, E.

1954   *The Metaphysical Foundations of Modern Physical Science.* Garden City, NY: Doubleday.

Caneva, K. L.

1990   Teleology with regrets. *Annals of Science* 47:291–300.

Canfield, J.

1964   Teleological explanation in biology. *British Journal for the Philosophy of Science* 14:285–295.

1965   Teleological explanation in biology: A reply. *British Journal for the Philosophy of Science* 15:327–331.

Caponi, G.

2004   Georges Cuvier ¿Un nombre olvidado en la historia de la fisiología? *Asclepio* 56:169–207.

Carothers, J. H.

1986   An experimental confirmation of morphological adaptation: Toe fringes in the sand-dwelling lizard *Uma scoparia. Evolution* 40:871–874.

Carpenter, W. B.

1854   *Principles of Comparative Physiology.* 4th ed. London: Churchill.

Carroll, S. B.

2005   *Endless Forms Most Beautiful: The New Science of Evo-Devo and the Making of the Animal Kingdom.* New York: Norton.

Chambers, R.

1844   *Vestiges of the Natural History of Creation.* London: Churchill.

Charlesworth, B., M. T. Morgan, and D. Charlesworth

1993   The effect of deleterious mutations on neutral molecular variation. *Genetics* 134:1289–1303.

Charlesworth, D., B. Charlesworth, and M. T. Morgan

1995   The pattern of neutral molecular variation under the background selection model. *Genetics* 141:1619–1632.

Chase, J. M., and M. A. Leibold

2003   *Ecological Niches: Linking Classical and Contemporary Approaches.* Chicago: University of Chicago Press.

Cheung, T.

2000   *Die Organisation des Lebendigen: Die Entstehung des biologischen Organismusbegriffs bei Cuvier, Leibniz und Kant.* Frankfurt: Campus.

Chevin, L., and F. Hospital

2006   The hitchhiking effect of an autosomal meiotic drive gene. *Genetics* 173:1829–1832.

Chiappe, L. M.

2007   *Glorified Dinosaurs: The Origin and Early Evolution of Birds.* Hoboken, NJ: Wiley.

Cicero

1961   *De Natura Deorum, Academica.* Trans. H. Rackham. Cambridge, MA: Harvard University Press.

Clarke, C. A., and P. M. Sheppard

1963   Frequencies of the melanic forms of the moth *Biston betularia* (L.) on Deeside and in adjacent areas. *Nature (London)* 198:1279–1282.

1964   Genetic control of the melanic form *insularia* of the moth *Biston betularia* (L.). *Nature (London)* 202:215–216.

1966   A local survey of the distribution of the industrial melanic forms in the moth *Biston betularia* and estimates of the selective values of these in an industrial environment. *Proceedings of the Royal Society of London B, Biological Sciences* 165:424–439.

Clarke, J. A., and K. Middleton

2006   Bird evolution. *Current Biology* 16:R350–R354.

Clutton-Brock, T. H.

1988a   Introduction. In T. H. Clutton-Brock (ed.), *Reproductive Success: Studies of Individual Variation in Contrasting Breeding Systems* (pp. 1–9). Chicago: University of Chicago Press.

1988b   Reproductive success. In T. H. Clutton-Brock (ed.), *Reproductive Success: Studies of Individual Variation in Contrasting Breeding Systems* (pp. 472–485). Chicago: University of Chicago Press.

1988c   ed. *Reproductive Success: Studies of Individual Variation in Contrasting Breeding Systems.* Chicago: University of Chicago Press.

Coddington, J. A.

1988   Cladistic tests of adaptational hypotheses. *Cladistics* 4:3–22.

Coleman, W.

1964   *Georges Cuvier, Zoologist: A Study in the History of Evolution Theory.* Cambridge, MA: Harvard University Press.

Coltman, D. W., J. A. Smith, D. R. Bancroft, J. Pilkington, A. D. C. MacColl, T. H. Clutton-Brock, and J. M. Pemberton

1999   Density-dependent variation in lifetime reproductive success and natural and sexual selection in Soay rams. *American Naturalist* 154:730–746.

Cook, L. M.

2000   Changing views on melanic moths. *Biological Journal of the Linnean Society* 69:431–441.

2003   The rise and fall of the *carbonaria* form of the peppered moth. *Quarterly Review of Biology* 78:399–417.

Cook, L. M., R. R. Askew, and J. A. Bishop

1970   Increasing frequency of the typical form of the peppered moth in Manchester. *Nature (London)* 227:1155.

Cooper, J. M.

1987   Hypothetical necessity and natural teleology. In A. Gotthelf and J. G. Lennox (eds.), *Philosophical Issues in Aristotle's Biology* (pp. 243–274). Cambridge: Cambridge University Press.

Cornford, F. M.

1937   *Plato's Cosmology: The Timaeus of Plato.* Indianapolis, IN: Bobbs-Merrill.

Corsi, P.

1988   *The Age of Lamarck: Evolutionary Theories in France 1790–1830.* Berkeley: University of California Press.

Coulson, T., T. G. Benton, P. Lundberg, S. R. X. Dall, B. E. Kendall, and J.-M. Gaillard

2006   Estimating individual contributions to population growth: Evolutionary fitness in ecological time. *Proceedings of the Royal Society of London B, Biological Sciences* 273:547–555.

Coyne, J. A.

2002   Evolution under pressure: A look at the controversy about industrial melanism in the peppered moth. *Nature (London)* 418:19–20.

Crabbe, G.

1840   *An Outline of a System of Natural Theology.* London: Pickering.

Crain, C. M., and M. D. Bertness

2006   Ecosystem engineering across environmental gradients: Implications for conservation and management. *BioScience* 56:211–218.

Creed, E. R., and D. R. Lees

1980   Pre-adult viability differences of melanic *Biston betularia* (L.) (Lepidoptera). *Biological Journal of the Linnean Society* 13:251–262.

Crocker, L. G.

1954   *The Embattled Philosopher: A Biography of Denis Diderot.* Lansing: Michigan State College Press.

1968   Diderot and eighteenth century French transformism. In B. Glass, O. Temkin, and W. L. Straus (eds.), *Forerunners of Darwin 1745–1859* (pp. 114–143). Baltimore: Johns Hopkins Press.

Croizat, L.
1962   Space, Time, Form: The Biological Synthesis. Caracas: Croizat.
Crombie, A.
1829   Natural Theology; or Essays on the Existence of Deity and of Providence, on the Immateriality of the Soul, and a Future State. London: Hunter.
Crow, J. F.
1958   Some possibilities for measuring selection intensities in man. Human Biology 30:1–13.
1962   Population genetics: Selection. In W. J. Burdette (ed.), Methodology in Human Genetics (pp. 53–75). San Francisco: Holden-Day.
1989a   Update to "Some Possibilities for Measuring Selection Intensities in Man." Human Biology 61:776–780.
1989b   Fitness variation in natural populations. In W. G. Hill, T. F. C. Mackay, and A. Robertson (eds.), Evolution and Animal Breeding: Reviews on Molecular and Quantitative Approaches in Honour of Alan Robertson (pp. 91–97). Wallingford, UK: CAB International.
Crow, J. F., and N. E. Morton
1955   Measurement of gene frequency drift in small populations. Evolution 9:202–214.
Crowther, J. G.
1960   Founders of British Science: John Wilkins, Robert Boyle, John Ray, Christopher Wren, Robert Hooke, Isaac Newton. London: Cresset.
Cummins, R.
1975   Functional analysis. Journal of Philosophy 72:741–765.
2002   Neo-teleology. In A. Ariew, R. Cummins and M. Perlman (eds.), Functions: New Essays in the Philosophy of Psychology and Biology (pp. 157–172). Oxford: Oxford University Press.
Cutter, A. D., J. D. Wasmuth, and M. L. Blaxter
2006   The evolution of biased codon and amino acid usage in nematode genomes. Molecular Biology and Evolution 23:2303–2315.
Cuvier, G.
1796a   Mémoire sur les espèces d'éléphans tant vivantes que fossiles, lu à la séance publique de l'Institut National le 15 germinal, an IV. Magasin encyclopédique, 2e année, 3:440–445. Translated in Rudwick 1997.
1796b   Notice sur le squelette d'une très-grande espèce de quadrupède inconnue jusqu'à présent, trouvé au Paraguay, et déposé au cabinet d'Histoire naturelle de Madrid, rédigée par G. Cuvier. Magasin encyclopédique, 2e année, 1:303–310. Translated in Rudwick 1997.
1798   Tableau élémentaire de l'histoire naturelle des animaux. Paris: Baudouin.
1800   Extrait d'un ouvrage sur les espèces de quadrupèdes dont on a trouvé les ossemens dans l'intérieur de la terre, addressé aux savants et aux amateurs des sciences: Imprimé par ordre de la classe des sciences mathématiques et physiques de l'Institut National, du 26 brumaire an 9. Journal de physique, de l'histoire naturelle, et des arts 52:253–267. Translated in Rudwick 1997.
1800–
1805   Leçons d'anatomie comparée. Paris: Baudouin.

1809   *Vorlesungen über vergleichende Anatomie.* Leipzig: Kummer.

1811   Aristote. In J. F. Michaud and L. G. Michaud (eds.), *Biographie universelle, ancienne et moderne, ou, Histoire par ordre alphabétique de la vie publique et privée de tous les hommes qui se sont fait remarquer par leurs écrits, leurs actions, leurs talents, leurs vertus ou leurs crimes: Ouvrage entièrement neuf,* vol. 2, pp. 456–464. Paris: Michaud.

1812   *Recherches sur les ossemens fossiles de quadrupèdes, où l'on rétablit les caractères de plusieurs espèces d'animaux que les révolutions du globe paroissent avoir détruites.* Paris: Deterville.

1817   *Le règne animal distribué d'après son organisation, pour servir de base à l'histoire naturelle des animaux et d'introduction à l'anatomie comparée.* Paris: Deterville.

1825   Nature. In F. Cuvier (ed.), *Dictionnaire des sciences naturelles* (vol. 34, pp. 261–268). Paris: Levrault.

1831   *A Discourse on the Revolutions of the Surface of the Globe, and the Changes Thereby Produced in the Animal Kingdom.* Philadelphia: Carey & Lea.

1841–
1845   *Histoire des sciences naturelles, depuis leur origine jusqu'à nos jours, chez tous les peuples connus, professée au Collége de France, par Georges Cuvier, complétée, rédigés, annotée et publiée par M. Magdeleine de Saint-Agy.* Paris: Fortin, Masson & Cie.

1861   *Recueil des éloges historiques lus dans les séances publiques de l'Institut de France.* Paris: Didot.

1997
(1798)   Extrait d'un Mémoire sur un animal dont on trouve les ossements dans la pierre à plâtre des environs de Paris, & qui paraît ne plus exister vivant aujourd'hui. In M. J. S. Rudwick (ed. and trans.), *Georges Cuvier, Fossil Bones, and Geological Catastrophes* (pp. 285–290; English trans., pp. 35–41). Chicago: University of Chicago Press. (First published here, but given in 1798.)

Cuvier, G., and A. Valenciennes
1828–
1849   *Histoire naturelle des poissons.* Paris: Levrault; Brussels: Librarie Parisienne.

Danforth, C. H.
1923   The frequency of mutation and the incidence of hereditary traits in man. *Eugenics, Genetics and the Family, Scientific Papers of the 2nd International Congress on Eugenics, N.Y., 1921,* 1:120–128.

Darwin, C.
1838   Notebook D. In P. H. Barrett, P. J. Gautrey, S. Herbert, D. Kohn, and S. Smith (eds.), 1987, *Charles Darwin's Notebooks, 1836–1844* (pp. 329–393). Ithaca, NY: Cornell University Press.

1839   *Journal of Researches into the Geology and Natural History of the Various Countries Visited by H. M. S. Beagle, under the Command of Captain Fitzroy, R.N. from 1832 to 1836.* London: Colburn.

1859   *On the Origin of Species by Means of Natural Selection, or the Preservation of Favoured Races in the Struggle for Life.* London: Murray.

1881   *The Formation of Vegetable Mould, through the Action of Worms, with Observations on Their Habits.* London: Murray.

1887   *The Life and Letters of Charles Darwin.* Ed. F. Darwin. New York: Appleton.

Davies, P. S.

2001   *Norms of Nature: Naturalism and the Nature of Functions.* Cambridge, MA: MIT Press.

Dawkins, R.

1976   *The Selfish Gene.* New York: Oxford University Press.

1986   *The Blind Watchmaker.* New York: Norton.

1989   The evolution of evolvability. In C. Langton (ed.), *Artificial Life* (pp. 201–220). Redwood City, CA: Addison-Wesley.

2004   Extended phenotype—but not *too* extended: A reply to Laland, Turner, and Jablonka. *Biology & Philosophy* 19:377–396.

2006   *The Selfish Gene: 30th Anniversary Edition.* Oxford: Oxford University Press.

de Beer, G.

1954   *Archaeopteryx Lithographica: A Study Based on the British Museum Specimen.* London: British Museum.

de Blainville, H. M. D.

1890   *Cuvier et Geoffroy Saint-Hilaire: Biographies scientifiques.* Paris: Baillière.

De la Beche, H. T.

1833   *A Geological Manual.* 3rd ed. London: Knight.

de Maillet, B.

1968   *Telliamed: Or Conversations between an Indian Philosopher and a French Missionary on the Diminution of the Sea.* Trans. A. V. Carozzi. Urbana: University of Illinois Press.

Derham, W.

1713   *Physico-Theology.* London: Innys.

Descartes, R.

1983   *Principles of Philosophy.* Trans. V. R. Miller and R. P. Miller. Dordrecht: Reidel.

1985   *The Philosophical Writings of Descartes, Vol. II.* Trans. J. Cottingham, R. Stoothoff, and D. Murdoch. Cambridge: Cambridge University Press.

Desmond, A.

1989   *The Politics of Evolution: Morphology, Medicine and Reform in Radical London.* Chicago: University of Chicago Press.

Desmond, A., and J. Moore

1991   *Darwin.* New York: Warner Books.

Deveny, A. J., and L. R. Fox

2006   Indirect interactions between browsers and seed predators affect the seed bank dynamics of a chaparral shrub. *Oecologia (Berlin)* 150:69–77.

d'Holbach, B.

2004   *The System of Nature; Or, the Laws of the Moral and Physical World.* Trans. S. Wilkinson. Whitefish, MT: Kessinger.

Dial, K. P.

2003   Wing-assisted incline running and the evolution of flight. *Science (Washington, DC)* 299:402–404.

Dickerson, R. E.

1971   The structure of cytochrome *c* and the rates of molecular evolution. *Journal of Molecular Evolution* 1:26–45.

Dobzhansky, T.

1955   A review of some fundamental concepts and problems of population genetics. *Cold Spring Harbor Symposia on Quantitative Biology* 20:1–15.

Doolittle, W. F., and C. Sapienza

1980   Selfish genes, the phenotype paradigm and genome evolution. *Nature (London)* 284:601–603.

Dover, G.

1982   Molecular drive: a cohesive mode of species evolution. *Nature (London)* 299:111–117.

2000   *Dear Mr. Darwin: Letters on the Evolution of Life and Human Nature*. Berkeley: University of California Press.

2002   Molecular drive. *Trends in Genetics* 18:587–589.

Downhower, J. F., L. S. Blumer, and L. Brown

1987   Opportunity for selection: An appropriate measure for evaluating variation in the potential for selection? *Evolution* 41:1395–1400.

Dykhuizen, D., and D. Hartl

1981   Evolution of competitive ability in *Escherichia coli*. *Evolution* 35:581–594.

Edwards, A. W. F.

1990   Fisher, $\overline{W}$, and the fundamental theorem. *Theoretical Population Biology* 38:276–284.

1994.   The fundamental theorem of natural selection. *Biological Review* 69:443–474.

2000.   Sewall Wright's equation $\Delta q = (q(1 - q)\partial w/\partial q)/2w$. *Theoretical Population Biology* 57:67–70.

2002   The fundamental theorem of natural selection. *Theoretical Population Biology* 61:335–337.

Eibl-Eiblesfeldt, I.

1971   *Love and Hate*. London: Methuen.

Eiseley, L.

1958   *Darwin's Century: Evolution and the Men Who Discovered It*. Garden City, NY: Doubleday.

Eldon, B., and J. Wakeley

2006   Coalescent processes when the distribution of offspring number among individuals is highly skewed. *Genetics* 172:2621–2633.

Elena, S. F., and R. E. Lenski

2003   Evolution experiments with microorganisms: The dynamics and genetic bases of adaptation. *Nature Reviews Genetics* 4:457–469.

Elton, C.

1927   *Animal Ecology*. London: Methuen.

Endler, J. A.

1986   *Natural Selection in the Wild*. Princeton, NJ: Princeton University Press.

Erséus, C., and M. Källersjö

2004   18S rDNA phylogeny of the Clitellata (Annelida). *Zoologica Scripta* 33:187–196.

Erwin, D. H.

2005   Seeds of diversity. *Science (Washington, DC)* 308:1752–1753.

Ewens, W. J.

1989   An interpretation and proof of the fundamental theorem of natural selection. *Theoretical Population Biology* 36:167–180.

Eyre-Walker, A.
2006   The genomic rate of adaptive evolution. *Trends in Ecology & Evolution* 21:569–575.
Falconer, D. S.
1964   *Introduction to Quantitative Genetics.* New York: Ronald.
Feduccia, A.
1999   *The Origin and Evolution of Birds.* 2nd ed. New Haven, CT: Yale University Press.
Feldman, M. W., and R. C. Lewontin
1975   The heritability hang-up. *Science (Washington, DC)* 190:1163–1168.
Feller, W.
1967   On fitness and the cost of natural selection. *Genetical Research* 9:1–15.
Fisher, D. C.
1985   Evolutionary morphology: Beyond the analogous, the anecdotal, and the ad hoc. *Paleobiology* 11:120–138.
Fisher, R. A.
1918   The correlation between relatives on the supposition of Mendelian inheritance. *Transactions of the Royal Society of Edinburgh* 52:399–433.
1921   Review of *The Relative Value of the Processes Causing Evolution* (A. L. and A. C. Hagedoorn). *Eugenics Review* 13:467–470.
1922a   On the dominance ratio. *Proceedings of the Royal Society of Edinburgh* 42:321–341.
1922b   On the mathematical foundations of theoretical statistics. *Philosophical Transactions of the Royal Society of London A, Mathematics, Physical and Engineering Sciences* 222:309–368.
1925   Theory of statistical estimation. *Proceedings of the Cambridge Philosophical Society* 22:700–725.
1930a   *The Genetical Theory of Natural Selection.* Oxford: Clarendon. Reprinted 1999 in a variorum edition, ed. J. H. Bennett, Oxford: Oxford University Press.
1930b   R. A. Fisher letter to E. B. Wilson, 2 June. Fisher Papers, Barr Smith Library, University of Adelaide, MSS 0013/Series 1. (Electronic copy available from R. A. Fisher Digital Archive, The University of Adelaide Digital Library, http://digital.library .adelaide.edu.au/coll/special/fisher/corres/wilsoneb/WilsonEB300602a.html.)
1939   Stage of development as a factor influencing the variance in the number of offspring, frequency of mutants and related quantities. *Annals of Eugenics* 9:406–408.
1941   Average excess and average effect of a gene substitution. *Annals of Eugenics* 11: 53–63.
1958a   *The Genetical Theory of Natural Selection.* 2nd rev. ed. New York: Dover.
1958b   The nature of probability. *Centennial Review* 2:261–274.
1959   Natural selection from the genetical standpoint. *Australian Journal of Science* 22:16–17. Reprinted in 1974 with omitted text restored in J. H. Bennett (ed.), *The Collected Papers of R. A. Fisher* (vol. 5, pp. 444–449). Adelaide: University of Adelaide.
Fisher, R. A., and E. B. Ford
1947   The spread of a gene in natural conditions in a colony of the moth *Panaxia dominula* L. *Heredity* 623:143–174.
Flourens, P.
1856   *Recueil des éloges historiques lus dans les séances publiques de l'Academie des sciences.* Paris: Garnier.

Ford, E. B.
1937   Problems of heredity in the Lepidoptera. *Biological Review* 12:461–503.
1940   Genetic research in the Lepidoptera. *Annals of Eugenics* 10:227–252.

Forrest, B., and P. E. Gross
2003   *Creationism's Trojan Horse: The Wedge of Intelligent Design.* New York: Oxford University Press.

Foucault, M.
1973   *The Order of Things: An Archaeology of the Human Sciences.* New York: Vintage Books.

Frank, S. A.
1998   *Foundations of Social Evolution.* Princeton, NJ: Princeton University Press.

Frank, S. A., and M. Slatkin
1992   Fisher's fundamental theorem of natural selection. *Trends in Ecology & Evolution* 7:92–95.

Frankel, O. H., and M. E. Soulé
1981   *Conservation and Evolution.* Cambridge: Cambridge University Press.

Franklin, I. R.
1980   Evolutionary change in small populations. In M. E. Soulé and B. A. Wilcox (eds.), *Conservation Biology: An Evolutionary-Ecological Perspective* (pp. 135–149). Sunderland, MA: Sinauer.

Freeman, S., and J. C. Herron
2007   *Evolutionary Analysis.* 4th ed. Upper Saddle River, NJ: Pearson/Prentice Hall.

Frumhoff, P. C., and H. K. Reeve
1994   Using phylogenies to test hypotheses of adaptation: A critique of some current proposals. *Evolution* 48:172–180.

Futuyma, D. J.
2005   *Evolution.* Sunderland, MA: Sinauer.

Galileo, G.
1954   *Dialogues Concerning Two New Sciences.* Trans. H. Crew and A. de Salvio. New York: Dover.

Gardner, M.
1986   WAP, SAP, PAP, and FAP. *New York Review of Books* 33(8):22–25.

Garner, J. P., G. K. Taylor, and A. L. R. Thomas
1999   On the origins of birds: The sequence of character acquisition in the evolution of avian flight. *Proceedings of the Royal Society of London B, Biological Sciences* 266:1259–1266.

Gatesy, S. M., and D. B. Baier
2005   The origin of the avian flight stroke: A kinematic and kinetic perspective. *Paleobiology* 31:382–399.

Gause, G. F.
1964   *The Struggle for Existence.* New York: Hafner.

Gavrilets, S.
2003   Models of speciation: What have we learned in 40 years? *Evolution* 57:2197–2215.
2004   *Fitness Landscapes and the Origin of Species.* Princeton, NJ: Princeton University Press.

Gay, P.

1967   *The Enlightenment: An Interpretation. The Rise of Modern Paganism*. New York: Knopf.

Gayon, J.

1998   *Darwinism's Struggle for Survival: Heredity and the Hypothesis of Natural Selection.* Cambridge: Cambridge University Press.

Ge, F., L.-S. Wang, and J. Kim

2005   The cobweb of life revealed by genome-scale estimates of horizontal gene transfer. *PLoS Biology* 3:1709–1718.

Gibbs, H. L., and P. R. Grant

1987   Oscillating selection on Darwin's finches. *Nature (London)* 327:511–513.

Gillespie, J. H.

1977   Natural selection for variances in offspring numbers: A new evolutionary principle. *American Naturalist* 111:1010–1014.

1991a   The burden of genetic load. *Science (Washington, DC)* 254:1049.

1991b   *The Causes of Molecular Evolution.* New York: Oxford University Press.

2000   Genetic drift in an infinite population: The pseudohitchhiking model. *Genetics* 155:909–919.

2001   Is the population size of a species relevant to its evolution? *Evolution* 55:2161–2169.

2004   *Population Genetics: A Concise Guide.* 2nd ed. Baltimore: Johns Hopkins University Press.

Gillispie, C. C.

1951   *Genesis and Geology.* Cambridge, MA: Harvard University Press.

2004   *Science and Polity in France: The Revolutionary and Napoleonic Years.* Princeton, NJ: Princeton University Press.

Glass, B.

1968   Maupertuis, pioneer of genetics and evolution. In B. Glass, O. Temkin, and W. L. Straus (eds.), *Forerunners of Darwin 1745–1859* (pp. 51–83). Baltimore: Johns Hopkins Press.

Godfrey-Smith, P.

1994   A modern history theory of functions. *Noûs* 28:344–362.

2007   Conditions for evolution by natural selection. *Journal of Philosophy* 104:489–516.

Goodman, L. A.

1960   On the exact variance of products. *Journal of the American Statistical Association* 55:708–713.

1962   The variance of the product of K random variables. *Journal of the American Statistical Association* 57:54–60.

Gotthelf, A.

1987   First principles in Aristotle's *Parts of Animals.* In A. Gotthelf and J. G. Lennox (eds.), *Philosophical Issues in Aristotle's Biology* (pp. 167–198). Cambridge: Cambridge University Press.

1999   Darwin on Aristotle. *Journal of the History of Biology* 32:3–30.

Gould, S. J.

1977   *Ontogeny and Phylogeny.* Cambridge, MA: Harvard University Press.

1989   A developmental constraint in *Cerion*, with comments on the definition and interpretation of constraint in evolution. *Evolution* 43:516–539.

2002.   *The Structure of Evolutionary Theory*. Cambridge, MA: Harvard University Press.

Gould, S. J., and R. C. Lewontin

1979   The spandrels of San Marco and the Panglossian paradigm: A critique of the adaptationist programme. *Proceedings of the Royal Society of London B, Biological Sciences* 205:581–598.

Gould, S. J., and E. S. Vrba

1982   Exaptation: A missing term in the science of form. *Paleobiology* 8:4–15.

Grafen, A.

1988   On the uses of data on lifetime reproductive success. In T. H. Clutton-Brock (ed.), *Reproductive Success: Studies of Individual Variation in Contrasting Breeding Systems* (pp. 454–471). Chicago: University of Chicago Press.

Grafen, A., and M. Ridley

2006   *Richard Dawkins: How a Scientist Changed the Way We Think*. New York: Oxford University Press.

Grant, B. S., A. D. Cook, C. A. Clarke, and D. F. Owen

1998   Geographic and temporal variation in the incidence of melanism in peppered moth populations in America and Britain. *Journal of Heredity* 89:465–471.

Grant, B. S., D. F. Owen, and C. A. Clarke

1996   Parallel rise and fall of melanic peppered moths in America and Britain. *Journal of Heredity* 87:351–357.

Grant, P. R.

1999   *Ecology and Evolution of Darwin's Finches*. Princeton, NJ: Princeton University Press.

Grant, P. R., and B. R. Grant

2000   Non-random fitness variation in two populations of Darwin's finches. *Proceedings of the Royal Society of London B, Biological Sciences* 267:131–138.

2002   Unpredictable evolution in a 30-year study of Darwin's finches. *Science (Washington, DC)* 296:707–711.

Gray, A.

1963   *Darwiniana*. Ed. A. Hunter Dupree. Cambridge, MA: Harvard University Press.

Green, J. J., and D. M. Newberry

2002   Reproductive investment and seedling survival of the mast-fruiting rain forest tree, *Microberlinia bisulcata* A. chev. *Plant Ecology* 162:169–183.

Grene, M., and D. Depew

2004   *The Philosophy of Biology: An Episodic History*. Cambridge: Cambridge University Press.

Griffiths, A. J. F., S. R. Wexler, R. C. Lewontin, and S. B. Carroll

2008   *An Introduction to Genetic Analysis*. 9th ed. New York: Freeman.

Griffiths, P. E.

1993   Functional analysis and proper functions. *British Journal for the Philosophy of Science* 44:409–422.

2005   Review of "Niche Construction." *Biology & Philosophy* 20:11–20.

Grinnell, J.

1917   The niche-relationships of the California thrasher. *Auk* 34:426–433.

Hagedoorn, A. L., and A. C. Hagedoorn

1921   *The Relative Value of the Processes Causing Evolution*. The Hague: Martinus Nijhoff.

Haldane, J. B. S.

1924   A mathematical theory of natural and artificial selection. Part I. *Transactions of the Cambridge Philosophical Society* 23:19–41.

1937   The effect of variation on fitness. *American Naturalist* 71:337–349.

1954   The statics of evolution. In J. Huxley, A. C. Hardy, and E. B. Ford (eds.), *Evolution as a Process* (pp. 109–121). London: Allen & Unwin.

1956a   The estimation of viabilities. *Journal of Genetics* 54:294–6.

1956b   The relation between density regulation and natural selection. *Proceedings of the Royal Society of London B, Biological Sciences* 145:303–306.

1956c   The theory of selection for melanism in Lepidoptera. *Proceedings of the Royal Society of London B, Biological Sciences* 145:303–306.

1957   The cost of natural selection. *Journal of Genetics* 55:511–524.

1961   More precise expressions for the cost of natural selection. *Journal of Genetics* 57:351–360.

Hall, B. G.

1982   Evolution on a Petri dish: The evolved β-galactosidase system as a model for studying acquisitive evolution in the laboratory. *Evolutionary Biology (New York)* 15:85–150.

Hankinson, R. J.

1998   *Cause and Explanation in Ancient Greek Thought.* Oxford: Clarendon.

Harris, M. P., S. Williamson, J. F. Fallon, H. Meinhardt, and R. O. Prum

2005   Molecular evidence for an activator-inhibitor mechanism in development of embryonic feather branching. *Proceedings of the National Academy of Sciences of the USA* 102:11734–11739.

Hartl, D. L., and D. E. Dykhuizen

1981   Potential for selection among nearly neutral allozymes of 6-phosphogluconate dehydrogenase in *Escherichia coli*. *Proceedings of the National Academy of Sciences of the USA* 78:6344–6348.

Hartl, D. L., D. E. Dykhuizen, and A. M. Dean

1985   Limits of adaptation: The evolution of selective neutrality. *Genetics* 111:655–674.

Harvey, P. H., and M. D. Pagel

1991   *The Comparative Method in Evolutionary Biology.* Oxford: Oxford University Press.

Hawking, S.

1988   *A Brief History of Time.* New York: Bantam Books.

Hedgecock, D.

1994   Does variance in reproductive success limit effective population sizes of marine organisms? In A. R. Beaumont (ed.), *Genetics and Evolution of Aquatic Organisms* (pp. 122–134). London: Chapman & Hall.

Hedrick, P.

2005   Large variance in reproductive success and the N_e/N ratio. *Evolution* 59:1596–1599.

Heffner, R. S., G. Koay, and H. E. Heffner

2006   Hearing in large *(Eidolon helvum)* and small *(Cynopterus brachyotis)* non-echolocating fruit bats. *Hearing Research* 221:17–25.

Heilmann, G.

1927   *The Origin of Birds.* New York: Appleton.

Hendrikse, J. L., T. E. Parsons, and B. Hallgrimsson
   2007   Evolvability as the proper focus of evolutionary developmental biology. *Evolution and Development* 9:393–401.
Herbert, S., ed.
   1980   *The Red Notebook of Charles Darwin.* Ithaca, NY: Cornell University Press.
Herre, E., C. A. Machado, and S. A. West
   2001   Selective regime and fig wasp sex ratios: Toward sorting rigor from pseudo-rigor in tests of adaptation. In S. H. Orzack and E. Sober (eds.), *Adaptationism and Optimality* (pp. 191–218). Cambridge: Cambridge University Press.
Hersch, E. I., and P. C. Phillips
   2004   Power and potential bias in field studies of natural selection. *Evolution* 58:479–485.
Heywood, J. S.
   2005   An exact form of the breeder's equation for the evolution of a quantitative trait under natural selection. *Evolution* 59:2287–2298.
Hicks, L. E.
   1883   *A Critique of Design-Arguments: A Historical Review and Free Examination of the Methods of Reasoning in Natural Theology.* New York: Scribner's.
Hill, R. W., G. A. Wyse, and M. Anderson
   2004   *Animal Physiology.* Sunderland, MA: Sinauer.
Hobbes, T.
   1973   *Leviathan.* New York: Dutton.
Hochachka, P. W., and G. N. Somero
   2002   *Biochemical Adaptation: Mechanism and Process in Physiological Evolution.* New York: Oxford University Press.
Hodge, M. J. S.
   1982   Darwin and the laws of the animate part of the terrestrial system (1835–1837): On the Lyellian origins of his zoonomical explanatory program. *Studies in the History of Biology* 6:1–106.
   1985   Darwin as a lifelong generation theorist. In D. Kohn (ed.), *The Darwinian Heritage* (pp. 207–243). Princeton, NJ: Princeton University Press.
Hoekstra, H. E., and J. A. Coyne
   2007   The locus of evolution: Evo devo and the genetics of adaptation. *Evolution* 61:995–1016.
Honkoop, P. J. C., and J. van der Meer
   1997   Reproductive output of *Macoma balthica* populations in relation to winter-temperature and intertidal-height mediated changes of body mass. *Marine Ecology Progress Series* 164:229–234.
Hooper, J.
   2002   *Of Moths and Men: The Untold Story of Science and the Peppered Moth.* New York: Norton.
Houle, D.
   1992   Comparing evolvability and variability of quantitative traits. *Genetics* 130:195–205.
Howard, R. D.
   1979   Estimating reproductive success in natural populations. *American Naturalist* 114:221–231.

Hubbell, S. P.

2001 *The Unified Neutral Theory of Biodiversity and Biogeography*. Princeton, NJ: Princeton University Press.

Huerta-Sanchez, E., R. Durrett, and C. D. Bustamante

2008 Population genetics of polymorphism and divergence under fluctuating selection. *Genetics* 178:325–337.

Hull, D. L.

1980 Individuality and selection. *Annual Review of Ecology and Systematics* 11:311–332.

Hume, D.

1947 *Dialogues Concerning Natural Religion*. Indianapolis: Bobbs-Merrill.

Huneman, P.

2006 Naturalising purpose: From comparative anatomy to the "adventure of reason." *Studies in History and Philosophy of Biological and Biomedical Sciences* 37:649–674.

ed.

2007 *Understanding Purpose: Kant and the Philosophy of Biology*. Rochester, NY: University of Rochester Press.

Hunt, J., L. F. Bussière, M. D. Jennions, and R. Brooks

2004 What is genetic quality? *Trends in Ecology & Evolution* 19:329–333.

Hutchinson, G. E.

1957 Concluding remarks. *Cold Spring Harbor Symposia on Quantitative Biology* 22:415–427.

1959 Homage to Santa Rosalia or why are there so many kinds of animals? *American Naturalist* 93:145–159.

1965 *The Ecological Theater and the Evolutionary Play*. New Haven, CT: Yale University Press.

1978 *An Introduction to Population Ecology*. New Haven, CT: Yale University Press.

Ikemura, T.

1985 Codon usage and tRNA content in unicellular and multicellular organisms. *Molecular Biology and Evolution* 2:13–34.

Inman, H., and W. Stephan

2003 Distinguishing the hitchhiking and background selection models. *Genetics* 165:2307–2312.

Inwood, B.

2001 *The Poem of Empedocles*. Rev. ed. Toronto: University of Toronto Press.

Isaak, M.

2007 *The Counter-Creationism Handbook*. Berkeley: University of California Press.

Jacob, F.

1977 Evolution and tinkering. *Science (Washington, DC)* 196:1161–1166.

Jacquard, A.

1983 Heritability: One word, three concepts. *Biometrics* 39:465–477.

Jamieson, B. G. M.

1988 On the phylogeny and higher classification of the Clitellata. *Cladistics* 4:367–410.

Jamieson, B. G. M., S. Tillier, A. Tillier, J.-L. Justine, E. Ling, S. James, K. McDonald, and A. F. Hugall

2002 Phylogeny of the Megascolecidae and Crassiclitellata (Annelida, Oligochaeta): Combined versus partitioned analysis using nuclear (28S) and mitochondrial (12S, 16S) rDNA. *Zoosystema* 24:707–734.

Jeffreys, A. J., and C. A. May
2004   Intense and highly localized gene conversion activity in human meiotic crossover hot spots. *Nature Genetics* 36:151–156.

Jeffreys, A. J., and R. Neumann
2005   Factors influencing recombination frequency and distribution in a human meiotic crossover hotspot. *Human Molecular Genetics* 14:2277–2287.

Johannesson, H., J. P. Townsend, C.-Y. Hung, G. T. Cole, and J. W. Taylor
2005   Concerted evolution in the repeats of an immunomodulating cell surface protein, SOWgp, of the human pathogenic fungi *Coccidiodes immitis* and *C. posadasii*. *Genetics* 171:109–117.

Joshi, A., M. H. Do, and L. D. Mueller
1999   Poisson distribution of male mating success in laboratory populations of *Drosophila melanogaster*. *Genetical Research* 73:239–249.

Jukes, T. H.
1977   Nearest-neighbor doublets in protein-coding regions of MS2 RNA. *Journal of Molecular Evolution* 9:299–303.

Kaneko, K.
2006   *Life: An Introduction to Complex Systems Biology*. New York: Springer.

Kant, I.
1969   *Universal Natural History and Theory of the Heavens*. Trans. W. Hastie. Ann Arbor: University of Michigan Press.
1987   *Critique of Judgment*. Trans. W. S. Pluhar. Indianapolis: Hackett.
1992   The only possible argument in support of a demonstration of the existence of God. In D. Walford and R. Meerbote (eds. and trans.), *Theoretical Philosophy, 1755–1770* (pp. 107–201). Cambridge: Cambridge University Press.

Karlin, S., and U. Lieberman
1974   Random temporal variation in selection intensities: Case of large population size. *Theoretical Population Biology* 6:355–382.

Kauffman, S.
1993   *The Origins of Order: Self-Organization and Selection in Evolution*. New York: Oxford University Press.

Kauffman, S., and S. Levin
1987   Towards a general theory of adaptive walks on rugged fitness landscapes. *Journal of Theoretical Biology* 128:11–45.

Keller, L., ed.
1999   *Levels of Selection in Evolution*. Princeton, NJ: Princeton University Press.

Keller, L. F., I. Biebach, and P. E. A. Hoeck
2007   The need for a better understanding of inbreeding effects on population growth. *Animal Conservation* 10:286–287.

Kellogg, V. L.
1908   *Darwinism To-day*. New York: Holt.

Kempthorne, O.
1997   Heritability: Uses and abuses. *Genetica (Dordrecht)* 99:109–112.

Kettlewell, H. B. D.
1955a   Recognition of appropriate backgrounds by the pale and black phases of Lepidoptera. *Nature (London)* 175:943–944.

1955b   Selection experiments on industrial melanism in the *Lepidoptera. Heredity* 9:323–342.

1956a   Further selection experiments on industrial melanism in the *Lepidoptera. Heredity* 10:287–301.

1956b   A résumé of investigations on the evolution of melanism in the Lepidoptera. *Proceedings of the Royal Society of London B, Biological Sciences* 145:297–303.

1958   A survey of the frequencies of *Biston betularia* (L.) (Lep.) and its melanic forms in Great Britain. *Heredity* 12:51–72.

1973   *The Evolution of Melanism: The Study of a Recurring Necessity.* Oxford: Oxford University Press.

Khelefi, A., J. Meunier, L. Duret, and D. Mouchiroud

2006   GC content evolution of the human and mouse genomes: Insights from the study of processed pseudogenes in regions of different recombination rates. *Journal of Molecular Evolution* 62:745–752.

Kidwell, M. G., and D. R. Lisch

2000   Transposable elements and host genome evolution. *Trends in Ecology & Evolution* 15:95–99.

2001   Perspective: Transposable elements, parasitic DNA, and genome evolution. *Evolution* 55:1–24.

Kimura, M.

1958   On the change of population fitness by natural selection. *Heredity* 12:145–167.

1968   Evolutionary rate at the molecular level. *Nature (London)* 217:624–626.

1980   A simple method for estimating evolutionary rates of base substitutions through comparative studies of nucleotide sequences. *Journal of Molecular Evolution* 16:111–120.

1983   *The Neutral Theory of Molecular Evolution.* Cambridge: Cambridge University Press.

1991   Recent development of the neutral theory viewed from the Wrightian tradition of theoretical population genetics. *Proceedings of the National Academy of Sciences of the USA* 88:5969–5973.

Kimura, M., and T. Ohta

1974   On some principles governing molecular evolution. *Proceedings of the National Academy of Sciences of the USA* 71:2848–2852.

King, J. L., and T. H. Jukes

1969   Non-Darwinian evolution. *Science (Washington, DC)* 164:788–798.

Kingsolver, J. G., H. E. Hoekstra, J. M. Hoekstra, D. Berrigan, S. N. Vignieri, C. E. Hill, A. Hoang, P. Gibert, and P. Beerli

2001   The strength of phenotypic selection in natural populations. *American Naturalist* 157:245–261.

Kirschner, M., and J. Gerhart

1998   Evolvability. *Proceedings of the National Academy of Sciences of the USA* 95:8420–8427.

Kitcher, P.

1993   Function and design. *Midwest Studies in Philosophy* 18:379–397.

Kohn, D.

1980   Theories to work by: Rejected theories, reproduction, and Darwin's path to natural selection. *Studies in the History of Biology* 4:67–170.

Konopka, A. K.
2006   *Systems Biology: Principles, Methods, and Concepts.* New York: CRC.

Koonin, E. V., K. S. Makarova, Y. I. Wolf, and L. Aravind
2002   Horizontal gene transfer and its role in the evolution of prokaryotes. In M. Syvanen and C. I. Kado (eds.), *Horizontal Gene Transfer* (pp. 277–304). San Diego, CA: Academic Press.

Kosakovsky Pond, S. L., and S. D. W. Frost
2005   Not so different after all: A comparison of methods for detecting amino acid sites under selection. *Molecular Biology and Evolution* 22:1208–1222.

Kottler, M. J.
1985   Charles Darwin and Alfred Russel Wallace: Two decades of debate over natural selection. In D. Kohn (ed.), *The Darwinian Heritage* (pp. 367–432). Princeton, NJ: Princeton University Press.

Kruijt, J. P., and G. J. de Vos
1988   Individual variation in reproductive success in male black grouse, *Tetrao tetrix* L. In T. Clutton-Brock (ed.), *Reproductive Success: Studies of Individual Variation in Contrasting Breeding Systems* (pp. 279–290). Chicago: University of Chicago Press.

Kruuk, L. E. B., T. H. Clutton-Brock, J. Slate, J. M. Pemberton, S. Brotherstone, and F. E. Guinness
2000   Heritability of fitness in a wild mammal population. *Proceedings of the National Academy of Sciences of the USA* 97:698–703.

Kruuk, L. E. B., J. Slate, J. M. Pemberton, S. Brotherstone, F. E. Guinness, and T. H. Clutton-Brock
2002   Antler size in red deer: Heritability and selection but no evolution. *Evolution* 56:1683–1695.

Kubitschek, H. E.
1970   *Introduction to Research with Continuous Cultures.* Englewood Cliffs, NJ: Prentice Hall.

Kurland, C. G.
2005   What tangled web: Barriers to rampant horizontal gene transfer. *BioEssays,* 27:741–747.

Kurland, C. G., B. Canback, and O. G. Berg
2003   Horizontal gene transfer: A critical view. *Proceedings of the National Academy of Sciences of the USA* 100:9658–9662.

Laforsch, C., W. Ngwa, W. Grill, and R. Tollrian
2004   An acoustic microscopy technique reveals hidden morphological defenses in *Daphnia. Proceedings of the National Academy of Sciences of the USA* 101:15911–15914.

Laland, K. N., and K. Sterelny
2006   Seven reasons (not) to neglect niche construction. *Evolution* 60:1751–1762.

Lande, R.
1976   Natural selection and random genetic drift in phenotypic evolution. *Evolution* 30:314–334.

Lande, R., and S. J. Arnold
1983   The measurement of selection on correlated characters. *Evolution* 37:1210–1226.

Larson, J. L.

1994   *Interpreting Nature: The Science of Living Forms from Linnaeus to Kant.* Baltimore: Johns Hopkins University Press.

Laubichler, M. D., and J. Maienschein, eds.

2007   *From Embryology to Evo-Devo: A History of Developmental Evolution.* Cambridge, MA: MIT Press.

Lauder, G. V.

1996   The argument from design. In G. V. Lauder and M. R. Rose (eds.), *Adaptation* (pp. 55–91). San Diego, CA: Academic Press.

Lauder, G. V., and M. R. Rose, eds.

1996   *Adaptation.* San Diego, CA: Academic Press.

Le Boeuf, B. J., and J. Reiter

1988   Lifetime reproductive success in northern elephant seals. In T. Clutton-Brock (ed.), *Reproductive Success: Studies of Individual Variation in Contrasting Breeding Systems* (pp. 344–362). Chicago: University of Chicago Press.

Le Guyader, H.

2004   *Geoffroy Saint-Hilaire: A Visionary Naturalist.* Chicago: University of Chicago Press.

Lee, S. B.

1833   *Memoirs of Baron Cuvier.* New York: Harper.

Lees, D. R.

1968   Genetic control of the melanic form *insularia* of the peppered moth *Biston betularia* (L.). *Nature (London)* 220:1249–1250.

Leibniz, G. W.

1951   *Leibniz: Selections.* Trans. P. P. Wiener. New York: Scribner's.

1991   *Discourse on Metaphysics and Other Essays.* Trans. D. Garber and R. Ariew. Indianapolis: Hackett.

Lenhoff, S. G., and H. M. Lenhoff

1986   *Hydra and the Birth of Experimental Biology—1744.* Pacific Grove, CA: Boxwood.

Lennox, J. G.

1992   Teleology. In E. F. Keller and E. A. Lloyd (eds.), *Keywords in Evolutionary Biology* (pp. 324–333). Cambridge, MA: Harvard University Press.

1993   Darwin *was* a teleologist. *Biology & Philosophy* 8:409–421.

1994   Teleology by another name: A reply to Ghiselin. *Biology & Philosophy* 9:493–495.

2001   *Aristotle's Philosophy of Biology: Studies in the Origin of Life Science.* Cambridge: Cambridge University Press.

Lenoir, T.

1989   *The Strategy of Life: Teleology and Mechanics in Nineteenth-Century German Biology.* Chicago: University of Chicago Press.

Lenski, R. E.

2004   Phenotypic and genomic evolution during a 20,000-generation experiment with the bacterium *Escherichia coli. Plant Breeding Review* 24:225–265.

Lenski, R. E., C. L. Winkworth, and M. A. Riley

2003   Rates of DNA sequence evolution in experimental populations of *Escherichia coli* during 20,000 generations. *Journal of Molecular Evolution* 56:498–508.

Leroi, A. M., M. R. Rose, and G. V. Lauder
1994   What does the comparative method reveal about adaptation? *American Naturalist* 143:381–402.

Lessard, S.
1997   Fisher's fundamental theorem of natural selection revisited. *Theoretical Population Biology* 52:119–136.

Letteney, M. J.
1999   Georges Cuvier, transcendental naturalist: A study of teleological explanation in biology. PhD diss., University of Notre Dame, Notre Dame, IN.

Levins, R.
1968   *Evolution in Changing Environments: Some Theoretical Explorations.* Princeton, NJ: Princeton University Press.

Lewontin, R. C.
1970   The units of selection. *Annual Review of Ecology and Systematics* 1:1–18.
1974   *The Genetic Basis of Evolutionary Change.* New York: Columbia University Press.
1978   Adaptation. *Scientific American* 239:212–230.
1983   Gene, organism, and environment. In D. S. Bendall (ed.), *Evolution from Molecules to Men* (pp. 273–285). Cambridge: Cambridge University Press.
1984   Adaptation. In E. Sober (ed.), *Conceptual Issues in Evolutionary Biology: An Anthology* (pp. 235–251). MIT Press, Cambridge, MA.
2000   *The Triple Helix.* Cambridge, MA: Harvard University Press.

Lewontin, R. C., and J. L. Hubby
1966   A molecular approach to the study of genic heterozygosity in natural populations. II. Amount of variation and degree of heterozygosity in natural populations of *Drosophila pseudoobscura. Genetics* 54:595–609.

Long, A. A.
1986   *Hellenistic Philosophy: Stoics, Epicureans, Sceptics.* 2nd ed. Berkeley: University of California Press.

Longrich, N.
2006   Structure and function of hindlimb feathers in *Archaeopteryx lithographica. Paleobiology* 32:417–431.

Lorenz, K.
1966   *On Aggression.* London: Methuen.

Lovejoy, A. O.
1936   *The Great Chain of Being.* New York: Harper & Row.

Lucretius
1995   *On the Nature of Things: De rerum natura.* Trans. A. M. Esolen. Baltimore: Johns Hopkins University Press.

Luke, C.
1986   Convergent evolution of lizard toe fringes. *Biological Journal of the Linnean Society* 27:1–16.

Lyell, C.
1830–
1833   *Principles of Geology.* London: Murray.

Lynch, M.

1990 The rate of morphological evolution in mammals from the standpoint of the neutral expectation. *American Naturalist* 136:727–741.

2007 The frailty of adaptive hypotheses for the origins of organismal complexity. *Proceedings of the National Academy of Sciences of the USA* 104:8597–8604.

Lynch, M., and W. Gabriel

1990 Mutation load and the survival of small populations. *Evolution* 44:1725–1737.

Lynch, M., and W. G. Hill

1986 Phenotypic evolution by neutral mutation. *Evolution* 40:915–935.

Lynch, M., and B. Walsh

1998 *Genetics and Analysis of Quantitative Traits*. Sunderland, MA: Sinauer.

MacArthur, R. H.

1969 The theory of the niche. In R. C. Lewontin (ed.), *Population Biology and Evolution* (pp. 159–176). Syracuse, NY: Syracuse University Press.

MacLean, R. C., and G. Bell

2002 Experimental adaptive radiation in *Pseudomonas*. *American Naturalist* 160:569–580.

Maharjan, R., S. Seeto, L. Notley-McRobb, and T. Ferenci

2006 Clonal adaptive radiation in a constant environment. *Science (Washington, DC)* 313:514–517.

Majerus, M. E. N.

1998 *Melanism: Evolution in Action*. Oxford: Oxford University Press.

Marchant, J.

1916 *Alfred Russel Wallace: Letters and Reminiscences*. New York: Harper.

Marchant, L.

1858 *Lettres de Georges Cuvier à C.M. Pfaff sur l'histoire naturelle, la politique et la littérature, 1788–1792*. Trans. L. Marchant. Paris: Masson.

Martin, P. S.

2005 *Twilight of the Mammoths: Ice Age Extinctions and the Rewilding of America*. Berkeley: University of California Press.

Martin, P. S., and R. G. Klein, eds.

1984 *Quaternary Extinctions: A Prehistoric Revolution*. Tucson: University of Arizona Press.

Mason, J. H.

1982 *The Irresistible Diderot*. London: Quartet Books.

May, R. M.

1973 *Stability and Complexity in Model Ecosystems*. Princeton, NJ: Princeton University Press.

Maynard Smith, J., R. M. Burian, S. Kauffman, P. Alberch, J. Campbell, B. Goodwin, R. Lande, D. Raup, and L. Wolpert

1985 Developmental constraints and evolution. *Quarterly Review of Biology* 60:265–287.

Maynard Smith, J., and J. Haigh

1974 The hitch-hiking effect of a favorable gene. *Genetical Research* 23:23–35.

Maynard Smith, J., and E. Szathmáry

1995 *The Major Transitions in Evolution*. Oxford: Freeman.

Mayr, E.

1962   Accident or design: The paradox of evolution. In G. W. Leeper (ed.), *The Evolution of Living Organisms* (pp. 1–14). Victoria: University of Melbourne Press.

1982   *The Growth of Biological Thought: Diversity, Evolution, and Inheritance.* Cambridge, MA: Harvard University Press.

McAdam, A. G., and S. Boutin

2003   Variation in viability selection among cohorts of juvenile red squirrels *(Tamiasciurus hudsonicus). Evolution* 57:1689–1697.

McClellan, C.

2001   The legacy of Georges Cuvier in Auguste Comte's natural philosophy. *Studies in the History and Philosophy of Science* 32:1–29.

McDonald, J. H., and M. Kreitman

1991   Adaptive protein evolution at the Adh locus in *Drosophila. Nature (London)* 351:652–654.

McGhee, G. R.

2007   *The Geometry of Evolution: Adaptive Landscapes and the Origin of Species.* Cambridge: Cambridge University Press.

McGill, B. J., B. J. Enquist, E. Weiher, and M. Westoby

2006   Rebuilding community ecology from functional traits. *Trends in Ecology & Evolution* 21:178–185.

McGowan, A., L. I. Wright, and J. Hunt

2007   Inbreeding and population dynamics: Implications for conservation strategies. *Animal Conservation* 10:284–285.

McLaughlin, P.

1990   *Kant's Critique of Teleology in Biological Explanation: Antinomy and Teleology.* Lewiston, NY: Mellen.

2001   *What Functions Explain: Functional Explanation and Self-Reproducing Systems.* Cambridge: Cambridge University Press.

Merilä, J., B. C. Sheldon, and L. E. B. Kruuk

2001   Explaining stasis: Microevolutionary studies in natural populations. *Genetica (Dordrecht)* 112–113:199–222.

Michod, R. E.

1999   *Darwinian Dynamics: Evolutionary Transitions in Fitness and Individuality.* Princeton, NJ: Princeton University Press.

Millikan, R.

1989   In defense of proper functions. *Philosophy of Science* 56:288–302.

Monod, J.

1971   *Chance and Necessity: An Essay on the Natural Philosophy of Modern Biology.* New York: Knopf.

Morrell, P. L., D. M. Toleno, K. F. Lundy, and M. T. Clegg

2006   Estimating the contribution of mutation, recombination and gene conversion in the generation of haplotypic diversity. *Genetics* 173:1705–1723.

Morrison, M.

2000   *Unifying Scientific Theories: Physical Concepts and Mathematical Structures.* Cambridge: Cambridge University Press.

Müller, G. B.

2007   Evo-devo: Extending the evolutionary synthesis. *Nature Reviews Genetics* 9:943–949.

Müller, G. B., and S. A. Newman, eds.

2003   *Origination of Organismal Form: Beyond the Gene in Developmental and Evolutionary Biology.* Cambridge, MA: MIT Press.

Muller, H. J.

1950   Our load of mutations. *American Journal of Human Genetics* 2:111–176.

Mulley, J. F., C.-H. Chiu, and P. W. H. Holland

2006   Breakup of a homeobox cluster after genome duplication in teleosts. *Proceedings of the National Academy of Sciences of the USA* 103:10369–10372.

Myers, S., L. Bottolo, C. Freeman, G. McVean, and P. Donnely

2005   A fine-scale map of recombination rates and hotspots across the human genome. *Science (Washington, DC)* 310:321–324.

Nagel, E.

1961   *The Structure of Science: Problems in the Logic of Scientific Explanation.* New York: Harcourt, Brace.

Neander, K.

1991   Functions as selected effects: The conceptual analyst's defense. *Philosophy of Science* 58:168–184.

Necşulea, A., and J. R. Lobry

2006   Revisiting the directional mutation pressure theory: The analysis of a particular genomic structure in *Leishmania major. Gene (Amsterdam)* 385:28–40.

Neff, B. D., and T. E. Pitcher

2005   Genetic quality and sexual selection: An integrated framework for good genes and compatible genes. *Molecular Ecology* 14:19–38.

Nei, M., and A. P. Rooney

2005   Concerted and birth-and-death evolution of multigene families. *Annual Review of Genetics* 39:121–152.

Nelson, W. A., T. A. Knott, and P. W. Carhart, eds.

1941   *Webster's New International Dictionary of the English Language.* Springfield, MA: Merriam.

Newton, I.

1952a   Mathematical principles of natural philosophy. In R. M. Hutchins (ed.), *Great Books of the Western World: Newton, Huygens* (vol. 34, pp. 1–372). Trans. A. Motte and F. Cajori. Chicago: Encyclopaedia Britannica.

1952b   Optics. In R. M. Hutchins (ed.), *Great Books of the Western World: Newton, Huygens* (vol. 34, pp. 375–544). Chicago: Encyclopaedia Britannica.

Novozhilov, A. S.

2005   Mathematical modeling of evolution of horizontally transferred genes. *Molecular Biology and Evolution* 22:1721–1732.

Nunney, L., and E. L. Schuenzel

2006   Detecting natural selection at the molecular level: A reexamination of some "classic" examples of adaptive evolution. *Journal of Molecular Evolution* 62: 176–195.

Nyhart, L. K.

1995   *Biology Takes Form: Animal Morphology and the German Universities, 1800–1900*. Chicago: University of Chicago Press.

Odling-Smee, F. J., K. N. Laland, and M. W. Feldman

2003   *Niche Construction: The Neglected Process in Evolution*. Princeton, NJ: Princeton University Press.

Ohta, T.

2002   Near-neutrality in evolution of genes and gene regulation. *Proceedings of the National Academy of Sciences of the USA* 99:16134–16137.

Okasha, S.

2006   *Evolution and the Levels of Selection*. Oxford: Oxford University Press.

Orgel, L. E., and F. H. C. Crick

1980   Selfish DNA: The ultimate parasite. *Nature (London)* 284:604–607.

Osborn, H. F.

1929   *From the Greeks to Darwin: The Development of the Evolution Idea Through Twenty-four Centuries*. 2nd ed. New York: Scribner's.

Osler, M. J.

1994   *Divine Will and the Mechanical Philosophy: Gassendi and Descartes on Contingency and Necessity in the Created World*. Cambridge: Cambridge University Press.

Ospovat, D.

1978   Perfect adaptation and teleological explanation: Approaches to the problem of the history of life in the mid–nineteenth century. *Studies in the History of Biology* 2:33–56.

1980   God and natural selection: The Darwinian idea of design. *Journal of the History of Biology* 13:169–194.

1981   *The Development of Darwin's Theory: Natural History, Natural Theology, and Natural Selection, 1838–1859*. Cambridge: Cambridge University Press.

Ostrom, J. H.

1973   The ancestry of birds. *Nature (London)* 242:136.

1974   *Archaeopteryx* and the origin of flight. *Quarterly Review of Biology* 49:27–47.

1976   *Archaeopteryx* and the origin of birds. *Biological Journal of the Linnean Society* 8:91–182.

1979   Bird flight: How did it begin? *American Scientist* 67:46–56.

Outram, D.

1978   The language of natural power: The "éloges" of Georges Cuvier and the public language of nineteenth century science. *History of Science* 16:153–178.

1984   *Georges Cuvier: Vocation, Science and Authority in Post-Revolutionary France*. Manchester: Manchester University Press.

1986   Uncertain legislator: Georges Cuvier's laws of nature in their intellectual context. *Journal of the History of Biology* 19:323–368.

Owen, R.

1992   The 1837 Hunterian Lectures. In P. R. Sloan (ed.), *The Hunterian Lectures in Comparative Anatomy, May and June, 1837* (pp. 73–302). Chicago: University of Chicago Press.

Padian, K.

2001a   Cross-testing adaptive hypotheses: Phylogenetic analysis and the origin of bird flight. *American Zoologist* 41:598–607.

2001b   Stages in the origin of bird flight: Beyond the arboreal-cursorial dichotomy. In J. Gauthier and L. F. Gall (eds.), *New Perspective on the Origin and Evolution of Birds* (pp. 255–272). New Haven, CT: Peabody Museum of Natural History, Yale University.

Padian, K., and L. M. Chiappe

1998   The origin and early evolution of birds. *Biological Review* 73:1–42.

Paley, W.

1802   *Natural Theology; Or, Evidences of the Existence and Attributes of the Deity, Collected from the Appearances of Nature.* London: Faulder.

1809   *Natural Theology; Or, Evidences of the Existence and Attributes of the Deity, Collected from the Appearances of Nature.* 12th ed. London: Faulder.

Palsson, B.

2006   *Systems Biology: Properties of Reconstructed Networks.* New York: Cambridge University Press.

Patrides, C. A., ed.

1980   *The Cambridge Platonists.* Cambridge: Cambridge University Press.

Paul, G. S.

2002   *Dinosaurs of the Air: The Evolution and Loss of Flight in Dinosaurs and Birds.* Baltimore: Johns Hopkins University Press.

Pausas, J. G., J. E. Keeley, and M. Verdú

2006   Inferring differential evolutionary processes of plant persistence traits in Northern Hemisphere Mediterranean fire-prone ecosystems. *Journal of Ecology* 94:31–39.

Pearson, M. B., ed.

2002   *A. R. Wallace's Malay Archipelago Journals and Notebook.* London: Linnean Society of London.

Perakh, M.

2004   *Unintelligent Design.* Amherst, NY: Prometheus Books.

Pickering, M.

1993   *Auguste Comte: An Intellectual Biography.* Vol. 1. Cambridge: Cambridge University Press.

Pigliucci, M., and J. Kaplan

2006   *Making Sense of Evolution: The Conceptual Foundations of Evolutionary Biology.* Chicago: University of Chicago Press.

Plato

1948   Phaedo. In S. Buchanan (ed.), *The Portable Plato* (pp. 191–278). Trans. B. Jowett. New York: Viking.

Poelwijk, F. J., D. J. Kiviet, D. M. Weinreich, and S. J. Tans

2007   Empirical fitness landscapes reveal accessible evolutionary paths. *Nature (London)* 445:383–386.

Powell, J. R., and E. N. Moriyama

1997   Evolution of codon usage bias in *Drosophila*. *Proceedings of the National Academy of Sciences of the USA* 94:7784–7790.

Price, G. R.

1970   Selection and covariance. *Nature (London)* 227:520–521.

1972a   Extension of covariance selection mathematics. *Annals of Human Genetics*, 35:485–490.

1972b   Fisher's "fundamental theorem" made clear. *Annals of Human Genetics*, 36:129–140.

1995   The nature of selection. *Journal of Theoretical Biology* 175:389–396.

Price, T. D., P. R. Grant, H. L. Gibbs, and P. T. Boag

1984   Recurrent patterns of natural selection in a population of Darwin's finches. *Nature (London)* 309:787–789.

Price, T. D., M. Kirkpatrick, and S. J. Arnold

1988   Directional selection and the evolution of breeding date in birds. *Science (Washington, DC)* 240:798–799.

Prigogine, I., and I. Stengers

1984   *Order Out of Chaos.* Boulder, CO: Shambhala.

Provine, W. B.

1985   Adaptation and mechanisms of evolution after Darwin: A study in persistent controversies. In D. Kohn (ed.), *The Darwinian Heritage* (pp. 825–866). Princeton, NJ: Princeton University Press.

1986   *Sewall Wright and Evolutionary Biology.* Chicago: University of Chicago Press.

Prum, R. O.

1999   Development and evolutionary origin of feathers. *Journal of Experimental Zoology Part B: Molecular and Developmental Evolution* 285:291–306.

2005   Evolution of the morphological innovations of feathers. *Journal of Experimental Zoology Part B: Molecular and Developmental Evolution* 304:570–579.

Prum, R. O., and A. H. Brush

2002   The evolutionary origin and diversification of feathers. *Quarterly Review of Biology* 77:261–295.

Prusinkiewicz, P., Y. Erasmus, B. Lane, L. D. Harder, and E. Coen

2007   Evolution and development of inflorescence architectures. *Science (Washington, DC)* 316:1452–1456.

Ptak, S. E., D. A. Hinds, K. Koehler, B. Nickel, N. Patil, D. G. Ballinger, M. Przeworski, K. A. Frazer, and S. Pääbo

2005   Fine-scale recombination patterns differ between chimpanzees and humans. *Nature Genetics* 37:429–434.

Queller, D. C., and J. E. Strassman

1988   Reproductive success and group nesting in the paper wasp, *Pollistes annularis*. In T. Clutton-Brock (ed.), *Reproductive Success: Studies of Individual Variation in Contrasting Breeding Systems* (pp. 76–96). Chicago: University of Chicago Press.

Raby, P.

2001   *Alfred Russel Wallace: A Life.* Princeton, NJ: Princeton University Press.

Raff, R. A.

1996   *The Shape of Life: Genes, Development, and the Evolution of Animal Form.* Chicago: University of Chicago Press.

Randall, J. H., Jr.

1960   *Aristotle.* New York: Columbia University Press.

Raven, C. E.

1986   *John Ray, Naturalist: His Life and Works.* 2nd ed. Cambridge: Cambridge University Press.

Ray, J.

1691   *The Wisdom of God Manifested in the Works of the Creation.* London: Smith.

Reed, D. H.

2005   Relationship between population size and fitness. *Conservation Biology* 19:563–568.

2007a   Extinction of island endemics: It is not inbreeding depression. *Animal Conservation* 10:145–146.

2007b   Natural selection and genetic diversity. *Heredity* 99:1–2.

Reed, D. H., and R. Frankham

2003   Correlation between fitness and genetic diversity. *Conservation Biology* 17:230–237.

Reed, D. H., A. C. Nicholas, and G. E. Stratton

2007a   Genetic quality of individuals impacts population dynamics. *Animal Conservation* 10:275–283.

2007b   Inbreeding levels and prey abundance interact to determine fecundity in natural populations of two species of wolf spider. *Conservation Genetics* 8:1061–1071.

Reeve, H. K., and P. W. Sherman

1993   Adaptation and the goals of evolutionary research. *Quarterly Review of Biology* 68:1–32.

Rehbock, P. F.

1983   *The Philosophical Naturalists: Themes in Early Nineteenth-Century British Biology.* Madison: University of Wisconsin Press.

Reill, P. H.

2005   *Vitalizing Nature in the Enlightenment.* Berkeley: University of California Press.

Reiss, J. O.

1989   On the fitness of "fitness": Determinism, teleology, and the conditions of existence. *American Zoologist* 29:93A.

2005   Natural selection and the conditions for existence: Representational vs. conditional teleology in biological explanation. *History and Philosophy of the Life Sciences* 27:249–280.

2007   Relative fitness, teleology, and the adaptive landscape. *Evolutionary Biology (New York)* 34:4–27.

Richards, R. J.

2002   *The Romantic Conception of Life.* Chicago: University of Chicago Press.

Richardson, M. K., and A. Chipman

2003   Developmental constraints in a comparative framework: A test case using variations in phalanx number during amniote evolution. *Journal of Experimental Zoology Part B: Molecular and Developmental Evolution* 296:8–22.

Ridley, M.

2004   *Evolution.* Malden, MA: Blackwell.

Robertson, A.

1968   The spectrum of genetic variation. In R. C. Lewontin (ed.), *Population Biology and Evolution* (pp. 5–16). Syracuse, NY: Syracuse University Press.

Roe, S. A.

1981   *Matter, Life, and Generation: Eighteenth Century Embryology and the Haller-Wolff Debate.* Cambridge: Cambridge University Press.

Roger, J.
  1997   *Buffon: A Life in Natural History.* Trans. S. L. Bonnefoi. Ithaca, NY: Cornell University Press.
Roget, P. M.
  1840   *Animal and Vegetable Physiology Considered with Reference to Natural Theology.* 3rd ed. London: Pickering.
Rose, K. E., T. H. Clutton-Brock, and F. E. Guinness
  1998   Cohort variation in male survival and lifetime breeding success in red deer. *Journal of Animal Ecology* 67:979–986.
Rosen, R.
  1985   *Anticipatory Systems: Philosophical, Mathematical and Methodological Foundations.* Oxford: Pergamon.
Rosenberg, M. S., S. Subramian, and S. Kumar
  2003   Patterns of transitional mutation biases within and among mammalian genomes. *Molecular Biology and Evolution* 20:988–993.
Rudd, M. K., G. Wray, and H. F. Willard
  2006   The evolutionary dynamics of α-satellite. *Genome Research* 16:88–96.
Rudge, D. W.
  2002   Cryptic designs on the peppered moth? *Revista Biologica Tropical* 50:1–7.
  2005   Did Kettlewell commit fraud? Re-examining the evidence. *Public Understanding of Science* 14:1–20.
Rudwick, M. J. S.
  1976   *The Meaning of Fossils: Episodes in the History of Palaeontology.* 2nd ed. New York: Elsevier.
  1997   *Georges Cuvier, Fossil Bones, and Geological Catastrophes: New Translations and Interpretations of the Primary Texts.* Chicago: University of Chicago Press.
  2005   *Bursting the Limits of Time: The Reconstruction of Geohistory in the Age of Revolution.* Chicago: University of Chicago Press.
  2008   *Worlds before Adam: The Reconstruction of Geohistory in the Age of Reform.* Chicago: University of Chicago Press.
Ruse, M.
  1971   Functional statements in biology. *Philosophy of Science* 38:87–95.
  1973   *The Philosophy of Biology.* London: Hutchinson.
  1996   *Monad to Man: The Concept of Progress in Evolutionary Biology.* Cambridge, MA: Harvard University Press.
  2003   *Darwin and Design: Does Evolution Have a Purpose?* Cambridge, MA: Harvard University Press.
Russell, E. S.
  1916   *Form and Function: A Contribution to the History of Animal Morphology.* London: Murray.
  1946   *The Directiveness of Organic Activities.* Cambridge: Cambridge University Press.
Saccheri, I., and I. Hanksi
  2006   Natural selection and population dynamics. *Trends in Ecology & Evolution* 21:341–347.

Salazar-Ciudad, I.

2006   Developmental constraints vs. variational properties: How pattern formation can help to understand evolution and development. *Journal of Experimental Zoology (Molecular Developmental Evolution)* 306B:107–125.

Salazar-Ciudad, I., and J. Jernvall

2002   A gene network model accounting for development and evolution of mammalian teeth. *Proceedings of the National Academy of Sciences of the USA* 99:8116–8120.

Sandler, L., and E. Novitski

1957   Meiotic drive as an evolutionary force. *American Naturalist* 91:105–110.

Savage, L. J.

1976   On rereading R. A. Fisher. *Annals of Statistics* 4:441–500.

Sawyer, S. L., and H. S. Malik

2006   Positive selection of yeast nonhomologous end-joining genes and a retrotransposon conflict hypothesis. *Proceedings of the National Academy of Sciences of the USA* 103:17614–17619.

Schemske, D. W., and P. Bierzychudek

2001   Evolution of flower color in the desert annual *Linanthus parryae*: Wright revisited. *Evolution* 55:1269–1282.

Schlosser, G., and G. P. Wagner

2004   *Modularity in Development and Evolution.* Chicago: University of Chicago Press.

Schmidt-Nielsen, K.

1979   *Animal Physiology: Adaptation and Environment.* 2nd ed. Cambridge: Cambridge University Press.

Schneider, D., and R. E. Lenski

2004   Dynamics of insertion sequence elements during experimental evolution of bacteria. *Research in Microbiology* 155:319–327.

Schwenk, K.

1995   A utilitarian approach to evolutionary constraint. *Zoology (Jena)*, 98:251–262.

Schwenk, K., and G. P. Wagner

2004   The relativism of constraints on phenotypic evolution. In M. Pigliucci and K. Preston (eds.), *Phenotypic Integration: Studying the Ecology and Evolution of Complex Phenotypes* (pp. 390–408). Oxford: Oxford University Press.

Scott, E. C.

2005   *Evolution vs. Creationism.* Berkeley: University of California Press.

Sedgwick, A.

1850   *A Discourse on the Studies of the University of Cambridge.* 5th ed. London: Parker.

Sedley, D.

2007   *Creationism and Its Critics in Antiquity.* Berkeley: University of California Press.

Seilacher, A.

1970   Arbeitsconzept zur Konstruktions-Morphologie. *Lethaia* 3:393–396.

1972   Divaricate patterns in pelecypod shells. *Lethaia* 5:325–343.

Semper, K.

1881   *The Natural Conditions of Existence as They Affect Animal Life.* London: Kegan Paul.

Senner, S. W.
1980   Inbreeding depression and the survival of zoo populations. In M. E. Soulé and B. A. Wilcox (eds.), *Conservation Biology: An Evolutionary-Ecological Perspective* (pp. 209–224). Sunderland, MA: Sinauer.

Shanks, N.
2004   *God, the Devil, and Darwin: A Critique of Intelligent Design Theory.* Oxford: Oxford University Press.

Sharma, S., and S. N. Raina
2005   Organization and evolution of highly repeated satellite DNA sequences in plant chromosomes. *Cytogenet. Genome Research* 109:15–26.

Shubin, N., and D. B. Wake
1996   Phylogeny, variation, and morphological integration. *American Zoologist* 36:51–60.

Shuster, S. M., and M. J. Wade
2003   *Mating Systems and Strategies.* Princeton, NJ: Princeton University Press.

Simberloff, D. S.
1978   Using island biogeographic distributions to determine if colonization is stochastic. *American Naturalist* 112:713–726.
1981   Santa Rosalia reconsidered: Size ratios and competition. *Evolution* 35:1206–1228.

Simon, A. L., E. A. Stone, and A. Sidow
2002   Inference of functional regions in proteins by quantification of evolutionary constraints. *Proceedings of the National Academy of Sciences of the USA* 99:2912–2917.

Simpson, G. G.
1944   *Tempo and Mode in Evolution.* New York: Columbia University Press.
1946   Fossil penguins. *Bulletin of the American Museum of Natural History* 87:1–99.
1953   *The Major Features of Evolution.* New York: Columbia University Press.

Sinervo, B., and A. L. Sinolo
1996   Testing adaptation using phenotypic manipulation. In G. V. Lauder and M. R. Rose (eds.), *Adaptation* (pp. 149–185). San Diego, CA: Academic Press.

Sloan, P. R.
1992   Introductory essay: On the edge of evolution. In P. R. Sloan (ed.), *The Hunterian Lectures in Comparative Anatomy* (pp. 3–72). Chicago: University of Chicago Press.
1997   Le Muséum de Paris vient à Londres. In C. Blanckaert, C. Cohen, P. Corsi, and J.-L. Fischer (eds.), *Le Muséum au premier siècle de son histoire* (pp. 607–634). Paris: Muséum national d'Histoire naturelle.
2006   Kant on the history of nature: The ambiguous heritage of the critical philosophy for natural history. *Studies in History and Philosophy of Biological and Biomedical Sciences* 37:627–648.

Smith, J. C.
1993   *Georges Cuvier: An Annotated Bibliography of His Published Works.* Washington, DC: Smithsonian Institution Press.

Smith, J. N. M.
1988   Determinants of lifetime reproductive success in the song sparrow. In T. Clutton-Brock (ed.), *Reproductive Success* (pp. 154–172). Chicago: University of Chicago Press.

Smith, N. K.

1947   Introduction, in D. Hume, *Dialogues Concerning Natural Religion* (pp. 1–75). Indianapolis: Bobbs-Merrill.

Smith, N. G. C., and A. Eyre-Walker

2002   Adaptive protein evolution in *Drosophila*. *Nature (London)* 415:1022–1024.

Sober, E.

1984   *The Nature of Selection: Evolutionary Theory in Philosophical Focus*. Cambridge, MA: MIT Press.

Sorhannus, U., and S. L. Kosakovsky Pond

2006   Evidence for positive selection on a sexual reproduction gene in the diatom genus *Thalassiosira* (Bacillariophyta). *Journal of Molecular Evolution* 63:231–239.

Soulé, M. E.

1980   Thresholds for survival: Maintaining fitness and evolutionary potential. In M. E. Soulé and B. A. Wilcox (eds.), *Conservation Biology: An Evolutionary-Ecological Perspective* (pp. 151–169). Sunderland, MA: Sinauer.

Soulé, M. E., and B. A. Wilcox, eds.

1980   *Conservation Biology: An Evolutionary-Ecological Perspective*. Sunderland, MA: Sinauer.

Spaemann, R., and R. Löw

2005   *Natürliche Ziele: Geschichte und Wiederentdeckung des teleologischen Denkens*. Stuttgart: Klett-Cotta.

Spanos, A.

2006   Where do statistical models come from? Revisiting the problem of specification. *IMS Lecture Notes—Monograph Series* 49:98–119.

Spencer, C. C. A., P. Deloukas, S. Hunt, J. Mullikan, S. Myers, B. Silverman, P. Donnely, D. Bentley, and G. McVean

2006   The influence of recombination on human genetic diversity. *PLoS Genetics* 2:1375–1385.

Spencer, H.

1864   *The Principles of Biology*. New York: Appleton.

Spink, J. S.

1960   *French Free-Thought from Gassendi to Voltaire*. London: Athlone, University of London.

Stauffer, R. C.

1975   *Charles Darwin's Natural Selection; Being the Second Part of His Big Species Book Written from 1836 to 1858*. Cambridge: Cambridge University Press.

Stebbins, R. C.

1944   Some aspects of the ecology of the iguanid genus *Uma*. *Ecological Monographs* 14:311–332.

Steinbeck, J., and E. F. Ricketts

1941   *The Sea of Cortez: A Leisurely Journal of Travel and Research*. New York: Viking.

Stevens, P. F.

1994   *The Development of Biological Systematics: Antoine-Laurent de Jussieu, Nature, and the Natural System*. New York: Columbia University Press.

Stewart, J., and J. Kemp
  1963   *Diderot, Interpreter of Nature: Selected Writings.* 2nd ed. Trans. J. Stewart and J. Kemp. New York: International.
Stoltzfus, A.
  1999   On the possibility of constructive neutral evolution. *Journal of Molecular Evolution* 49:169–181.
Stone, J. L., and A. F. Motten
  2002   Anther-stigma separation is associated with inbreeding depression in *Datura stramonium*, a predominantly self-fertilizing annual. *Evolution* 56:2187–2195.
Suzuki, Y., and T. Gojobori
  1999   A method for detecting positive selection at single amino acid sites. *Molecular Biology and Evolution* 16:1315–1328.
Swainson, W.
  1834   *A Preliminary Discourse on the Study of Natural History.* London: Longman, Rees, Orme, Brown, Green, & Longman, & J. Taylor.
Syvanen, M., and C. I. Kado, eds.
  2002   *Horizontal Gene Transfer.* 2nd ed. San Diego, CA: Academic Press.
Tabak, H. F., D. Hoepfner, A. v.d. Zand, H. J. Geuze, I. Braakman, and M. A. Huynen
  2006   Formation of peroxisomes: Present and past. *Biochimica et Biophysica Acta* 1763:1647–1654.
Tam, V. H., A. Louie, M. R. Deziel, W. Liu, R. Leary, and G. L. Drusano
  2005   Bacterial-population responses to drug-selective pressure: Examination of Garenoxacin's effect on *Pseudomonas aeruginosa. Journal of Infectious Diseases* 192:420–428.
Taquet, P.
  2006   *Georges Cuvier: Naissance d'un génie.* Paris: Odile Jacob.
Taquet, P., and K. Padian
  2004   The earliest known restoration of a pterosaur and the philosophical origins of Cuvier's *Ossemens Fossiles. Comptes Rendus Palevol* 3:157–175.
Taylor, A. E.
  1928   *A Commentary on Plato's Timaeus.* Oxford: Clarendon.
Taylor, D. R., and P. K. Ingvarsson
  2003   Common features of segregation distortion in plants and animals. *Genetica (Dordrecht)* 117:27–35.
Terborgh, J., and B. Winter
  1980   Some causes of extinction. In M. E. Soulé and B. A. Wilcox (eds.), *Conservation Biology: An Evolutionary-Ecological Perspective* (pp. 119–133). Sunderland, MA: Sinauer.
Terrall, M.
  2002   *The Man Who Flattened the Earth: Maupertuis and the Sciences in the Enlightenment.* Chicago: University of Chicago Press.
Thompson, D. W.
  1942   *On Growth and Form.* Cambridge: Cambridge University Press.
Tulloch, J.
  1874   *Rational Theology and Christian Philosophy in England in the 17th Century.* 2nd ed. Edinburgh: Blackwood.

Turner, J. S.

2000   *The Extended Organism: The Physiology of Animal-Built Structures*. Cambridge, MA: Harvard University Press.

2004   Extended phenotypes and extended organisms. *Biology & Philosophy* 19:327–352.

2007   *The Tinkerer's Accomplice: How Design Emerges from Life Itself*. Cambridge, MA: Harvard University Press.

Uhland, R.

1953   *Geschichte der hohen Karlsschule in Stuttgart*. Stuttgart: Kohlhammer.

Urrutia, A. O., and L. D. Hurst.

2001   Codon usage bias covaries with expression breadth and the rate of synonymous evolution in humans, but this is not evidence for selection. *Genetics* 159:1191–1199.

Van der Meer, J. M.

2005   Reading nature in the light of scripture: The case of Georges Cuvier (1769–1832). In K. van Berkel and A. Vanderjagt (eds.), *The Book of Nature in Modern Times* (pp. 183–195). Leuven, Belgium: Peeters.

Van Valen, L.

1973   Festschrift. *Science (Washington, DC)* 180:488.

Vartanian, A.

1950   Trembley's Polyp, La Mettrie, and Eighteenth-Century French Materialism. *Journal of the History of Ideas* 11:259–286.

1953   *Diderot and Descartes: A Study of Scientific Naturalism in the Enlightenment*. Princeton, NJ: Princeton University Press.

Vogler, D. W., and S. Kalisz

2001   Sex among the flowers: The distribution of plant mating systems. *Evolution* 55:202–204.

Vrba, E. S., and S. J. Gould

1986   The hierarchical expansion of sorting and selection: Sorting and selection cannot be equated. *Paleobiology* 12:217–228.

Wade, M. J.

1979   Sexual selection and variance in reproductive success. *American Naturalist* 114:742–747.

Wagner, A.

2005   *Robustness and Evolvability in Living Systems*. Princeton, NJ: Princeton University Press.

Wagner, G. P., and L. Altenberg

1996   Complex adaptations and the evolution of evolvability. *Evolution* 50:967–976.

Wagner, G. P., and K. Schwenk

2000   Evolutionarily stable configurations: Functional integration and the evolution of phenotypic stability. *Evolutionary Biology (New York)* 31:155–217.

Wake, D., G. Roth, and M. Wake

1983   On the problem of stasis in organismal evolution. *Journal of Theoretical Biology* 101:211–224.

Wake, D. B.

1991   Homoplasy: The result of natural selection, or evidence of design limitations? *American Naturalist* 138:543–567.

Wallace, A. R.

1858   On the tendency of varieties to depart indefinitely from the original type. *Journal of the Proceedings of the Linnean Society of London, Zoology* 3:53–62.

1889   *Darwinism: An Exposition of the Theory of Natural Selection, with Some of Its Applications.* London: Macmillan.

1891   *Natural Selection and Tropical Nature.* New ed. London: Macmillan.

1908a   *My Life: A Record of Events and Opinions.* New ed. London: Chapman & Hall.

1908b   The present position of Darwinism. *Contemporary Review* 94:129–141.

Wallace, B.

1968a   Polymorphism, population size, and genetic load. In R. C. Lewontin (ed.), *Population Biology and Evolution* (pp. 87–108). Syracuse, NY: Syracuse University Press.

1968b   *Topics in Population Genetics.* New York: Norton.

1970   *Genetic Load: Its Biological and Conceptual Aspects.* Englewood Cliffs, NJ: Prentice Hall.

1975   Hard and soft selection revisited. *Evolution* 29:465–473.

1991   *Fifty Years of Genetic Load: An Odyssey.* Ithaca, NY: Cornell University Press.

Walsh, D. M.

1996   Fitness and function. *British Journal for the Philosophy of Science* 47:553–574.

Walsh, D. M., and A. Ariew

1996   A taxonomy of functions. *Canadian Journal of Philosophy* 26:493–514.

Warringer, J., and A. Blomberg

2006   Evolutionary constraints on yeast protein size. *BMC Evolutionary Biology* 6:61.

Weatherhead, P. J., and P. T. Boag

1997   Genetic estimates of annual and lifetime reproductive success in male red-winged blackbirds. *Ecology (New York)* 78:884–896.

Welch, J. J.

2006   Estimating the genomewide rate of adaptive protein evolution in *Drosophila. Genetics* 173:821–837.

Wells, J.

2000   *Icons of Evolution: Science or Myth?* Washington, DC: Regnery.

West-Eberhard, M. J.

2003   *Developmental Plasticity and Evolution.* Oxford: Oxford University Press.

Whewell, W.

1857   *The Philosophy of the Inductive Sciences, Founded upon Their History.* New ed. London: Parker.

1859   *History of the Inductive Sciences.* 3rd ed. New York: Appleton.

Whyte, L. L.

1965   *Internal Factors in Evolution.* New York: Braziller.

Wilcox, B. A.

1980   Insular ecology and conservation. In M. E. Soulé and B. A. Wilcox (eds.), *Conservation Biology: An Evolutionary-Ecological Perspective* (pp. 95–117). Sunderland, MA: Sinauer.

Willey, B.

1934   *The Seventeenth Century Background.* London: Chatto & Windus.

1940   *The Eighteenth Century Background.* New York: Columbia University Press.

Williams, G. C.

1966   *Adaptation and Natural Selection: A Critique of Some Current Evolutionary Thought.* Princeton, NJ: Princeton University Press.

1992   *Natural Selection: Domains, Levels, and Challenges.* Oxford: Oxford University Press.

Wilson, D. J., and G. McVean

2006   Estimating diversifying selection and functional constraint in the presence of recombination. *Genetics* 172:1411–1425.

Wilson, D. S.

1980   *The Natural Selection of Populations and Communities.* Menlo Park, CA: Benjamin/Cummings.

Wimsatt, W. C.

1972   Teleology and the logical structure of function statements. *Studies in History and Philosophy of Science* 3:1–80.

Wolff, C.

1963   *Preliminary Discourse on Philosophy in General.* Trans. R. J. Blackwell. Indianapolis: Bobbs-Merrill.

Woodfield, A.

1998   Teleology. In E. Craig (ed.), *Routledge encyclopedia of philosophy* (vol. 9, pp. 295–297). London: Routledge.

Woods, R., D. Schneider, C. L. Winkworth, M. A. Riley, and R. E. Lenski

2006   Tests of parallel molecular evolution in a long-term experiment with *Escherichia coli. Proceedings of the National Academy of Sciences of the USA* 103:9107–9112.

Worster, D.

1994   *Nature's Economy: A History of Ecological Ideas.* 2nd ed. Cambridge: Cambridge University Press.

Wouters, A. G.

1995   Viability explanation. *Biology & Philosophy* 10:435–457.

1999   Explanation without a cause. PhD diss., Utrecht University.

2003a   Four notions of biological function. *Studies in History and Philosophy of Biological and Biomedical Sciences* 34:633–668.

2003b   Philosophers on functions (review of *Functions*, 2002, A. Ariew et al., eds.). *Acta Biotheoretica* 51:223–235.

2005a   The function debate in philosophy. *Acta Biotheoretica* 53:123–151.

2005b   The functional perspective of organismal biology. In T. A. C. Reydon and L. Hemerik (eds.), *Current Themes in Theoretical Biology: A Dutch Perspective* (pp. 33–69). Dordrecht: Springer.

2007   Design explanation: Determining the constraints on what can be alive. *Erkenntnis,* 67:65–80.

Wright, L.

1973   Functions. *Philosophical Review* 82:139–168.

Wright, M. R., ed.

2000   *Reason and Necessity: Essays on Plato's Timaeus.* London: Duckworth and the Classical Press of Wales.

Wright, S.

1923   Mendelian analysis of pure breeds of livestock II: The Duchess family of shorthorns as bred by Thomas Bates. *Journal of Heredity* 14:405–422.

1931   Evolution in Mendelian populations. *Genetics* 16:97–159.

1932   The roles of mutation, inbreeding, crossbreeding, and selection in evolution. *Proceedings of the Sixth International Congress of Genetics* 1:356–366.

1935a   The analysis of variance and the correlations between relatives with respect to deviations from an optimum. *Journal of Genetics* 30:243–256.

1935b   Evolution in populations in approximate equilibrium. *Journal of Genetics* 30:257–266.

1937   The distribution of gene frequencies in populations. *Proceedings of the National Academy of Sciences of the USA* 23:307–320.

1942   Statistical genetics and evolution. *Bulletin of the American Mathematical Society* 48:223–246.

1948   On the roles of directed and random changes in gene frequency in the genetics of populations. *Evolution* 2:279–294.

1949   Adaptation and selection. In G. L. Jepson, G. G. Simpson, and E. Mayr (eds.), *Genetics, Paleontology, and Evolution* (pp. 365–389). Princeton, NJ: Princeton University Press.

1956a   Classification of the factors of evolution. *Cold Spring Harbor Symposia on Quantitative Biology* 20:16–24.

1956b   Modes of selection. *American Naturalist* 90:5–24.

1969   *Evolution and the Genetics of Populations. Vol. 2. The Theory of Gene Frequencies.* Chicago: University of Chicago Press.

1977   *Evolution and the Genetics of Populations. Vol. 3. Experimental Results and Evolutionary Deductions.* Chicago: University of Chicago Press.

1978   The relation of livestock breeding to theories of evolution. *Journal of Animal Science* 46:1192–1200.

1986   *Evolution: Selected Papers.* Ed. W. Provine. Chicago: University of Chicago Press.

1988   Surfaces of selective value revisited. *American Naturalist* 131:115–123.

Wynne-Edwards, V. C.

1962   *Animal Dispersion in Relation to Social Behaviour.* Edinburgh: Oliver & Boyd.

Xenophon

1965   *Recollections of Socrates and Socrates' Defense before the Jury.* Trans. A. S. Benjamin. Indianapolis: Bobbs-Merrill.

Xu, X.

2006   Scales, feathers and dinosaurs. *Nature (London)* 440:287–288.

Zhang, J.

2006   Parallel adaptive origins of digestive RNases in Asian and African leaf monkeys. *Nature Genetics* 38:819–823.

Zhou, Z.

2004   The origin and early evolution of birds: Discoveries, disputes, and perspectives from fossil evidence. *Naturwissenschaften* 91:455–471.

INDEX

Abel, Johann Friedrich, 104
Abel, O., 299
Abzhanov, A., 328
adaptation, term, 3–5, 20, 25, 257–258,
    264–265, 301
adaptation (process), 3–6, 25, 258
  adaptedness and, 258, 264–265
  Bock on, 301
  to a changing environment, 258
  conservation biology and, 347
  constraints and, 294
  Darwin on, 129–141, 147
  definition, 258
  vs. drift, 242
  evolutionary, 339–340
    Bock on, 301
  Fisher on, 184
  of individual features, 264–267
  molecular evolution and, 241–242
  natural selection and, 264–267
  origin of bird flight and, 297–311
  performance and, 265
  physiology and, 339–341
  sand-dwelling lizards and, 265–267
  term, 258, 264–265, 301
adaptationism, 309–310
adaptationist program, 294–295

adaptations. *See* adapted features
adapted features, 4, 259–260, 262,
    264–265
  vs. adaptedness, 265–267
  Bock on, 301–303
  current utility and, 265
  Darwin on, 129–141, 142, 147
  definition, 264
  explanation and, 130–132, 262
    in British natural theology, 123
  exquisite, 356
  eyes as, 263
  Fisher on, 185
  functions and, 4, 260–267, 261 (table)
  Futuyma on, 4
  Gould and Lewontin on, 264,
    294–295
  group, Fisher on, 185
  lizard toe fringes as, 262
  multilevel selection and, 324–326
  neutral evolution of, 252
  Odling-Smee on, 336
  phylogenetic analysis and, 264–265
  physiology and, 339
  as products of natural selection, 20, 225,
    264, 267
  survival and, 262

Composition: Michael Bass Associates
Text and display: Minion Pro (Open Type)
Printer and binder: Thomson-Shore